INTRODUCTION TO BIOSTATISTICS

INTRODUCTION TO

BIOSTATISTICS

SECOND EDITION

Robert R. Sokal and F. James Rohlf
STATE UNIVERSITY OF NEW YORK AT STONY BROOK

W. H. FREEMAN AND COMPANY / NEW YORK

Library of Congress Cataloging-in-Publication Data

Sokal, Robert R.
  Introduction to biostatistics.

  Bibliography: p.
  Includes index.
  1. Biometry.   I. Rohlf, F. James, 1936–
II. Title.
QH323.5.S633 1987   574′.072   86-31838
ISBN 0-716-71805-7

Printed in the United States of America

    4 5 6 7 8 9   VB   9 9 8 7 6 5 4 3 2

to Julie and Pat

Contents

Preface

The favorable reception that the first edition of this book received from teachers and students encouraged us to prepare a second edition. In this revised edition, we provide a thorough foundation in biological statistics for the undergraduate student who has a minimal knowledge of mathematics. We intend *Introduction to Biostatistics* to be used in comprehensive biostatistics courses, but it can also be adapted for short courses in medical and professional schools; thus, we include examples from the health-related sciences.

We have extracted most of this text from the more-inclusive second edition of our own *Biometry*. We believe that the proven pedagogic features of that book, such as its informal style, will be valuable here.

We have modified some of the features from *Biometry*; for example, in *Introduction to Biostatistics* we provide detailed outlines for statistical computations but we place less emphasis on the computations themselves. Why? Students in many undergraduate courses are not motivated to and have few opportunities to perform lengthy computations with biological research material; also, such computations can easily be made on electronic calculators and microcomputers. Thus, we rely on the course instructor to advise students on the best computational procedures to follow.

We present material in a sequence that progresses from descriptive statistics to fundamental distributions and the testing of elementary statistical hypotheses; we then proceed immediately to the analysis of variance and the familiar *t* test (which is treated as a special case of the analysis of variance and relegated to several sections of the book). We do this deliberately for two reasons: (1) since today's biologists all need a thorough foundation in the analysis of variance, students should become acquainted with the subject early in the course; and (2) if analysis of variance is understood early, the need to use the t distribution is reduced. (One would still want to use it for the setting of confidence limits and in a few other special situations.) All t tests can be carried out directly as analyses of variance, and the amount of computation of these analyses of variance is generally equivalent to that of t tests.

This larger second edition includes the Kolgorov-Smirnov two-sample test, nonparametric regression, stem-and-leaf diagrams, hanging histograms, and the Bonferroni method of multiple comparisons. We have rewritten the chapter on the analysis of frequencies in terms of the G statistic rather than χ^2, because the former has been shown to have more desirable statistical properties. Also, because of the availability of logarithm functions on calculators, the computation of the G statistic is now easier than that of the earlier chi-square test. Thus, we reorient the chapter to emphasize log-likelihood-ratio tests. We have also added new homework exercises.

We call special, double-numbered tables "boxes." They can be used as convenient guides for computation because they show the computational methods for solving various types of biostatistical problems. They usually contain all the steps necessary to solve a problem—from the initial setup to the final result. Thus, students familiar with material in the book can use them as quick summary reminders of a technique.

We found in teaching this course that we wanted students to be able to refer to the material now in these boxes. We discovered that we could not cover even half as much of our subject if we had to put this material on the blackboard during the lecture, and so we made up and distributed boxes and asked students to refer to them during the lecture. Instructors who use this book may wish to use the boxes in a similar manner.

We emphasize the practical applications of statistics to biology in this book; thus, we deliberately keep discussions of statistical theory to a minimum. Derivations are given for some formulas, but these are consigned to Appendix A1, where they should be studied and reworked by the student. Statistical tables to which the reader can refer when working through the methods discussed in this book are found in Appendix A2.

We are grateful to K. R. Gabriel, R. C. Lewontin, and M. Kabay for their extensive comments on the second edition of *Biometry* and to M. D. Morgan, E. Russek-Cohen, and M. Singh for comments on an early draft of this book. We also appreciate the work of our secretaries, Resa Chapey and Cheryl Daly, with preparing the manuscripts, and of Donna DiGiovanni, Patricia Rohlf, and Barbara Thomson with proofreading.

Robert R. Sokal

F. James Rohlf

INTRODUCTION TO BIOSTATISTICS

CHAPTER 1

Introduction

This chapter sets the stage for your study of biostatistics. In Section 1.1, we define the field itself. We then cast a necessarily brief glance at its historical development in Section 1.2. Then in Section 1.3 we conclude the chapter with a discussion of the attitudes that the person trained in statistics brings to biological research.

1.1 Some definitions

We shall define *biostatistics* as *the application of statistical methods to the solution of biological problems*. The biological problems of this definition are those arising in the basic biological sciences as well as in such applied areas as the health-related sciences and the agricultural sciences. Biostatistics is also called *biological statistics* or *biometry*.

The definition of biostatistics leaves us somewhat up in the air—"statistics" has not been defined. *Statistics* is a science well known by name even to the layman. The number of definitions you can find for it is limited only by the number of books you wish to consult. We might define statistics in its modern sense as *the scientific study of numerical data based on natural phenomena*. All parts of this definition are important and deserve emphasis:

Scientific study: Statistics must meet the commonly accepted criteria of validity of scientific evidence. We must always be objective in presentation and evaluation of data and adhere to the general ethical code of scientific methodology, or we may find that the old saying that "figures never lie, only statisticians do" applies to us.

Data: Statistics generally deals with populations or groups of individuals; hence it deals with *quantities* of information, not with a single *datum*. Thus, the measurement of a single animal or the response from a single biochemical test will generally not be of interest.

Numerical: Unless data of a study can be quantified in one way or another, they will not be amenable to statistical analysis. Numerical data can be measurements (the length or width of a structure or the amount of a chemical in a body fluid, for example) or counts (such as the number of bristles or teeth).

Natural phenomena: We use this term in a wide sense to mean not only all those events in animate and inanimate nature that take place outside the control of human beings, but also those evoked by scientists and partly under their control, as in experiments. Different biologists will concern themselves with different levels of natural phenomena; other kinds of scientists, with yet different ones. But all would agree that the chirping of crickets, the number of peas in a pod, and the age of a woman at menopause are natural phenomena. The heartbeat of rats in response to adrenalin, the mutation rate in maize after irradiation, or the incidence or morbidity in patients treated with a vaccine may still be considered natural, even though scientists have interfered with the phenomenon through their intervention. The average biologist would not consider the number of stereo sets bought by persons in different states in a given year to be a natural phenomenon. Sociologists or human ecologists, however, might so consider it and deem it worthy of study. The qualification "natural phenomena" is included in the definition of statistics mostly to make certain that the phenomena studied are not arbitrary ones that are entirely under the will and control of the researcher, such as the number of animals employed in an experiment.

The word "statistics" is also used in another, though related, way. It can be the plural of the noun *statistic*, which refers to any one of many computed or estimated statistical quantities, such as the mean, the standard deviation, or the correlation coefficient. Each one of these is a statistic.

1.2 The development of biostatistics

Modern statistics appears to have developed from two sources as far back as the seventeenth century. The first source was political science; a form of statistics developed as a quantitive description of the various aspects of the affairs of a government or state (hence the term "statistics"). This subject also became known as political arithmetic. Taxes and insurance caused people to become interested in problems of censuses, longevity, and mortality. Such considerations assumed increasing importance, especially in England as the country prospered during the development of its empire. John Graunt (1620–1674) and William Petty (1623–1687) were early students of vital statistics, and others followed in their footsteps.

At about the same time, the second source of modern statistics developed: the mathematical theory of probability engendered by the interest in games of chance among the leisure classes of the time. Important contributions to this theory were made by Blaise Pascal (1623–1662) and Pierre de Fermat (1601–1665), both Frenchmen. Jacques Bernoulli (1654–1705), a Swiss, laid the foundation of modern probability theory in *Ars Conjectandi*. Abraham de Moivre (1667–1754), a Frenchman living in England, was the first to combine the statistics of his day with probability theory in working out annuity values and to approximate the important normal distribution through the expansion of the binomial.

A later stimulus for the development of statistics came from the science of astronomy, in which many individual observations had to be digested into a coherent theory. Many of the famous astronomers and mathematicians of the eighteenth century, such as Pierre Simon Laplace (1749–1827) in France and Karl Friedrich Gauss (1777–1855) in Germany, were among the leaders in this field. The latter's lasting contribution to statistics is the development of the method of least squares.

Perhaps the earliest important figure in biostatistic thought was Adolphe Quetelet (1796–1874), a Belgian astronomer and mathematician, who in his work combined the theory and practical methods of statistics and applied them to problems of biology, medicine, and sociology. Francis Galton (1822–1911), a cousin of Charles Darwin, has been called the father of biostatistics and eugenics. The inadequacy of Darwin's genetic theories stimulated Galton to try to solve the problems of heredity. Galton's major contribution to biology was his application of statistical methodology to the analysis of biological variation, particularly through the analysis of variability and through his study of regression and correlation in biological measurements. His hope of unraveling the laws of genetics through these procedures was in vain. He started with the most difficult material and with the wrong assumptions. However, his methodology has become the foundation for the application of statistics to biology.

Karl Pearson (1857–1936), at University College, London, became interested in the application of statistical methods to biology, particularly in the demonstration of natural selection. Pearson's interest came about through the influence of W. F. R. Weldon (1860–1906), a zoologist at the same institution. Weldon, incidentally, is credited with coining the term "biometry" for the type of studies he and Pearson pursued. Pearson continued in the tradition of Galton and laid the foundation for much of descriptive and correlational statistics.

The dominant figure in statistics and biometry in the twentieth century has been Ronald A. Fisher (1890–1962). His many contributions to statistical theory will become obvious even to the cursory reader of this book.

Statistics today is a broad and extremely active field whose applications touch almost every science and even the humanities. New applications for statistics are constantly being found, and no one can predict from what branch of statistics new applications to biology will be made.

1.3 The statistical frame of mind

A brief perusal of almost any biological journal reveals how pervasive the use of statistics has become in the biological sciences. Why has there been such a marked increase in the use of statistics in biology? Apparently, because biologists have found that the interplay of biological causal and response variables does not fit the classic mold of nineteenth-century physical science. In that century, biologists such as Robert Mayer, Hermann von Helmholtz, and others tried to demonstrate that biological processes were nothing but physicochemical phenomena. In so doing, they helped create the impression that the experimental methods and natural philosophy that had led to such dramatic progress in the physical sciences should be imitated fully in biology.

Many biologists, even to this day, have retained the tradition of strictly mechanistic and deterministic concepts of thinking (while physicists, interestingly enough, as their science has become more refined, have begun to resort to statistical approaches). In biology, most phenomena are affected by many causal factors, uncontrollable in their variation and often unidentifiable. Statistics is needed to measure such variable phenomena, to determine the error of measurement, and to ascertain the reality of minute but important differences.

A misunderstanding of these principles and relationships has given rise to the attitude of some biologists that if differences induced by an experiment, or observed by nature, are not clear on plain inspection (and therefore are in need of statistical analysis), they are not worth investigating. There are few legitimate fields of inquiry, however, in which, from the nature of the phenomena studied, statistical investigation is unnecessary.

Statistical thinking is not really different from ordinary disciplined scientific thinking, in which we try to quantify our observations. In statistics we express our degree of belief or disbelief as a probability rather than as a vague, general statement. For example, a statement that individuals of species A are larger than those of species B or that women suffer more often from disease X than do men is of a kind commonly made by biological and medical scientists. Such statements can and should be more precisely expressed in quantitative form.

In many ways the human mind is a remarkable statistical machine, absorbing many facts from the outside world, digesting these, and regurgitating them in simple summary form. From our experience we know certain events to occur frequently, others rarely. "Man smoking cigarette" is a frequently observed event, "Man slipping on banana peel," rare. We know from experience that Japanese are on the average shorter than Englishmen and that Egyptians are on the average darker than Swedes. We associate thunder with lightning almost always, flies with garbage cans in the summer frequently, but snow with the southern Californian desert extremely rarely. All such knowledge comes to us as a result of experience, both our own and that of others, which we learn about by direct communication or through reading. All these facts have been processed by that remarkable computer, the human brain, which furnishes an abstract. This abstract is constantly under revision, and though occasionally faulty and biased, it is on the whole astonishingly sound; it is our knowledge of the moment.

Although statistics arose to satisfy the needs of scientific research, the development of its methodology in turn affected the sciences in which statistics is applied. Thus, through positive feedback, statistics, created to serve the needs of natural science, has itself affected the content and methods of the biological sciences. To cite an example: Analysis of variance has had a tremendous effect in influencing the types of experiments researchers carry out. The whole field of quantitative genetics, one of whose problems is the separation of environmental from genetic effects, depends upon the analysis of variance for its realization, and many of the concepts of quantitative genetics have been directly built around the designs inherent in the analysis of variance.

CHAPTER 2

Data in Biostatistics

In Section 2.1 we explain the statistical meaning of the terms "sample" and "population," which we shall be using throughout this book. Then, in Section 2.2, we come to the types of observations that we obtain from biological research material; we shall see how these correspond to the different kinds of variables upon which we perform the various computations in the rest of this book. In Section 2.3 we discuss the degree of accuracy necessary for recording data and the procedure for rounding off figures. We shall then be ready to consider in Section 2.4 certain kinds of derived data frequently used in biological science— among them ratios and indices—and the peculiar problems of accuracy and distribution they present us. Knowing how to arrange data in frequency distributions is important because such arrangements give an overall impression of the general pattern of the variation present in a sample and also facilitate further computational procedures. Frequency distributions, as well as the presentation of numerical data, are discussed in Section 2.5. In Section 2.6 we briefly describe the computational handling of data.

2.1 Samples and populations

We shall now define a number of important terms necessary for an under-standing of biological data. The *data* in biostatistics are generally based on *individual observations*. They are *observations or measurements taken on the smallest sampling unit*. These smallest sampling units frequently, but not neces-sarily, are also individuals in the ordinary biological sense. If we measure weight in 100 rats, then the weight of each rat is an individual observation; the hundred rat weights together represent the *sample of observations*, defined as *a collection of individual observations selected by a specified procedure*. In this instance, one individual observation (an *item*) is based on one individual in a biological sense—that is, one rat. However, if we had studied weight in a single rat over a period of time, the sample of individual observations would be the weights recorded on one rat at successive times. If we wish to measure temperature in a study of ant colonies, where each colony is a basic sampling unit, each temperature reading for one colony is an individual observation, and the sample of observations is the temperatures for all the colonies considered. If we consider an estimate of the DNA content of a single mammalian sperm cell to be an individual observation, the sample of observations may be the estimates of DNA content of all the sperm cells studied in one individual mammal.

We have carefully avoided so far specifying what particular variable was being studied, because the terms "individual observation" and "sample of ob-servations" as used above define only the structure but not the nature of the data in a study. The *actual property* measured by the individual observations is the *character*, or *variable*. The more common term employed in general sta-tistics is "variable." However, in biology the word "character" is frequently used synonymously. More than one variable can be measured on each smallest sampling unit. Thus, in a group of 25 mice we might measure the blood *p*H and the erythrocyte count. Each mouse (a biological individual) is the smallest sampling unit, blood *p*H and red cell count would be the two variables studied, the *p*H readings and cell counts are individual observations, and two samples of 25 observations (on *p*H and on erythrocyte count) would result. Or we might speak of a *bivariate sample* of 25 observations, each referring to a *p*H reading paired with an erythrocyte count.

Next we define *population*. The biological definition of this term is well known. It refers to all the individuals of a given species (perhaps of a given life-history stage or sex) found in a circumscribed area at a given time. In statistics, population always means *the totality of individual observations about which inferences are to be made, existing anywhere in the world or at least within a definitely specified sampling area limited in space and time*. If you take five men and study the number of leucocytes in their peripheral blood and you are prepared to draw conclusions about all men from this sample of five, then the population from which the sample has been drawn represents the leucocyte counts of all extant males of the species *Homo sapiens*. If, on the other hand, you restrict yourself to a more narrowly specified sample, such as five male

Chinese, aged 20, and you are restricting your conclusions to this particular group, then the population from which you are sampling will be leucocyte numbers of all Chinese males of age 20.

A common misuse of statistical methods is to fail to define the statistical population about which inferences can be made. A report on the analysis of a sample from a restricted population should not imply that the results hold in general. The population in this statistical sense is sometimes referred to as the *universe*.

A population may represent variables of a concrete collection of objects or creatures, such as the tail lengths of all the white mice in the world, the leucocyte counts of all the Chinese men in the world of age 20, or the DNA content of all the hamster sperm cells in existence; or it may represent the outcomes of experiments, such as all the heartbeat frequencies produced in guinea pigs by injections of adrenalin. In cases of the first kind the population is generally finite. Although in practice it would be impossible to collect, count, and examine all hamster sperm cells, all Chinese men of age 20, or all white mice in the world, these populations are in fact finite. Certain smaller populations, such as all the whooping cranes in North America or all the recorded cases of a rare but easily diagnosed disease X, may well lie within reach of a total census. By contrast, an experiment can be repeated an infinite number of times (at least in theory). A given experiment, such as the administration of adrenalin to guinea pigs, could be repeated as long as the experimenter could obtain material and his or her health and patience held out. The sample of experiments actually performed is a sample from an infinite number that *could* be performed.

Some of the statistical methods to be developed later make a distinction between sampling from finite and from infinite populations. However, though populations are theoretically finite in most applications in biology, they are generally so much larger than samples drawn from them that they can be considered de facto infinite-sized populations.

2.2 Variables in biostatistics

Each biological discipline has its own set of variables, which may include conventional morphological measurements; concentrations of chemicals in body fluids; rates of certain biological processes; frequencies of certain events, as in genetics, epidemiology, and radiation biology; physical readings of optical or electronic machinery used in biological research; and many more.

We have already referred to biological variables in a general way, but we have not yet defined them. We shall define a *variable* as *a property with respect to which individuals in a sample differ in some ascertainable way*. If the property does not differ within a sample at hand or at least among the samples being studied, it cannot be of statistical interest. Length, height, weight, number of teeth, vitamin C content, and genotypes are examples of variables in ordinary, genetically and phenotypically diverse groups of organisms. Warm-bloodedness in a group of mammals is not, since mammals are all alike in this regard, although body temperature of individual mammals would, of course, be a variable.

We can divide variables as follows:

Variables

Measurement variables
    Continuous variables
    Discontinuous variables
Ranked variables
Attributes

Measurement variables are *those measurements and counts that are expressed numerically.* Measurement variables are of two kinds. The first kind consists of *continuous variables*, which at least theoretically can assume an infinite number of values between any two fixed points. For example, between the two length measurements 1.5 and 1.6 cm there are an infinite number of lengths that could be measured if one were so inclined and had a precise enough method of calibration. Any given reading of a continuous variable, such as a length of 1.57 mm, is therefore an approximation to the exact reading, which in practice is unknowable. Many of the variables studied in biology are continuous variables. Examples are lengths, areas, volumes, weights, angles, temperatures, periods of time, percentages, concentrations, and rates.

Contrasted with continuous variables are the *discontinuous variables*, also known as *meristic* or *discrete variables*. These are variables that have only certain fixed numerical values, with no intermediate values possible in between. Thus the number of segments in a certain insect appendage may be 4 or 5 or 6 but never $5\frac{1}{2}$ or 4.3. Examples of discontinuous variables are numbers of a given structure (such as segments, bristles, teeth, or glands), numbers of offspring, numbers of colonies of microorganisms or animals, or numbers of plants in a given quadrat.

Some variables cannot be measured but at least can be ordered or ranked by their magnitude. Thus, in an experiment one might record the rank order of emergence of ten pupae without specifying the exact time at which each pupa emerged. In such cases we code the data as a *ranked variable*, the order of emergence. Special methods for dealing with such variables have been developed, and several are furnished in this book. By expressing a variable as a series of ranks, such as 1, 2, 3, 4, 5, we do not imply that the difference in magnitude between, say, ranks 1 and 2 is identical to or even proportional to the difference between ranks 2 and 3.

Variables that cannot be measured but must be expressed qualitatively are called *attributes*, or *nominal variables*. These are all properties, such as black or white, pregnant or not pregnant, dead or alive, male or female. When such attributes are combined with frequencies, they can be treated statistically. Of 80 mice, we may, for instance, state that four were black, two agouti, and the rest gray. When attributes are combined with frequencies into tables suitable for statistical analysis, they are referred to as *enumeration data*. Thus the enumeration data on color in mice would be arranged as follows:

Color	Frequency
Black	4
Agouti	2
Gray	74
Total number of mice	80

In some cases attributes can be changed into measurement variables if this is desired. Thus colors can be changed into wavelengths or color-chart values. Certain other attributes that can be ranked or ordered can be coded to become ranked variables. For example, three attributes referring to a structure as "poorly developed," "well developed," and "hypertrophied" could be coded 1, 2, and 3.

A term that has not yet been explained is *variate*. In this book we shall use it as a single reading, score, or observation of a given variable. Thus, if we have measurements of the length of the tails of five mice, tail length will be a continuous variable, and each of the five readings of length will be a variate. In this text we identify variables by capital letters, the most common symbol being Y. Thus Y may stand for tail length of mice. A variate will refer to a given length measurement; Y_i is the measurement of tail length of the ith mouse, and Y_4 is the measurement of tail length of the fourth mouse in our sample.

2.3 Accuracy and precision of data

"Accuracy" and "precision" are used synonymously in everyday speech, but in statistics we define them more rigorously. *Accuracy* is *the closeness of a measured or computed value to its true value*. *Precision* is *the closeness of repeated measurements*. A biased but sensitive scale might yield inaccurate but precise weight. By chance, an insensitive scale might result in an accurate reading, which would, however, be imprecise, since a repeated weighing would be unlikely to yield an equally accurate weight. Unless there is bias in a measuring instrument, precision will lead to accuracy. We need therefore mainly be concerned with the former.

Precise variates are usually, but not necessarily, whole numbers. Thus, when we count four eggs in a nest, there is no doubt about the exact number of eggs in the nest if we have counted correctly; it is 4, not 3 or 5, and clearly it could not be 4 plus or minus a fractional part. Meristic, or discontinuous, variables are generally measured as exact numbers. Seemingly, continuous variables derived from meristic ones can under certain conditions also be exact numbers. For instance, ratios between exact numbers are themselves also exact. If in a colony of animals there are 18 females and 12 males, the ratio of females to males (a continuous variable) is 1.5, an exact number.

Most continuous variables, however, are approximate. We mean by this that the exact value of the single measurement, the variate, is unknown and probably unknowable. The last digit of the measurement stated should imply precision; that is, it should indicate the limits on the measurement scale between which we believe the true measurement to lie. Thus, a length measurement of 12.3 mm implies that the true length of the structure lies somewhere between 12.25 and 12.35 mm. Exactly where between these *implied limits* the real length is we do not know. But where would a true measurement of 12.25 fall? Would it not equally likely fall in either of the two classes 12.2 and 12.3—clearly an unsatisfactory state of affairs? Such an argument is correct, but when we record a number as either 12.2 or 12.3, we imply that the decision whether to put it into the higher or lower class has already been taken. This decision was not taken arbitrarily, but presumably was based on the best available measurement. If the scale of measurement is so precise that a value of 12.25 would clearly have been recognized, then the measurement should have been recorded originally to four significant figures. *Implied limits, therefore, always carry one more figure beyond the last significant one measured by the observer.*

Hence, it follows that if we record the measurement as 12.32, we are implying that the true value lies between 12.315 and 12.325. Unless this is what we mean, there would be no point in adding the last decimal figure to our original measurements. If we do add another figure, we must imply an increase in precision. We see, therefore, that accuracy and precision in numbers are not absolute concepts, but are relative. Assuming there is no bias, a number becomes increasingly more accurate as we are able to write more significant figures for it (increase its precision). To illustrate this concept of the relativity of accuracy, consider the following three numbers:

	Implied limits
193	192.5–193.5
192.8	192.75–192.85
192.76	192.755–192.765

We may imagine these numbers to be recorded measurements of the same structure. Let us assume that we had extramundane knowledge that the true length of the given structure was 192.758 units. If that were so, the three measurements would increase in accuracy from the top down, as the interval between their implied limits decreased. You will note that the implied limits of the topmost measurement are wider than those of the one below it, which in turn are wider than those of the third measurement.

Meristic variates, though ordinarily exact, may be recorded approximately when large numbers are involved. Thus when counts are reported to the nearest thousand, a count of 36,000 insects in a cubic meter of soil, for example, implies that the true number varies somewhere from 35,500 to 36,500 insects.

To how many significant figures should we record measurements? If we array the sample by order of magnitude from the smallest individual to the largest one, an easy rule to remember is that *the number of unit steps from the smallest to the largest measurement in an array should usually be between 30* and *300*. Thus, if we are measuring a series of shells to the nearest millimeter and the largest is 8 mm and the smallest is 4 mm wide, there are only four unit steps between the largest and the smallest measurement. Hence, we should measure our shells to one more significant decimal place. Then the two extreme measurements might be 8.2 mm and 4.1 mm, with 41 unit steps between them (counting the last significant digit as the unit); this would be an adequate number of unit steps. The reason for such a rule is that an error of 1 in the last significant digit of a reading of 4 mm would constitute an inadmissible error of 25%, but an error of 1 in the last digit of 4.1 is less than 2.5%. Similarly, if we measured the height of the tallest of a series of plants as 173.2 cm and that of the shortest of these plants as 26.6 cm, the difference between these limits would comprise 1466 unit steps (of 0.1 cm), which are far too many. It would therefore be advisable to record the heights to the nearest centimeter, as follows: 173 cm for the tallest and 27 cm for the shortest. This would yield 146 unit steps. Using the rule we have stated for the number of unit steps, we shall record two or three digits for most measurements.

The last digit should always be significant; that is, it should imply a range for the true measurement of from half a "unit step" below to half a "unit step" above the recorded score, as illustrated earlier. This applies to all digits, zero included. Zeros should therefore not be written at the end of approximate numbers to the right of the decimal point unless they are meant to be significant digits. Thus 7.80 must imply the limits 7.795 to 7.805. If 7.75 to 7.85 is implied, the measurement should be recorded as 7.8.

When the number of significant digits is to be reduced, we carry out the process of *rounding off* numbers. The rules for rounding off are very simple. A digit to be rounded off is not changed if it is followed by a digit less than 5. If the digit to be rounded off is followed by a digit greater than 5 or by 5 followed by other nonzero digits, it is increased by 1. When the digit to be rounded off is followed by a 5 standing alone or a 5 followed by zeros, it is unchanged if it is even but increased by 1 if it is odd. The reason for this last rule is that when such numbers are summed in a long series, we should have as many digits raised as are being lowered, on the average; these changes should therefore balance out. Practice the above rules by rounding off the following numbers to the indicated number of significant digits:

Number	Significant digits desired	Answer
26.58	2	27
133.7137	5	133.71
0.03725	3	0.0372
0.03715	3	0.0372
18,316	2	18,000
17.3476	3	17.3

Most pocket calculators or larger computers round off their displays using a different rule: they increase the preceding digit when the following digit is a 5 standing alone or with trailing zeros. However, since most of the machines usable for statistics also retain eight or ten significant figures internally, the accumulation of rounding errors is minimized. Incidentally, if two calculators give answers with slight differences in the final (least significant) digits, suspect a different number of significant digits in memory as a cause of the disagreement.

2.4 Derived variables

The majority of variables in biometric work are observations recorded as direct measurements or counts of biological material or as readings that are the output of various types of instruments. However, there is an important class of variables in biological research that we may call the *derived* or *computed variables*. These are generally based on two or more independently measured variables whose relations are expressed in a certain way. We are referring to ratios, percentages, concentrations, indices, rates, and the like.

A *ratio* expresses as a single value the relation that two variables have, one to the other. In its simplest form, a ratio is expressed as in 64:24, which may represent the number of wild-type versus mutant individuals, the number of males versus females, a count of parasitized individuals versus those not parasitized, and so on. These examples imply ratios based on counts. A ratio based on a continuous variable might be similarly expressed as 1.2:1.8, which may represent the ratio of width to length in a sclerite of an insect or the ratio between the concentrations of two minerals contained in water or soil. Ratios may also be expressed as fractions; thus, the two ratios above could be expressed as $\frac{64}{24}$ and $\frac{1.2}{1.8}$. However, for computational purposes it is more useful to express the ratio as a quotient. The two ratios cited would therefore be 2.666... and 0.666..., respectively. These are pure numbers, not expressed in measurement units of any kind. It is this form for ratios that we shall consider further. *Percentages* are also a type of ratio. Ratios, percentages, and concentrations are basic quantities in much biological research, widely used and generally familiar.

An *index* is *the ratio of the value of one variable to the value of a so-called standard one*. A well-known example of an index in this sense is the cephalic index in physical anthropology. Conceived in the wide sense, an index could be the average of two measurements—either simply, such as $\frac{1}{2}$(length of A + length of B), or in weighted fashion, such as $\frac{1}{3}[(2 \times \text{length of } A) + \text{length of } B]$.

Rates are important in many experimental fields of biology. The amount of a substance liberated per unit weight or volume of biological material, weight gain per unit time, reproductive rates per unit population size and time (birth rates), and death rates would fall in this category.

The use of ratios and percentages is deeply ingrained in scientific thought. Often ratios may be the only meaningful way to interpret and understand certain types of biological problems. If the biological process being investigated operates on the ratio of the variables studied, one must examine this ratio to understand the process. Thus, Sinnott and Hammond (1935) found that inheritance of the shapes of squashes of the species *Cucurbita pepo* could be interpreted through a form index based on a length-width ratio, but not through the independent dimensions of shape. By similar methods of investigation, we should be able to find selection affecting body proportions to exist in the evolution of almost any organism.

There are several disadvantages to using ratios. First, they are relatively inaccurate. Let us return to the ratio $\frac{1.2}{1.8}$ mentioned above and recall from the previous section that a measurement of 1.2 implies a true range of measurement of the variable from 1.15 to 1.25; similarly, a measurement of 1.8 implies a range from 1.75 to 1.85. We realize, therefore, that the true ratio may vary anywhere from $\frac{1.15}{1.85}$ to $\frac{1.25}{1.75}$, or from 0.622 to 0.714. We note a possible maximal error of 4.2% if 1.2 is an original measurement: $(1.25 - 1.2)/1.2$; the corresponding maximal error for the ratio is 7.0%: $(0.714 - 0.667)/0.667$. Furthermore, the best estimate of a ratio is not usually the midpoint between its possible ranges. Thus, in our example the midpoint between the implied limits is 0.668 and the ratio based on $\frac{1.2}{1.8}$ is 0.666 . . . ; while this is only a slight difference, the discrepancy may be greater in other instances.

A second disadvantage to ratios and percentages is that they may not be approximately normally distributed (see Chapter 5) as required by many statistical tests. This difficulty can frequently be overcome by transformation of the variable (as discussed in Chapter 10). A third disadvantage of ratios is that in using them one loses information about the relationships between the two variables except for the information about the ratio itself.

2.5 Frequency distributions

If we were to sample a population of birth weights of infants, we could represent each sampled measurement by a point along an axis denoting magnitude of birth weight. This is illustrated in Figure 2.1A, for a sample of 25 birth weights. If we sample repeatedly from the population and obtain 100 birth weights, we shall probably have to place some of these points on top of other points in order to record them all correctly (Figure 2.1B). As we continue sampling additional hundreds and thousands of birth weights (Figure 2.1C and D), the assemblage of points will continue to increase in size but will assume a fairly definite shape. The outline of the mound of points approximates the distribution of the variable. Remember that a continuous variable such as birth weight can assume an infinity of values between any two points on the abscissa. The refinement of our measurements will determine how fine the number of recorded divisions between any two points along the axis will be.

The distribution of a variable is of considerable biological interest. If we find that the distribution is asymmetrical and drawn out in one direction, it tells us that there is, perhaps, selection that causes organisms to fall preferentially in one of the tails of the distribution, or possibly that the scale of measurement

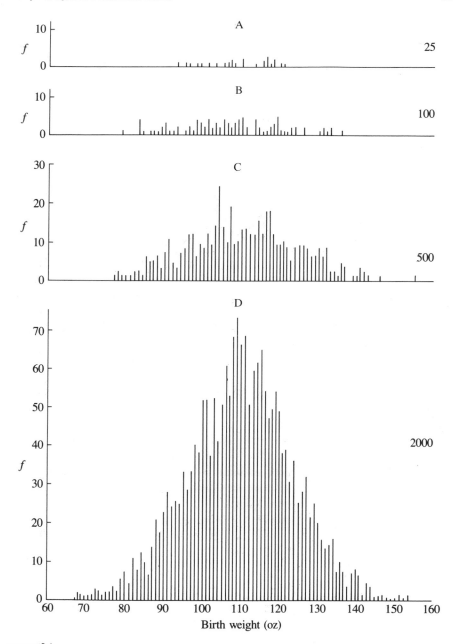

FIGURE 2.1
Sampling from a population of birth weights of infants (a continuous variable). A. A sample of 25.
B. A sample of 100. C. A sample of 500. D. A sample of 2000.

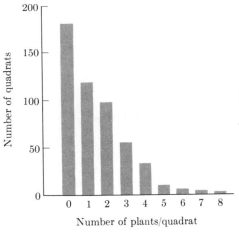

FIGURE 2.2

Bar diagram. Frequency of the sedge *Carex flacca* in 500 quadrats. Data from Table 2.2; orginally from Archibald (1950).

chosen is such as to bring about a distortion of the distribution. If, in a sample of immature insects, we discover that the measurements are bimodally distributed (with two peaks), this would indicate that the population is dimorphic. This means that different species or races may have become intermingled in our sample. Or the dimorphism could have arisen from the presence of both sexes or of different instars.

There are several characteristic shapes of frequency distributions. The most common is the symmetrical bell shape (approximated by the bottom graph in Figure 2.1), which is the shape of the normal frequency distribution discussed in Chapter 5. There are also skewed distributions (drawn out more at one tail than the other), L-shaped distributions as in Figure 2.2, U-shaped distributions, and others, all of which impart significant information about the relationships they represent. We shall have more to say about the implications of various types of distributions in later chapters and sections.

After researchers have obtained data in a given study, they must arrange the data in a form suitable for computation and interpretation. We may assume that variates are randomly ordered initially or are in the order in which the measurements have been taken. A simple arrangement would be an *array* of the data by order of magnitude. Thus, for example, the variates 7, 6, 5, 7, 8, 9, 6, 7, 4, 6, 7 could be arrayed in order of decreasing magnitude as follows: 9, 8, 7, 7, 7, 7, 6, 6, 6, 5, 4. Where there are some variates of the same value, such as the 6's and 7's in this fictitious example, a time-saving device might immediately have occurred to you—namely, to list a frequency for each of the recurring variates; thus: 9, 8, 7(4 ×), 6(3 ×), 5, 4. Such a shorthand notation is one way to represent a *frequency distribution*, which is simply an arrangement of the classes of variates with the frequency of each class indicated. Conventionally, a frequency distribution is stated in tabular form; for our example, this is done as follows:

Variable Y	Frequency f
9	1
8	1
7	4
6	3
5	1
4	1

The above is an example of a *quantitative frequency distribution*, since Y is clearly a measurement variable. However, arrays and frequency distributions need not be limited to such variables. We can make frequency distributions of attributes, called *qualitative frequency distributions*. In these, the various classes are listed in some logical or arbitrary order. For example, in genetics we might have a qualitative frequency distribution as follows:

Phenotype	f
A−	86
aa	32

This tells us that there are two classes of individuals, those identifed by the A − phenotype, of which 86 were found, and those comprising the homozygote recessive *aa*, of which 32 were seen in the sample.

An example of a more extensive qualitative frequency distribution is given in Table 2.1, which shows the distribution of melanoma (a type of skin cancer) over body regions in men and women. This table tells us that the trunk and limbs are the most frequent sites for melanomas and that the buccal cavity, the rest of the gastrointestinal tract, and the genital tract are rarely afflicted by this

TABLE 2.1

Two qualitative frequency distributions. Number of cases of skin cancer (melanoma) distributed over body regions of 4599 men and 4786 women.

Anatomic site	Observed frequency Men f	Women f
Head and neck	949	645
Trunk and limbs	3243	3645
Buccal cavity	8	11
Rest of gastrointestinal tract	5	21
Genital tract	12	93
Eye	382	371
Total cases	4599	4786

Source: Data from Lee (1982).

TABLE **2.2**

A meristic frequency distribution.
Number of plants of the sedge *Carex
flacca* found in 500 quadrats.

No. of plants per quadrat Y	Observed frequency f
0	181
1	118
2	97
3	54
4	32
5	9
6	5
7	3
8	1
Total	500

Source: Data from Archibald (1950).

type of cancer. We often encounter other examples of qualitative frequency distributions in ecology in the form of tables, or species lists, of the inhabitants of a sampled ecological area. Such tables catalog the inhabitants by species or at a higher taxonomic level and record the number of specimens observed for each. The arrangement of such tables is usually alphabetical, or it may follow a special convention, as in some botanical species lists.

A quantitative frequency distribution based on meristic variates is shown in Table 2.2. This is an example from plant ecology: the number of plants per quadrat sampled is listed at the left in the variable column; the observed frequency is shown at the right.

Quantitative frequency distributions based on a continuous variable are the most commonly employed frequency distributions; you should become thoroughly familiar with them. An example is shown in Box 2.1. It is based on 25 femur lengths measured in an aphid population. The 25 readings are shown at the top of Box 2.1 in the order in which they were obtained as measurements. (They could have been arrayed according to their magnitude.) The data are next set up in a frequency distribution. The variates increase in magnitude by unit steps of 0.1. The frequency distribution is prepared by entering each variate in turn on the scale and indicating a count by a conventional tally mark. When all of the items have been tallied in the corresponding class, the tallies are converted into numerals indicating frequencies in the next column. Their sum is indicated by Σf.

What have we achieved in summarizing our data? The original 25 variates are now represented by only 15 classes. We find that variates 3.6, 3.8, and 4.3 have the highest frequencies. However, we also note that there are several classes, such as 3.4 or 3.7, that are not represented by a single aphid. This gives the entire frequency distribution a drawn-out and scattered appearance. The reason for this is that we have only 25 aphids, too few to put into a frequency distribution with 15 classes. To obtain a more cohesive and smooth-looking distribution, we have to condense our data into fewer classes. This process is known as *grouping of classes* of frequency distributions; it is illustrated in Box 2.1 and described in the following paragraphs.

We should realize that grouping individual variates into classes of wider range is only an extension of the same process that took place when we obtained the initial measurement. Thus, as we have seen in Section 2.3, when we measure an aphid and record its femur length as 3.3 units, we imply thereby that the true measurement lies between 3.25 and 3.35 units, but that we were unable to measure to the second decimal place. In recording the measurement initially as 3.3 units, we estimated that it fell within this range. Had we estimated that it exceeded the value of 3.35, for example, we would have given it the next higher score, 3.4. Therefore, all the measurements between 3.25 and 3.35 were in fact grouped into the class identified by the *class mark* 3.3. Our *class interval* was 0.1 units. If we now wish to make wider class intervals, we are doing nothing but extending the range within which measurements are placed into one class.

Reference to Box 2.1 will make this process clear. We group the data twice in order to impress upon the reader the flexibility of the process. In the first example of grouping, the class interval has been doubled in width; that is, it has been made to equal 0.2 units. If we start at the lower end, the implied class limits will now be from 3.25 to 3.45, the limits for the next class from 3.45 to 3.65, and so forth.

Our next task is to find the class marks. This was quite simple in the frequency distribution shown at the left side of Box 2.1, in which the original measurements were used as class marks. However, now we are using a class interval twice as wide as before, and the class marks are calculated by taking the midpoint of the new class intervals. Thus, to find the class mark of the first class, we take the midpoint between 3.25 and 3.45, which turns out to be 3.35. We note that the class mark has one more decimal place than the original measurements. We should not now be led to believe that we have suddenly achieved greater precision. Whenever we designate a class interval whose last *significant* digit is even (0.2 in this case), the class mark will carry one more decimal place than the original measurements. On the right side of the table in Box 2.1 the data are grouped once again, using a class interval of 0.3. Because of the odd last significant digit, the class mark now shows as many decimal places as the original variates, the midpoint between 3.25 and 3.55 being 3.4.

Once the implied class limits and the class mark for the first class have been correctly found, the others can be written down by inspection without any special computation. Simply add the class interval repeatedly to each of the values. Thus, starting with the lower limit 3.25, by adding 0.2 we obtain 3.45, 3.65, 3.85, and so forth; similarly, for the class marks, we obtain 3.35, 3.55, 3.75, and so forth. It should be obvious that the wider the class intervals, the more compact the data become but also the less precise. However, looking at

20

BOX 2.1

Preparation of frequency distribution and grouping into fewer classes with wider class intervals.

Twenty-five femur lengths of the aphid *Pemphigus*. Measurements are in mm $\times 10^{-1}$.

Original measurements

3.8	3.6	3.5	4.3
3.3	4.3	3.9	3.8
3.9	4.4	3.8	3.6
4.1	4.4	4.7	3.8
4.4	4.1	3.6	3.9
		4.5	
		3.6	
		4.2	

Original frequency distribution

Implied limits	Y	Tally marks	f
3.25–3.35	3.3	|	1
3.35–3.45	3.4		0
3.45–3.55	3.5	|	1
3.55–3.65	3.6	||||	4
3.65–3.75	3.7		0
3.75–3.85	3.8	||||	4
3.85–3.95	3.9	|||	3
3.95–4.05	4.0		0
4.05–4.15	4.1	||	2
4.15–4.25	4.2	|	1

Grouping into 8 classes of interval 0.2

Implied limits	Class mark	Tally marks	f
3.25–3.45	3.35	|	1
3.45–3.65	3.55	|||||	5
3.65–3.85	3.75	||||	4
3.85–4.05	3.95	|||	3
4.05–4.25	4.15	|||	3

Grouping into 5 classes of interval 0.3

Implied limits	Class mark	Tally marks	f
3.25–3.55	3.4	||	2
3.55–3.85	3.7	||||\ |||	8
3.85–4.15	4.0	||||	5
4.15–4.45	4.3	||||\ |||	8

4.25–4.35	4.3	||||	4	4.25–4.45	4.35	⫫||	7
4.35–4.45	4.4	|||	3				
4.45–4.55	4.5	|	1	4.45–4.65	4.55	|	1	4.45–4.75	4.6	||	2
4.55–4.65	4.6		0				
4.65–4.75	4.7	|	1	4.65–4.85	4.75	|	1
$\sum f$			$\overline{25}$				$\overline{25}$				$\overline{25}$

Source: Data from R. R. Sokal.

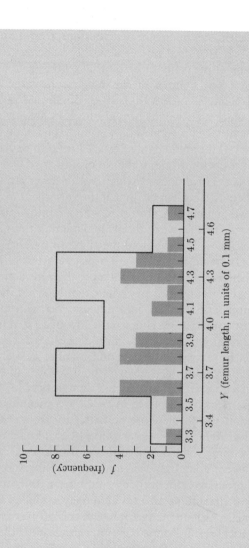

Histogram of the original frequency distribution shown above and of the grouped distribution with 5 classes. Line below abscissa shows class marks for the grouped frequency distribution. Shaded bars represent original frequency distribution; hollow bars represent grouped distribution.

For a detailed account of the process of grouping, see Section 2.5.

the frequency distribution of aphid femur lengths in Box 2.1, we notice that the initial rather chaotic structure is being simplified by grouping. When we group the frequency distribution into five classes with a class interval of 0.3 units, it becomes notably bimodal (that is, it possesses two peaks of frequencies).

In setting up frequency distributions, from 12 to 20 classes should be established. This rule need not be slavishly adhered to, but it should be employed with some of the common sense that comes from experience in handling statistical data. The number of classes depends largely on the size of the sample studied. Samples of less than 40 or 50 should rarely be given as many as 12 classes, since that would provide too few frequencies per class. On the other hand, samples of several thousand may profitably be grouped into more than 20 classes. If the aphid data of Box 2.1 need to be grouped, they should probably not be grouped into more than 6 classes.

If the original data provide us with fewer classes than we think we should have, then nothing can be done if the variable is meristic, since this is the nature of the data in question. However, with a continuous variable a scarcity of classes would indicate that we probably had not made our measurements with sufficient precision. If we had followed the rules on number of significant digits for measurements stated in Section 2.3, this could not have happened.

Whenever we come up with more than the desired number of classes, grouping should be undertaken. When the data are meristic, the implied limits of continuous variables are meaningless. Yet with many meristic variables, such as a bristle number varying from a low of 13 to a high of 81, it would probably be wise to group the variates into classes, each containing several counts. This can best be done by using an odd number as a class interval so that the class mark representing the data will be a whole rather than a fractional number. Thus, if we were to group the bristle numbers 13, 14, 15, and 16 into one class, the class mark would have to be 14.5, a meaningless value in terms of bristle number. It would therefore be better to use a class ranging over 3 bristles or 5 bristles, giving the integral value 14 or 15 as a class mark.

Grouping data into frequency distributions was necessary when computations were done by pencil and paper. Nowadays even thousands of variates can be processed efficiently by computer without prior grouping. However, frequency distributions are still extremely useful as a tool for data analysis. This is especially true in an age in which it is all too easy for a researcher to obtain a numerical result from a computer program without ever really examining the data for outliers or for other ways in which the sample may not conform to the assumptions of the statistical methods.

Rather than using tally marks to set up a frequency distribution, as was done in Box 2.1, we can employ Tukey's *stem-and-leaf display*. This technique is an improvement, since it not only results in a frequency distribution of the variates of a sample but also permits easy checking of the variates and ordering them into an array (neither of which is possible with tally marks). This technique will therefore be useful in computing the median of a sample (see Section 3.3) and in computing various tests that require ordered arrays of the sample variates (see Sections 10.3 and 12.5).

To learn how to construct a stem-and-leaf display, let us look ahead to Table 3.1 in the next chapter, which lists 15 blood neutrophil counts. The unordered measurements are as follows: 4.9, 4.6, 5.5, 9.1, 16.3, 12.7, 6.4, 7.1, 2.3, 3.6, 18.0, 3.7, 7.3, 4.4, and 9.8. To prepare a stem-and-leaf display, we scan the variates in the sample to discover the lowest and highest leading digit or digits. Next, we write down the entire range of leading digits in unit increments to the left of a vertical line (the "stem"), as shown in the accompanying illustration. We then put the next digit of the first variate (a "leaf") at that level of the stem corresponding to its leading digit(s). The first observation in our sample is 4.9. We therefore place a 9 next to the 4. The next variate is 4.6. It is entered by finding the stem level for the leading digit 4 and recording a 6 next to the 9 that is already there. Similarly, for the third variate, 5.5, we record a 5 next to the leading digit 5. We continue in this way until all 15 variates have been entered (as "leaves") in sequence along the appropriate leading digits of the stem. The completed array is the equivalent of a frequency distribution and has the appearance of a histogram or bar diagram (see the illustration). Moreover, it permits the efficient ordering of the variates. Thus, from the completed array it becomes obvious that the appropriate ordering of the 15 variates is 2.3, 3.6, 3.7, 4.4, 4.6, 4.9, 5.5, 6.4, 7.1, 7.3, 9.1, 9.8, 12.7, 16.3, 18.0. The median can easily be read off the stem-and-leaf display. It is clearly 6.4. For very large samples, stem-and-leaf displays may become awkward. In such cases a conventional frequency distribution as in Box 2.1 would be preferable.

Step 1		Step 2		...	Step 7		...	Completed array (Step 15)	
2		2			2			2	3
3		3			3			3	67
4	9	4	96		4	96		4	964
5		5			5	5		5	5
6		6			6	4		6	4
7		7			7			7	13
8		8			8			8	
9		9			9	1		9	18
10		10			10			10	
11		11			11			11	
12		12			12	7		12	7
13		13			13			13	
14		14			14			14	
15		15			15			15	
16		16			16	3		16	3
17		17			17			17	
18		18			18			18	0

When the shape of a frequency distribution is of particular interest, we may wish to present the distribution in graphic form when discussing the results. This is generally done by means of frequency diagrams, of which there are two common types. For a distribution of meristic data we employ a *bar diagram*, as shown in Figure 2.2 for the sedge data of Table 2.2. The abscissa represents

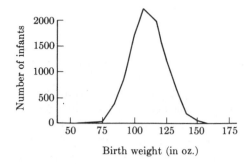

FIGURE 2.3
Frequency polygon. Birth weights of 9465
males infants. Chinese third-class patients in
Singapore, 1950 and 1951. Data from Millis
and Seng (1954).

the variable (in our case, the number of plants per quadrat), and the ordinate
represents the frequencies. The important point about such a diagram is that
the bars do not touch each other, which indicates that the variable is not continuous. By contrast, continuous variables, such as the frequency distribution
of the femur lengths of aphid stem mothers, are graphed as a *histogram*. In a
histogram the width of each bar along the abscissa represents a class interval
of the frequency distribution and the bars touch each other to show that the
actual limits of the classes are contiguous. The midpoint of the bar corresponds
to the class mark. At the bottom of Box 2.1 are shown histograms of the frequency distribution of the aphid data, ungrouped and grouped. The height of
each bar represents the frequency of the corresponding class.

To illustrate that histograms are appropriate approximations to the continuous distributions found in nature, we may take a histogram and make the
class intervals more narrow, producing more classes. The histogram would then
clearly have a closer fit to a continuous distribution. We can continue this process until the class intervals become infinitesimal in width. At this point the
histogram becomes the continuous distribution of the variable.

Occasionally the class intervals of a grouped continuous frequency distribution are unequal. For instance, in a frequency distribution of ages we might
have more detail on the different ages of young individuals and less accurate
identification of the ages of old individuals. In such cases, the class intervals
for the older age groups would be wider, those for the younger age groups, narrower. In representing such data, the bars of the histogram are drawn with
different widths.

Figure 2.3 shows another graphical mode of representation of a frequency
distribution of a continuous variable (in this case, birth weight in infants). As
we shall see later the shapes of distributions seen in such frequency polygons
can reveal much about the biological situations affecting the given variable.

2.6 The handling of data

Data must be handled skillfully and expeditiously so that statistics can be prac-
ticed successfully. Readers should therefore acquaint themselves with the various techniques available for carrying out statistical computations.

In this book we ignore "pencil-and-paper" short-cut methods for computations, found in earlier textbooks of statistics, since we assume that the student has access to a calculator or a computer. Some statistical methods are very easy to use because special tables exist that provide answers for standard statistical problems; thus, almost no computation is involved. An example is Finney's table, a 2-by-2 contingency table containing small frequencies that is used for the test of independence (Pearson and Hartley, 1958, Table 38). For small problems, Finney's table can be used in place of Fisher's method of finding exact probabilities, which is very tedious. Other statistical techniques are so easy to carry out that no mechanical aids are needed. Some are inherently simple, such as the sign test (Section 10.3). Other methods are only approximate but can often serve the purpose adequately; for example, we may sometimes substitute an easy-to-evaluate median (defined in Section 3.3) for the mean (described in Sections 3.1 and 3.2) which requires computation.

We can use many new types of equipment to perform statistical computations—many more than we could have when *Introduction to Biostatistics* was first published. The once-standard electrically driven mechanical desk calculator has completely disappeared. Many new electronic devices, from small pocket calculators to larger desk-top computers, have replaced it. Such devices are so diverse that we will not try to survey the field here. Even if we did, the rate of advance in this area would be so rapid that whatever we might say would soon become obsolete.

We cannot really draw the line between the more sophisticated electronic calculators, on the one hand, and digital computers. There is no abrupt increase in capabilities between the more versatile programmable calculators and the simpler microcomputers, just as there is none as we progress from microcomputers to minicomputers and so on up to the large computers that one associates with the central computation center of a large university or research laboratory. All can perform computations automatically and be controlled by a set of detailed instructions prepared by the user. Most of these devices, including programmable small calculators, are adequate for all of the computations described in this book, even for large sets of data.

The material in this book consists of relatively standard statistical computations, most of which have probably already been programmed in any given computer installation. "BIOM," a package of FORTRAN programs covering the topics in this book, is available at nominal cost.*

The use of modern data processing procedures has one inherent danger. One can all too easily either feed in erroneous data or choose an inappropriate program. Users must select programs carefully to ensure that those programs perform the desired computations, give numerically reliable results, and are as free from error as possible. When using a program for the first time, one should test it using data from textbooks with which one is familiar. Some programs

---

* For information or to order, write to Applied Biostatistics, Inc., 3 Heritage Lane, Setauket, NY 11733. These programs have been used on the IBM PC and compatible microcomputers as well as on a variety of mainframe computers with few, if any, changes needed.

are notorious because the programmer has failed to guard against excessive rounding errors or other problems. Users of a program should carefully check the data being analyzed so that typing errors are not present. In addition, programs should help users identify and remove bad data values and should provide them with transformations so that they can make sure that their data satisfy the assumptions of various analyses.

Exercises

2.1 Round the following numbers to three significant figures: 106.55, 0.06819, 3.0495, 7815.01, 2.9149, and 20.1500. What are the implied limits before and after rounding? Round these same numbers to one decimal place.
ANS. For the first value: 107; 106.545–106.555; 106.5–107.5; 106.6

2.2 Differentiate between the following pairs of terms and give an example of each. (a) Statistical and biological populations. (b) Variate and individual. (c) Accuracy and precision (repeatability). (d) Class interval and class mark. (e) Bar diagram and histogram. (f) Abscissa and ordinate.

2.3 Given 200 measurements ranging from 1.32 to 2.95 mm, how would you group them into a frequency distribution? Give class limits as well as class marks.

2.4 Group the following 40 measurements of interorbital width of a sample of domestic pigeons into a frequency distribution and draw its histogram (data from Olson and Miller, 1958). Measurements are in millimeters.

12.2	12.9	11.8	11.9	11.6	11.1	12.3	12.2	11.8	11.8
10.7	11.5	11.3	11.2	11.6	11.9	13.3	11.2	10.5	11.1
12.1	11.9	10.4	10.7	10.8	11.0	11.9	10.2	10.9	11.6
10.8	11.6	10.4	10.7	12.0	12.4	11.7	11.8	11.3	11.1

2.5 How precisely should you measure the wing length of a species of mosquitoes in a study of geographic variation if the smallest specimen has a length of about 2.8 mm and the largest a length of about 3.5 mm?

2.6 Transform the 40 measurements in Exercise 2.4 into common logarithms (use a table or calculator) and make a frequency distribution of these transformed variates. Comment on the resulting change in the pattern of the frequency distribution from that found before.

2.7 For the data of Tables 2.1 and 2.2 identify the individual observations, samples, populations, and variables.

2.8 Make a stem-and-leaf display of the data given in Exercise 2.4.

2.9 The distribution of ages of striped bass captured by hook and line from the East River and the Hudson River during 1980 were reported as follows (Young, 1981):

Age	f
1	13
2	49
3	96
4	28
5	16
6	8

Show this distribution in the form of a bar diagram.

CHAPTER 3

Descriptive Statistics

An early and fundamental stage in any science is the descriptive stage. Until phenomena can be accurately described, an analysis of their causes is premature. The question "What?" comes before "How?" Unless we know something about the usual distribution of the sugar content of blood in a population of guinea pigs, as well as its fluctuations from day to day and within days, we shall be unable to ascertain the effect of a given dose of a drug upon this variable. In a sizable sample it would be tedious to obtain our knowledge of the material by contemplating each individual observation. We need some form of summary to permit us to deal with the data in manageable form, as well as to be able to share our findings with others in scientific talks and publications. A histogram or bar diagram of the frequency distribution would be one type of summary. However, for most purposes, a numerical summary is needed to describe concisely, yet accurately, the properties of the observed frequency distribution. Quantities providing such a summary are called *descriptive statistics*. This chapter will introduce you to some of them and show how they are computed.

Two kinds of descriptive statistics will be discussed in this chapter: statistics of location and statistics of dispersion. The *statistics of location* (also known as *measures of central tendency*) describe the *position of a sample along a given dimension representing a variable.* For example, after we measure the length of the animals within a sample, we will then want to know whether the animals are closer, say, to 2 cm or to 20 cm. To express a representative value for the sample of observations—for the length of the animals—we use a statistic of location. But statistics of location will not describe the shape of a frequency distribution. The shape may be long or very narrow, may be humped or U-shaped, may contain two humps, or may be markedly asymmetrical. Quantitative measures of such aspects of frequency distributions are required. To this end we need to define and study the *statistics of dispersion.*

The arithmetic mean, described in Section 3.1, is undoubtedly the most important single statistic of location, but others (the geometric mean, the harmonic mean, the median, and the mode) are briefly mentioned in Sections 3.2, 3.3, and 3.4. A simple statistic of dispersion (the range) is briefly noted in Section 3.5, and the standard deviation, the most common statistic for describing dispersion, is explained in Section 3.6. Our first encounter with contrasts between sample statistics and population parameters occurs in Section 3.7, in connection with statistics of location and dispersion. In Section 3.8 there is a description of practical methods for computing the mean and standard deviation. The coefficient of variation (a statistic that permits us to compare the relative amount of dispersion in different samples) is explained in the last section (Section 3.9).

The techniques that will be at your disposal after you have mastered this chapter will not be very powerful in solving biological problems, but they will be indispensable tools for any further work in biostatistics. Other descriptive statistics, of both location and dispersion, will be taken up in later chapters.

An important note: We shall first encounter the use of logarithms in this chapter. To avoid confusion, common logarithms have been consistently abbreviated as log, and natural logarithms as ln. Thus, log x means $\log_{10} x$ and ln x means $\log_e x$.

3.1 The arithmetic mean

The most common statistic of location is familiar to everyone. It is the *arithmetic mean,* commonly called the *mean* or *average.* The mean is calculated by summing all the individual observations or items of a sample and dividing this sum by the number of items in the sample. For instance, as the result of a gas analysis in a respirometer an investigator obtains the following four readings of oxygen percentages and sums them:

$$
\begin{array}{r}
14.9 \\
10.8 \\
12.3 \\
23.3 \\
\hline
\text{Sum} = 61.3
\end{array}
$$

The investigator calculates the mean oxygen percentage as the sum of the four items divided by the number of items. Thus the average oxygen percentage is

$$\text{Mean} = \frac{61.3}{4} = 15.325\%$$

Calculating a mean presents us with the opportunity for learning statistical symbolism. We have already seen (Section 2.2) that an individual observation is symbolized by Y_i, which stands for the ith observation in the sample. Four observations could be written symbolically as follows:

$$Y_1, Y_2, Y_3, Y_4$$

We shall define n, the *sample size*, as the number of items in a sample. In this particular instance, the sample size n is 4. Thus, in a large sample, we can symbolize the array from the first to the nth item as follows:

$$Y_1, Y_2, \ldots, Y_n$$

When we wish to sum items, we use the following notation:

$$\sum_{i=1}^{i=n} Y_i = Y_1 + Y_2 + \cdots + Y_n$$

The capital Greek sigma, Σ, simply means the sum of the items indicated. The $i = 1$ means that the items should be summed, starting with the first one and ending with the nth one, as indicated by the $i = n$ above the Σ. The subscript and superscript are necessary to indicate how many items should be summed. The "$i =$ " in the superscript is usually omitted as superfluous. For instance, if we had wished to sum only the first three items, we would have written $\Sigma_{i=1}^3 Y_i$. On the other hand, had we wished to sum all of them except the first one, we would have written $\Sigma_{i=2}^n Y_i$. With some exceptions (which will appear in later chapters), it is desirable to omit subscripts and superscripts, which generally add to the apparent complexity of the formula and, when they are unnecessary, distract the student's attention from the important relations expressed by the formula. Below are seen increasing simplifications of the complete summation notation shown at the extreme left:

$$\sum_{i=1}^{i=n} Y_i = \sum_{i=1}^{n} Y_i = \sum_{i} Y_i = \sum^{n} Y = \sum Y$$

The third symbol might be interpreted as meaning, "Sum the Y_i's over all available values of i." This is a frequently used notation, although we shall not employ it in this book. The next, with n as a superscript, tells us to sum n items of Y; note that the i subscript of the Y has been dropped as unnecessary. Finally, the simplest notation is shown at the right. It merely says sum the Y's. This will be the form we shall use most frequently: if a summation sign precedes a variable, the summation will be understood to be over n items (all the items in the sample) unless subscripts or superscripts specifically tell us otherwise.

We shall use the symbol $\bar{Y}$ for the arithmetic mean of the variable Y. Its formula is written as follows:

$$\bar{Y} = \frac{\sum Y}{n} = \frac{1}{n}\sum Y \tag{3.1}$$

This formula tells us, "Sum all the (n) items and divide the sum by n."

The *mean of a sample* represents *the center of the observations in the sample.* If you were to draw a histogram of an observed frequency distribution on a sheet of cardboard and then cut out the histogram and lay it flat against a blackboard, supporting it with a pencil beneath, chances are that it would be out of balance, toppling to either the left or the right. If you moved the supporting pencil point to a position about which the histogram would exactly balance, this point of balance would correspond to the arithmetic mean.

We often must compute averages of means or of other statistics that may differ in their reliabilities because they are based on different sample sizes. At other times we may wish the individual items to be averaged to have different weights or amounts of influence. In all such cases we compute a *weighted average*. A general formula for calculating the weighted average of a set of values Y_i is as follows:

$$\bar{Y}_w = \frac{\sum\limits_{i}^{n} w_i Y_i}{\sum\limits^{n} w_i} \tag{3.2}$$

where n variates, each weighted by a factor w_i, are being averaged. The values of Y_i in such cases are unlikely to represent variates. They are more likely to be sample means $\bar{Y}_i$ or some other statistics of different reliabilities.

The simplest case in which this arises is when the Y_i are not individual variates but are means. Thus, if the following three means are based on differing sample sizes, as shown,

$\bar{Y}_i$	n_i
3.85	12
5.21	25
4.70	8

their weighted average will be

$$\bar{Y}_w = \frac{(12)(3.85) + (25)(5.21) + (8)(4.70)}{12 + 25 + 8} = \frac{214.05}{45} = 4.76$$

Note that in this example, computation of the weighted mean is exactly equivalent to adding up all the original measurements and dividing the sum by the total number of the measurements. Thus, the sample with 25 observations, having the highest mean, will influence the weighted average in proportion to its size.

3.2 Other means

We shall see in Chapters 10 and 11 that variables are sometimes transformed into their logarithms or reciprocals. If we calculate the means of such transformed variables and then change the means back into the original scale, these means will not be the same as if we had computed the arithmetic means of the original variables. The resulting means have received special names in statistics. The back-transformed mean of the logarithmically transformed variables is called the *geometric mean*. It is computed as follows:

$$GM_Y = \text{antilog} \frac{1}{n} \sum \log Y \tag{3.3}$$

which indicates that the geometric mean GM_Y is the antilogarithm of the mean of the logarithms of variable Y. Since addition of logarithms is equivalent to multiplication of their antilogarithms, there is another way of representing this quantity; it is

$$GM_Y = \sqrt[n]{Y_1 Y_2 Y_3 \cdots Y_n} \tag{3.4}$$

The geometric mean permits us to become familiar with another operator symbol: capital pi, Π, which may be read as "product." Just as Σ symbolizes summation of the items that follow it, so Π symbolizes the multiplication of the items that follow it. The subscripts and superscripts have exactly the same meaning as in the summation case. Thus, Expression (3.4) for the geometric mean can be rewritten more compactly as follows:

$$GM_Y = \sqrt[n]{\prod_{i=1}^{n} Y_i} \tag{3.4a}$$

The computation of the geometric mean by Expression (3.4a) is quite tedious. In practice, the geometric mean has to be computed by transforming the variates into logarithms.

The reciprocal of the arithmetic mean of reciprocals is called the *harmonic mean*. If we symbolize it by H_Y, the formula for the harmonic mean can be written in concise form (without subscripts and superscripts) as

$$\frac{1}{H_Y} = \frac{1}{n} \sum \frac{1}{Y} \tag{3.5}$$

You may wish to convince yourself that the geometric mean and the harmonic mean of the four oxygen percentages are 14.65% and 14.09%, respectively. Unless the individual items do not vary, the geometric mean is always less than the arithmetic mean, and the harmonic mean is always less than the geometric mean.

Some beginners in statistics have difficulty in accepting the fact that measures of location or central tendency other than the arithmetic mean are permissible or even desirable. They feel that the arithmetic mean is the "logical" average, and that any other mean would be a distortion. This whole problem relates to the proper scale of measurement for representing data; this scale is not always the linear scale familiar to everyone, but is sometimes by preference a logarithmic or reciprocal scale. If you have doubts about this question, we shall try to allay them in Chapter 10, where we discuss the reasons for transforming variables.

3.3 The median

The *median M* is a statistic of location occasionally useful in biological research. It is defined as *that value of the variable* (in an ordered array) *that has an equal number of items on either side of it*. Thus, the median divides a frequency distribution into two halves. In the following sample of five measurements,

$$14, 15, 16, 19, 23$$

$M = 16$, since the third observation has an equal number of observations on both sides of it. We can visualize the median easily if we think of an array from largest to smallest—for example, a row of men lined up by their heights. The median individual will then be that man having an equal number of men on his right and left sides. His height will be the median height of the sample considered. This quantity is easily evaluated from a sample array with an odd number of individuals. When the number in the sample is even, the median is conventionally calculated as the midpoint between the $(n/2)$th and the $[(n/2) + 1]$th variate. Thus, for the sample of four measurements

$$14, 15, 16, 19$$

the median would be the midpoint between the second and third items, or 15.5.

Whenever any one value of a variate occurs more than once, problems may develop in locating the median. Computation of the median item becomes more involved because all the members of a given class in which the median item is located will have the same class mark. The median then is the $(n/2)$th variate in the frequency distribution. It is usually computed as that point between the class limits of the median class where the median individual would be located (assuming the individuals in the class were evenly distributed).

The median is just one of a family of statistics dividing a frequency distribution into equal areas. It divides the distribution into two halves. The three *quartiles* cut the distribution at the 25, 50, and 75% points—that is, at points dividing the distribution into first, second, third, and fourth quarters by area (and frequencies). The second quartile is, of course, the median. (There are also quintiles, deciles, and percentiles, dividing the distribution into 5, 10, and 100 equal portions, respectively.)

Medians are most often used for distributions that do not conform to the standard probability models, so that nonparametric methods (see Chapter 10) must be used. Sometimes the median is a more representative measure of location than the arithmetic mean. Such instances almost always involve asymmetric distributions. An often quoted example from economics would be a suitable measure of location for the "typical" salary of an employee of a corporation. The very high salaries of the few senior executives would shift the arithmetic mean, the center of gravity, toward a completely unrepresentative value. The median, on the other hand, would be little affected by a few high salaries; it would give the particular point on the salary scale above which lie 50% of the salaries in the corporation, the other half being lower than this figure.

In biology an example of the preferred application of a median over the arithmetic mean may be in populations showing skewed distribution, such as weights. Thus a median weight of American males 50 years old may be a more meaningful statistic than the average weight. The median is also of importance in cases where it may be difficult or impossible to obtain and measure all the items of a sample. For example, suppose an animal behaviorist is studying the time it takes for a sample of animals to perform a certain behavioral step. The variable he is measuring is the time from the beginning of the experiment until each individual has performed. What he wants to obtain is an average time of performance. Such an average time, however, can be calculated only after records have been obtained on all the individuals. It may take a long time for the slowest animals to complete their performance, longer than the observer wishes to spend. (Some of them may never respond appropriately, making the computation of a mean impossible.) Therefore, a convenient statistic of location to describe these animals may be the median time of performance. Thus, so long as the observer knows what the total sample size is, he need not have measurements for the right-hand tail of his distribution. Similar examples would be the responses to a drug or poison in a group of individuals (the median lethal or effective dose, LD_{50} or ED_{50}) or the median time for a mutation to appear in a number of lines of a species.

3.4 The mode

The *mode* refers to *the value represented by the greatest number of individuals.* When seen on a frequency distribution, the mode is the value of the variable at which the curve peaks. In grouped frequency distributions the mode as a point has little meaning. It usually suffices to identify the modal class. In biology, the mode does not have many applications.

Distributions having two peaks (equal or unequal in height) are called *bimodal*; those with more than two peaks are *multimodal*. In those rare distributions that are **U**-shaped, we refer to the low point at the middle of the distribution as an *antimode*.

In evaluating the relative merits of the arithmetic mean, the median, and the mode, a number of considerations have to be kept in mind. The mean is generally preferred in statistics, since it has a smaller standard error than other statistics of location (see Section 6.2), it is easier to work with mathematically, and it has an additional desirable property (explained in Section 6.1): it will tend to be normally distributed even if the original data are not. The mean is

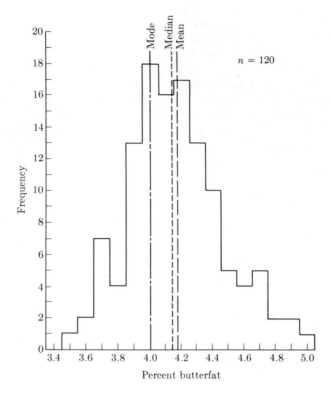

FIGURE 3.1

An asymmetrical frequency distribution (skewed to the right) showing location of the mean, median, and mode. Percent butterfat in 120 samples of milk (from a Canadian cattle breeders' record book).

markedly affected by outlying observations; the median and mode are not. The mean is generally more sensitive to changes in the shape of a frequency distribution, and if it is desired to have a statistic reflecting such changes, the mean may be preferred.

In symmetrical, unimodal distributions the mean, the median, and the mode are all identical. A prime example of this is the well-known normal distribution of Chapter 5. In a typical asymmetrical distribution, such as the one shown in Figure 3.1, the relative positions of the mode, median, and mean are generally these: the mean is closest to the drawn-out tail of the distribution, the mode is farthest, and the median is between these. An easy way to remember this sequence is to recall that they occur in alphabetical order from the longer tail of the distribution.

3.5 The range

We now turn to measures of dispersion. Figure 3.2 demonstrates that radically different-looking distributions may possess the identical arithmetic mean. It is therefore obvious that other ways of characterizing distributions must be found.

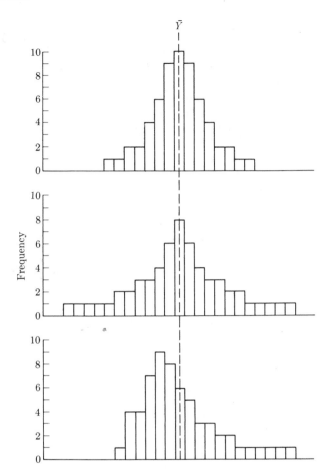

FIGURE 3.2
Three frequency distributions having identical means and sample sizes but differing in dispersion pattern.

One simple measure of dispersion is the *range*, which is defined as *the difference between the largest and the smallest items in a sample.* Thus, the range of the four oxygen percentages listed earlier (Section 3.1) is

$$\text{Range} = 23.3 - 10.8 = 12.5\%$$

and the range of the aphid femur lengths (Box 2.1) is

$$\text{Range} = 4.7 - 3.3 = 1.4 \text{ units of } 0.1 \text{ mm}$$

Since the range is a measure of the span of the variates along the scale of the variable, it is in the same units as the original measurements. The range is clearly affected by even a single outlying value and for this reason is only a rough estimate of the dispersion of all the items in the sample.

3.6 The standard deviation

We desire that a measure of dispersion take all items of a distribution into consideration, weighting each item by its distance from the center of the distribution. We shall now try to construct such a statistic. In Table 3.1 we show a sample of 15 blood neutrophil counts from patients with tumors. Column (1) shows the variates in the order in which they were reported. The computation of the mean is shown below the table. The mean neutrophil count turns out to be 7.713.

The distance of each variate from the mean is computed as the following deviation:

$$y = Y - \bar{Y}$$

Each individual deviation, or *deviate*, is by convention computed as the individual observation minus the mean, $Y - \bar{Y}$, rather than the reverse, $\bar{Y} - Y$. Deviates are symbolized by lowercase letters corresponding to the capital letters of the variables. Column (2) in Table 3.1 gives the deviates computed in this manner.

We now wish to calculate an average deviation that will sum all the deviates and divide them by the number of deviates in the sample. But note that when

TABLE **3.1**

The standard deviation. Long method, not recommended for hand or calculator computations but shown here to illustrate the meaning of the standard deviation. The data are blood neutrophil counts (divided by 1000) per microliter, in 15 patients with nonhematological tumors.

	(1) Y	(2) $y = Y - \bar{Y}$	(3) y^2
	4.9	−2.81	7.9148
	4.6	−3.11	9.6928
	5.5	−2.21	4.8988
	9.1	1.39	1.9228
	16.3	8.59	73.7308
	12.7	4.99	24.8668
	6.4	−1.31	1.7248
	7.1	−0.61	0.3762
	2.3	−5.41	29.3042
	3.6	−4.11	16.9195
	18.0	10.29	105.8155
	3.7	−4.01	16.1068
	7.3	−0.41	0.1708
	4.4	−3.31	10.9782
	9.8	2.09	4.3542
Total	115.7	0.05	308.7770

$$\text{Mean} \quad \bar{Y} = \frac{\sum Y}{n} = \frac{115.7}{15} = 7.713$$

Source: Liu et al. (1983)

we sum our deviates, negative and positive deviates cancel out, as is shown by the sum at the bottom of column (2); this sum appears to be unequal to zero only because of a rounding error. Deviations from the arithmetic mean always sum to zero because the mean is the center of gravity. Consequently, an average based on the sum of deviations would also always equal zero. You are urged to study Appendix A1.1, which demonstrates that the sum of deviations around the mean of a sample is equal to zero.

Squaring the deviates gives us column (3) of Table 3.1 and enables us to reach a result other than zero. (Squaring the deviates also holds other mathematical advantages, which we shall take up in Sections 7.5 and 11.3.) The sum of the squared deviates (in this case, 308.7770) is a very important quantity in statistics. It is called the *sum of squares* and is identified symbolically as Σy^2. Another common symbol for the sum of squares is *SS*.

The next step is to obtain the average of the n squared deviations. The resulting quantity is known as the *variance,* or the *mean square*:

$$\text{Variance} = \frac{\Sigma y^2}{n} = \frac{308.7770}{15} = 20.5851$$

The variance is a measure of fundamental importance in statistics, and we shall employ it throughout this book. At the moment, we need only remember that because of the squaring of the deviations, the variance is expressed in squared units. To undo the effect of the squaring, we now take the positive square root of the variance and obtain the *standard deviation*:

$$\text{Standard deviation} = +\sqrt{\frac{\Sigma y^2}{n}} = 4.5371$$

Thus, standard deviation is again expressed in the original units of measurement, since it is a square root of the squared units of the variance.

An important note: The technique just learned and illustrated in Table 3.1 is not the simplest for direct computation of a variance and standard deviation. However, it is often used in computer programs, where accuracy of computations is an important consideration. Alternative and simpler computational methods are given in Section 3.8.

The observant reader may have noticed that we have avoided assigning any symbol to either the variance or the standard deviation. We shall explain why in the next section.

3.7 Sample statistics and parameters

Up to now we have calculated statistics from samples without giving too much thought to what these statistics represent. When correctly calculated, a mean and standard deviation will always be absolutely true measures of location and dispersion for the samples on which they are based. Thus, the true mean of the four oxygen percentage readings in Section 3.1 is 15.325%. The standard deviation of the 15 neutrophil counts is 4.537. However, only rarely in biology (or in statistics in general) are we interested in measures of location and dispersion only as descriptive summaries of the samples we have studied. Almost always we are interested in the *populations* from which the samples have been taken. What we want to know is not the mean of the particular four oxygen precentages, but rather the true oxgyen percentage of the universe of readings from which the four readings have been sampled. Similarly, we would like to know the true mean neutrophil count of the population of patients with nonhematological tumors, not merely the mean of the 15 individuals measured. When studying dispersion we generally wish to learn the true standard deviations of the populations and not those of the samples. These population statistics, however, are unknown and (generally speaking) are unknowable. Who would be able to collect all the patients with this particular disease and measure their neutrophil counts? Thus we need to use *sample statistics* as estimators of *population statistics* or *parameters*.

It is conventional in statistics to use Greek letters for population parameters and Roman letters for sample statistics. Thus, the sample mean $\bar{Y}$ estimates μ, the parametric mean of the population. Similarly, a sample variance, symbolized by s^2, estimates a parametric variance, symbolized by σ^2. Such estimators should be *unbiased*. By this we mean that samples (regardless of the sample size) taken from a population with a known parameter should give sample statistics that, when averaged, will give the parametric value. An estimator that does not do so is called *biased*.

The sample mean $\bar{Y}$ is an unbiased estimator of the parametric mean μ. However, the sample variance as computed in Section 3.6 is not unbiased. On the average, it will underestimate the magnitude of the population variance σ^2. To overcome this bias, mathematical statisticians have shown that when sums of squares are divided by $n - 1$ rather than by n the resulting sample variances will be unbiased estimators of the population variance. For this reason, it is customary to compute variances by dividing the sum of squares by $n - 1$. The formula for the standard deviation is therefore customarily given as follows:

$$s = + \sqrt{\frac{\sum y^2}{n - 1}} \tag{3.6}$$

In the neutrophil-count data the standard deviation would thus be computed as

$$s = \sqrt{\frac{308.7770}{14}} = 4.6963$$

We note that this value is slightly larger than our previous estimate of 4.537. Of course, the greater the sample size, the less difference there will be between division by n and by $n - 1$. However, regardless of sample size, it is good practice to divide a sum of squares by $n - 1$ when computing a variance or standard deviation. It may be assumed that when the symbol s^2 is encountered, it refers to a variance obtained by division of the sum of squares by the *degrees of freedom*, as the quantity $n - 1$ is generally referred to.

Division of the sum of squares by n is appropriate only when the interest of the investigator is limited to the sample at hand and to its variance and standard deviation as descriptive statistics of the *sample*. This would be in contrast to using these as estimates of the population parameters. There are also the rare cases in which the investigator possesses data on the entire population; in such cases division by n is perfectly justified, because then the investigator is not estimating a parameter but is in fact evaluating it. Thus the variance of the wing lengths of all adult whooping cranes would be a parametric value; similarly, if the heights of all winners of the Nobel Prize in physics had been measured, their variance would be a parameter since it would be based on the entire population.

3.8 Practical methods for computing mean and standard deviation

Three steps are necessary for computing the standard deviation: (1) find Σy^2, the sum of squares; (2) divide by $n - 1$ to give the variance; and (3) take the square root of the variance to obtain the standard deviation. The procedure used to compute the sum of squares in Section 3.6 can be expressed by the following formula:

$$\Sigma y^2 = \Sigma (Y - \bar{Y})^2 \tag{3.7}$$

This formulation explains most clearly the meaning of the sum of squares, although it may be inconvenient for computation by hand or calculator, since one must first compute the mean before one can square and sum the deviations. A quicker computational formula for this quantity is

$$\Sigma y^2 = Y^2 = \Sigma Y^2 - \frac{(\Sigma Y)^2}{n} \tag{3.8}$$

Let us see exactly what this formula represents. The first term on the right side of the equation, ΣY^2, is the sum of all individual Y's, each squared, as follows:

$$\Sigma Y^2 = Y_1^2 + Y_2^2 + Y_3^2 + \cdots + Y_n^2$$

When referred to by name, ΣY^2 should be called the "sum of Y squared" and should be carefully distinguished from Σy^2, "the sum of squares of Y." These names are unfortunate, but they are too well established to think of amending them. The other quantity in Expression (3.8) is $(\Sigma Y)^2/n$. It is often called the *correction term* (CT). The numerator of this term is the square of the sum of the Y's; that is, all the Y's are first summed, and this sum is then squared. In general, this quantity is different from ΣY^2, which first squares the Y's and then sums them. These two terms are identical only if all the Y's are equal. If you are not certain about this, you can convince yourself of this fact by calculating these two quantities for a few numbers.

The disadvantage of Expression (3.8) is that the quantities ΣY^2 and $(\Sigma Y)^2/n$ may both be quite large, so that accuracy may be lost in computing their difference unless one takes the precaution of carrying sufficient significant figures.

Why is Expression (3.8) identical with Expression (3.7)? The proof of this identity is very simple and is given in Appendix A1.2. You are urged to work through it to build up your confidence in handling statistical symbols and formulas.

It is sometimes possible to simplify computations by recoding variates into simpler form. We shall use the term *additive coding* for the addition or subtraction of a constant (since subtraction is only addition of a negative number). We shall similarly use *multiplicative coding* to refer to the multiplication or division by a constant (since division is multiplication by the reciprocal of the divisor). We shall use the term *combination coding* to mean the application of both additive and multiplicative coding to the same set of data. In Appendix A1.3 we examine the consequences of the three types of coding in the computation of means, variances, and standard deviations.

For the case of *means*, the formula for combination coding and decoding is the most generally applicable one. If the coded variable is $Y_c = D(Y + C)$, then

$$\bar{Y} = \frac{\bar{Y}_c}{D} - C$$

where C is an additive code and D is a multiplicative code.

On considering the effects of coding variates on the values of *variances and standard deviations*, we find that additive codes have no effect on the sums of squares, variances, or standard deviations. The mathematical proof is given in Appendix A1.3, but we can see this intuitively, because an additive code has no effect on the distance of an item from its mean. The distance from an item of 15 to its mean of 10 would be 5. If we were to code the variates by subtracting a constant of 10, the item would now be 5 and the mean zero. The difference between them would still be 5. Thus, if only additive coding is employed, the only statistic in need of decoding is the mean. But multiplicative coding does have an effect on sums of squares, variances, and standard deviations. The standard deviations have to be divided by the multiplicative code, just as had to be done for the mean. However, the sums of squares or variances have to be divided by the multiplicative codes squared, because they are squared terms, and the multiplicative factor becomes squared during the operations. In combination coding the additive code can be ignored.

When the data are unordered, the computation of the mean and standard deviation proceeds as in Box 3.1, which is based on the unordered neutrophil-count data shown in Table 3.1. We chose not to apply coding to these data, since it would not have simplified the computations appreciably.

When the data are arrayed in a frequency distribution, the computations can be made much simpler. When computing the statistics, you can often avoid the need for manual entry of large numbers of individual variates if you first set up a frequency distribution. Sometimes the data will come to you already in the form of a frequency distribution, having been grouped by the researcher.

The computation of $\bar{Y}$ and s from a frequency distribution is illustrated in Box 3.2. The data are the birth weights of male Chinese children, first encountered in Figure 2.3. The calculation is simplified by coding to remove the awkward class marks. This is done by subtracting 59.5, the lowest class mark of the array.

BOX 3.1
Calculation of $\bar{Y}$ and s from unordered data.
Neutrophil counts, unordered as shown in Table 3.1.

Computation

$n = 15$

$\sum Y = 115.7$

$\bar{Y} = \dfrac{1}{n}\sum Y = 7.713$

$\sum Y^2 = 1201.21$

$\sum y^2 = \sum Y^2 - \dfrac{(\sum Y)^2}{n}$

$\quad = 1201.21 - \dfrac{(115.7)^2}{15}$

$\quad = 308.7773$

$s^2 = \dfrac{\sum y^2}{n-1} = \dfrac{308.7773}{14}$

$\quad = 22.056$

$s = \sqrt{22.056} = 4.696$

The resulting class marks are values such as 0, 8, 16, 24, 32, and so on. They are then divided by 8, which changes them to 0, 1, 2, 3, 4, and so on, which is the desired format. The details of the computation can be learned from the box.

When checking the results of calculations, it is frequently useful to have an approximate method for estimating statistics so that gross errors in computation can be detected. A simple method for estimating the mean is to average the largest and smallest observation to obtain the so-called *midrange*. For the neutrophil counts of Table 3.1, this value is $(2.3 + 18.0)/2 = 10.15$ (not a very good estimate). Standard deviations can be estimated from ranges by appropriate division of the range, as follows:

For samples of	Divide the range by
10	3
30	4
100	5
500	6
1000	$6\frac{1}{2}$

BOX 3.2
Calculation of $\bar{Y}$, s, and V from a frequency distribution.

Birth weights of male Chinese in ounces.

(1) Class mark Y	(2) f	(3) Coded class mark Y_c
59.5	2	0
67.5	6	1
75.5	39	2
83.5	385	3
91.5	888	4
99.5	1729	5
107.5	2240	6
115.5	2007	7
123.5	1233	8
131.5	641	9
139.5	201	10
147.5	74	11
155.5	14	12
163.5	5	13
171.5	1	14
	$9465 = n$	

Source: Millis and Seng (1954).

Computation	*Coding and decoding*
$\sum fY_c = 59{,}629$	Code: $Y_c = \dfrac{Y - 59.5}{8}$
$\bar{Y}_c = 6.300$	To decode $\bar{Y}_c$: $\bar{Y} = 8\bar{Y}_c + 59.5$
$\sum fY_c^2 = 402{,}987$	$= 50.4 + 59.5$
$CT = \dfrac{(\sum fY_c)^2}{n} = 375{,}659.550$	$= 109.9$ oz
$\sum fy_c^2 = \sum fY_c^2 - CT = 27{,}327.450$	
$s_c^2 = \dfrac{\sum fy_c^2}{n - 1} = 2.888$	
$s_c = 1.6991$	To decode s_c: $s = 8s_c = 13.593$ oz

$$V = \frac{s}{\bar{Y}} \times 100 = \frac{13.593}{109.9} \times 100 = 12.369\%$$

The range of the neutrophil counts is 15.7. When this value is divided by 4, we get an estimate for the standard deviation of 3.925, which compares with the calculated value of 4.696 in Box 3.1. However, when we estimate mean and standard deviation of the aphid femur lengths of Box 2.1 in this manner, we obtain 4.0 and 0.35, respectively. These are good estimates of the actual values of 4.004 and 0.3656, the sample mean and standard deviation.

3.9 The coefficient of variation

Having obtained the standard deviation as a measure of the amount of variation in the data, you may be led to ask, "Now what?" At this stage in our comprehension of statistical theory, nothing really useful comes of the computations we have carried out. However, the skills just learned are basic to all later statistical work. So far, the only use that we might have for the standard deviation is as an estimate of the amount of variation in a population. Thus, we may wish to compare the magnitudes of the standard deviations of similar populations and see whether population A is more or less variable than population B.

When populations differ appreciably in their means, the direct comparison of their variances or standard deviations is less useful, since larger organisms usually vary more than smaller one. For instance, the standard deviation of the tail lengths of elephants is obviously much greater than the entire tail length of a mouse. To compare the relative amounts of variation in populations having different means, the *coefficient of variation*, symbolized by V (or occasionally CV), has been developed. This is simply the standard deviation expressed as a percentage of the mean. Its formula is

$$V = \frac{s \times 100}{\bar{Y}} \tag{3.9}$$

For example, the coefficient of variation of the birth weights in Box 3.2 is 12.37%, as shown at the bottom of that box. The coefficient of variation is independent of the unit of measurement and is expressed as a percentage.

Coefficients of variation are used when one wishes to compare the variation of two populations without considering the magnitude of their means. (It is probably of little interest to discover whether the birth weights of the Chinese children are more or less variable than the femur lengths of the aphid stem mothers. However, we can calculate V for the latter as $(0.3656 \times 100)/4.004 = 9.13\%$, which would suggest that the birth weights are more variable.) Often, we shall wish to test whether a given biological sample is more variable for one character than for another. Thus, for a sample of rats, is body weight more variable than blood sugar content? A second, frequent type of comparison, especially in systematics, is among different populations for the same character. Thus, we may have measured wing length in samples of birds from several localities. We wish to know whether any one of these populations is more variable than the others. An answer to this question can be obtained by examining the coefficients of variation of wing length in these samples.

Exercises

3.1   Find $\bar{Y}$, s, V, and the median for the following data (mg of glycine per mg of creatinine in the urine of 37 chimpanzees; from Gartler, Firschein, and Dobzhansky, 1956). ANS. $\bar{Y} = 0.115$, $s = 0.10404$.

.008	.018	.056	.055	.135	.052	.077	.026	.440	.300
.025	.036	.043	.100	.120	.110	.100	.350	.100	.300
.011	.060	.070	.050	.080	.110	.110	.120	.133	.100
.100	.155	.370	.019	.100	.100	.116			

3.2   Find the mean, standard deviation, and coefficient of variation for the pigeon data given in Exercise 2.4. Group the data into ten classes, recompute $\bar{Y}$ and s, and compare them with the results obtained from ungrouped data. Compute the median for the grouped data.

3.3   The following are percentages of butterfat from 120 registered three-year-old Ayrshire cows selected at random from a Canadian stock record book.
(a) Calculate $\bar{Y}$, s, and V directly from the data.
(b) Group the data in a frequency distribution and again calculate $\bar{Y}$, s, and V. Compare the results with those of (a). How much precision has been lost by grouping? Also calculate the median.

4.32	4.24	4.29	4.00
3.96	4.48	3.89	4.02
3.74	4.42	4.20	3.87
4.10	4.00	4.33	3.81
4.33	4.16	3.88	4.81
4.23	4.67	3.74	4.25
4.28	4.03	4.42	4.09
4.15	4.29	4.27	4.38
4.49	4.05	3.97	4.32
4.67	4.11	4.24	5.00
4.60	4.38	3.72	3.99
4.00	4.46	4.82	3.91
4.71	3.96	3.66	4.10
4.38	4.16	3.77	4.40
4.06	4.08	3.66	4.70
3.97	3.97	4.20	4.41
4.31	3.70	3.83	4.24
4.30	4.17	3.97	4.20
4.51	3.86	4.36	4.18
4.24	4.05	4.05	3.56
3.94	3.89	4.58	3.99
4.17	3.82	3.70	4.33
4.06	3.89	4.07	3.58
3.93	4.20	3.89	4.60
4.38	4.14	4.66	3.97
4.22	3.47	3.92	4.91
3.95	4.38	4.12	4.52
4.35	3.91	4.10	4.09
4.09	4.34	4.09	4.88
4.28	3.98	3.86	4.58

3.4   What effect would adding a constant 5.2 to all observations have upon the numerical values of the following statistics: $\bar{Y}$, s, V, average deviation, median,

3.5   mode, range? What would be the effect of adding 5.2 and then multiplying the sums by 8.0? Would it make any difference in the above statistics if we multiplied by 8.0 first and then added 5.2?

3.5   Estimate μ and σ using the midrange and the range (see Section 3.8) for the data in Exercises 3.1, 3.2, and 3.3. How well do these estimates agree with the estimates given by $\bar{Y}$ and s? ANS. Estimates of μ and σ for Exercise 3.2 are 0.224 and 0.1014.

3.6   Show that the equation for the variance can also be written as

$$s^2 = \frac{\sum Y^2 - n\bar{Y}^2}{n-1}$$

3.7   Using the striped bass age distribution given in Exercise 2.9, compute the following statistics: $\bar{Y}$, s^2, s, V, median, and mode. ANS. $\bar{Y} = 3.043$, $s^2 = 1.2661$, $s = 1.125$, $V = 36.98\%$, median $= 2.948$, mode $= 3$.

3.8   Use a calculator and compare the results of using Equations 3.7 and 3.8 to compute s^2 for the following artificial data sets:
(a) 1, 2, 3, 4, 5
(b) 9001, 9002, 9003, 9004, 9005
(c) 90001, 90002, 90003, 90004, 90005
(d) 900001, 900002, 900003, 900004, 900005
Compare your results with those of one or more computer programs. What is the correct answer? Explain your results.

CHAPTER 4

Introduction to Probability Distributions: The Binomial and Poisson Distributions

In Section 2.5 we first encountered frequency distributions. For example, Table 2.2 shows a distribution for a meristic, or discrete (discontinuous), variable, the number of sedge plants per quadrat. Examples of distributions for continuous variables are the femur lengths of aphids in Box 2.1 and the human birth weights in Box 3.2. Each of these distributions informs us about the absolute frequency of any given class and permits us to compute the relative frequencies of any class of variable. Thus, most of the quadrats contained either no sedges or one or two plants. In the 139.5-oz class of birth weights, we find only 201 out of the total of 9465 babies recorded; that is, approximately only 2.1% of the infants are in that birth weight class.

We realize, of course, that these frequency distributions are only samples from given populations. The birth weights, for example, represent a population of male Chinese infants from a given geographical area. But if we knew our sample to be representative of that population, we could make all sorts of predictions based upon the sample frequency distribution. For instance, we could say that approximately 2.1% of male Chinese babies born in this population should weigh between 135.5 and 143.5 oz at birth. Similarly, we might say that the probability that the weight at birth of any one baby in this population will be in the 139.5-oz birth class is quite low. If all of the 9465 weights were mixed up in a hat and a single one pulled out, the probability that we would pull out one of the 201 in the 139.5-oz class would be very low indeed—only 2.1%. It would be much more probable that we would sample an infant of 107.5 or 115.5 oz, since the infants in these classes are represented by frequencies 2240 and 2007, respectively. Finally, if we were to sample from an unknown population of babies and find that the very first individual sampled had a birth weight of 170 oz, we would probably reject any hypothesis that the unknown population was the same as that sampled in Box 3.2. We would arrive at this conclusion because in the distribution in Box 3.2 only one out of almost 10,000 infants had a birth weight that high. Though it is possible that we could have sampled from the population of male Chinese babies and obtained a birth weight of 170 oz, the probability that the first individual sampled would have such a value is very low indeed. It seems much more reasonable to suppose that the unknown population from which we are sampling has a larger mean that the one sampled in Box 3.2.

We have used this empirical frequency distribution to make certain predictions (with what frequency a given event will occur) or to make judgments and decisions (is it likely that an infant of a given birth weight belongs to this population?). In many cases in biology, however, we shall make such predictions not from empirical distributions, but on the basis of theoretical considerations that in our judgment are pertinent. We may feel that the data should be distributed in a certain way because of basic assumptions about the nature of the forces acting on the example at hand. If our actually observed data do not conform sufficiently to the values expected on the basis of these assumptions, we shall have serious doubts about our assumptions. This is a common use of frequency distributions in biology. The assumptions being tested generally lead to a theoretical frequency distribution known also as a *probability distribution.* This may be a simple two-valued distribution, such as the 3:1 ratio in a Mendelian cross; or it may be a more complicated function, as it would be if we were trying to predict the number of plants in a quadrat. If we find that the observed data do not fit the expectations on the basis of theory, we are often led to the discovery of some biological mechanism causing this deviation from expectation. The phenomena of linkage in genetics, of preferential mating between different phenotypes in animal behavior, of congregation of animals at certain favored places or, conversely, their territorial dispersion are cases in point. We shall thus make use of probability theory to test our assumptions about the laws of occurrence of certain biological phenomena. We should point out to the reader, however, that probability theory underlies the entire structure of statistics, since, owing to the nonmathematical orientation of this book, this may not be entirely obvious.

In this chapter we shall first discuss probability, in Section 4.1, but only to the extent necessary for comprehension of the sections that follow at the intended level of mathematical sophistication. Next, in Section 4.2, we shall take up the binomial frequency distribution, which is not only important in certain types of studies, such as genetics, but also fundamental to an understanding of the various kinds of probability distributions to be discussed in this book.

The Poisson distribution, which follows in Section 4.3, is of wide applicability in biology, especially for tests of randomness of occurrence of certain events. Both the binomial and Poisson distributions are discrete probability distributions. The most common continuous probability distribution is the normal frequency distribution, discussed in Chapter 5.

4.1 Probability, random sampling, and hypothesis testing

We shall start this discussion with an example that is not biometrical or biological in the strict sense. We have often found it pedagogically effective to introduce new concepts through situations thoroughly familiar to the student, even if the example is not relevant to the general subject matter of biostatistics.

Let us betake ourselves to Matchless University, a state institution somewhere between the Appalachians and the Rockies. Looking at its enrollment figures, we notice the following breakdown of the student body: 70% of the students are American undergraduates (AU) and 26% are American graduate students (AG); the remaining 4% are from abroad. Of these, 1% are foreign undergraduates (FU) and 3% are foreign graduate students (FG). In much of our work we shall use proportions rather than percentages as a useful convention. Thus the enrollment consists of 0.70 AU's, 0.26 AG's, 0.01 FU's, and 0.03 FG's. The total student body, corresponding to 100%, is therefore represented by the figure 1.0.

If we were to assemble all the students and sample 100 of them at random, we would intuitively expect that, on the average, 3 would be foreign graduate students. The actual outcome might vary. There might not be a single FG student among the 100 sampled, or there might be quite a few more than 3. The ratio of the number of foreign graduate students sampled divided by the total number of students sampled might therefore vary from zero to considerably greater than 0.03. If we increased our sample size to 500 or 1000, it is less likely that the ratio would fluctuate widely around 0.03. The greater the sample taken, the closer the ratio of FG students sampled to the total students sampled will approach 0.03. In fact, the *probability* of sampling a foreign student can be defined as the limit as sample size keeps increasing of the ratio of foreign students to the total number of students sampled. Thus, we may formally summarize the situation by stating that the probability that a student at Matchless University will be a foreign graduate student is $P[FG] = 0.03$. Similarly, the probability of sampling a foreign undergraduate is $P[FU] = 0.01$, that of sampling an American undergraduate is $P[AU] = 0.70$, and that for American graduate students, $P[AG] = 0.26$.

Now let us imagine the following experiment: We try to sample a student at random from among the student body at Matchless University. This is not as easy a task as might be imagined. If we wanted to do this operation physically, we would have to set up a collection or trapping station somewhere on campus. And to make certain that the sample was truly random with respect to the entire student population, we would have to know the ecology of students on campus very thoroughly. We should try to locate our trap at some station where each student had an equal probability of passing. Few, if any, such places can be found in a university. The student union facilities are likely to be frequented more by independent and foreign students, less by those living in organized houses and dormitories. Fewer foreign and graduate students might be found along fraternity row. Clearly, we would not wish to place our trap near the International Club or House, because our probability of sampling a foreign student would be greatly enhanced. In front of the bursar's window we might sample students paying tuition. But those on scholarships might not be found there. We do not know whether the proportion of scholarships among foreign or graduate students is the same as or different from that among the American or undergraduate students. Athletic events, political rallies, dances, and the like would all draw a differential spectrum of the student body; indeed, no easy solution seems in sight. The time of sampling is equally important, in the seasonal as well as the diurnal cycle.

Those among the readers who are interested in sampling organisms from nature will already have perceived parallel problems in their work. If we were to sample only students wearing turbans or saris, their probability of being foreign students would be almost 1. We could no longer speak of a random sample. In the familiar ecosystem of the university these violations of proper sampling procedure are obvious to all of us, but they are not nearly so obvious in real biological instances where we are unfamiliar with the true nature of the environment. How should we proceed to obtain a random sample of leaves from a tree, of insects from a field, or of mutations in a culture? In sampling at random, we are attempting to permit the frequencies of various events occurring in nature to be reproduced unalteredly in our records; that is, we hope that on the average the frequencies of these events in our sample will be the same as they are in the natural situation. Another way of saying this is that in a random sample every individual in the population being sampled has an equal probability of being included in the sample.

We might go about obtaining a random sample by using records representing the student body, such as the student directory, selecting a page from it at random and a name at random from the page. Or we could assign an an arbitrary number to each student, write each on a chip or disk, put these in a large container, stir well, and then pull out a number.

Imagine now that we sample a single student physically by the trapping method, after carefully planning the placement of the trap in such a way as to make sampling random. What are the possible outcomes? Clearly, the student could be either an AU, AG, FU or FG. The *set* of these four possible outcomes exhausts the possibilities of this experiment. This set, which we can represent as {AU, AG, FU, FG} is called the *sample space*. Any single trial of the experiment described above would result in only one of the four possible outcomes (elements) in the set. A single element in a sample space is called a *simple event*. It is distinguished from an *event*, which is any subset of the sample space. Thus, in the sample space defined above {AU}, {AG}, {FU}, and {FG} are each simple events. The following sampling results are some of the possible events: {AU, AG, FU}, {AU, AG, FG}, {AG, FG}, {AU, FG}, ... By the definition of "event," simple events as well as the entire sample space are also events. The meaning of these events should be clarified. Thus {AU, AG, FU} implies being either an American or an undergraduate, or both.

Given the sampling space described above, the event **A** = {AU, AG} encompasses all possible outcomes in the space yielding an American student. Similarly, the event **B** = {AG, FG} summarizes the possibilities for obtaining a graduate student. The *intersection* of events **A** and **B**, written **A** ∩ **B**, describes only those events that are shared by **A** and **B**. Clearly only AG qualifies, as can be seen below:

$$\mathbf{A} = \{AU, AG\}$$

$$\mathbf{B} = \quad \{AG, FG\}$$

Thus, **A** ∩ **B** is that event in the sample space giving rise to the sampling of an American graduate student. When the intersection of two events is empty, as in **B** ∩ **C**, where **C** = {AU, FU}, events **B** and **C** are mutually exclusive. Thus there is no common element in these two events in the sampling space.

We may also define events that are *unions* of two other events in the sample space. Thus **A** ∪ **B** indicates that **A** or **B** or both **A** and **B** occur. As defined above, **A** ∪ **B** would describe all students who are either American students, graduate students, or American graduate students.

Why are we concerned with defining sample spaces and events? Because these concepts lead us to useful definitions and operations regarding the probability of various outcomes. If we can assign a number p, where $0 \le p \le 1$, to each simple event in a sample space such that the sum of these p's over all simple events in the space equals unity, then the space becomes a (finite) *probability space*. In our example above, the following numbers were associated with the appropriate simple events in the sample space:

$$\{AU, \ AG, \ FU, \ FG\}$$
$$\{0.70, 0.26, 0.01, 0.03\}$$

Given this probability space, we are now able to make statements regarding the probability of given events. For example, what is the probability that a student sampled at random will be an American graduate student? Clearly, it is $P[\{AG\}] = 0.26$. What is the probability that a student is either American or a graduate student? In terms of the events defined earlier, this is

$$P[\mathbf{A} \cup \mathbf{B}] = P[\{AU, AG\}] + P[\{AG, FG\}] - P[\{AG\}]$$
$$= 0.96 + 0.29 - 0.26$$
$$= 0.99$$

We subtract $P[\{AG\}]$ from the sum on the right-hand side of the equation because if we did not do so it would be included twice, once in $P[A]$ and once in $P[B]$, and would lead to the absurd result of a probability greater than 1.

Now let us assume that we have sampled our single student from the student body of Matchless University and that student turns out to be a foreign graduate student. What can we conclude from this? By chance alone, this result would happen 0.03, or 3%, of the time—not very frequently. The assumption that we have sampled at random should probably be rejected, since if we accept the hypothesis of random sampling, the outcome of the experiment is improbable. Please note that we said *improbable*, not *impossible*. It is obvious that we could have chanced upon an FG as the very first one to be sampled. However, it is not very likely. The probability is 0.97 that a single student sampled will be a non-FG. If we could be certain that our sampling method was random (as when drawing student numbers out of a container), we would have to decide that an improbable event has occurred. The decisions of this paragraph are all based on our definite knowledge that the proportion of students at Matchless University is indeed as specified by the probability space. If we were uncertain about this, we would be led to assume a higher proportion of foreign graduate students as a consequence of the outcome of our sampling experiment.

We shall now extend our experiment and sample two students rather than just one. What are the possible outcomes of this sampling experiment? The new sampling space can best be depicted by a diagram (Figure 4.1) that shows the set of the 16 possible simple events as points in a lattice. The simple events are the following possible combinations. Ignoring which student was sampled first, they are (AU, AU), (AU, AG), (AU, FU), (AU, FG), (AG, AG), (AG, FU), (AG, FG), (FU, FU), (FU, FG), and (FG, FG).

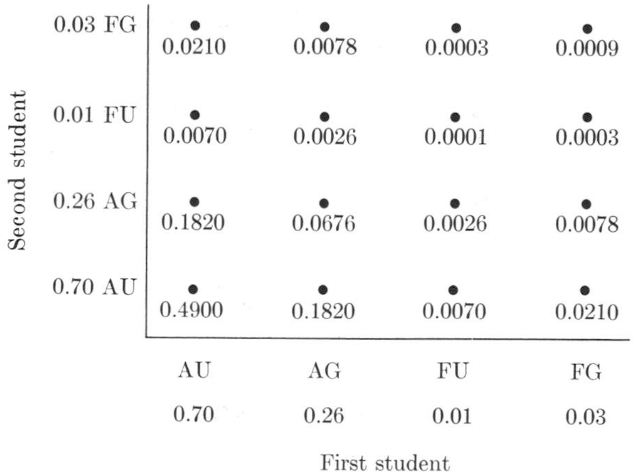

Second student	AU	AG	FU	FG
0.03 FG	0.0210	0.0078	0.0003	0.0009
0.01 FU	0.0070	0.0026	0.0001	0.0003
0.26 AG	0.1820	0.0676	0.0026	0.0078
0.70 AU	0.4900	0.1820	0.0070	0.0210
First student	AU 0.70	AG 0.26	FU 0.01	FG 0.03

FIGURE 4.1
Sample space for sampling two students from Matchless University.

What are the expected probabilities of these outcomes? We know the expected outcomes for sampling one student from the former probability space, but what will be the probability space corresponding to the new sampling space of 16 elements? Now the nature of the sampling procedure becomes quite important. We may sample with or without *replacement*: we may return the first student sampled to the population (that is, *replace* the first student), or we may keep him or her out of the pool of the individuals to be sampled. If we do not replace the first individual sampled, the probability of sampling a foreign graduate student will no longer be exactly 0.03. This is easily seen. Let us assume that Matchless University has 10,000 students. Then, since 3% are foreign graduate students, there must be 300 FG students at the university. After sampling a foreign graduate student first, this number is reduced to 299 out of 9999 students. Consequently, the probability of sampling an FG student now becomes $299/9999 = 0.0299$, a slightly lower probability than the value of 0.03 for sampling the first FG student. If, on the other hand, we return the original foreign student to the student population and make certain that the population is thoroughly randomized before being sampled again (that is, give the student a chance to lose him- or herself among the campus crowd or, in drawing student numbers out of a container, mix up the disks with the numbers on them), the probability of sampling a second FG student is the same as before—0.03. In fact, if we keep on replacing the sampled individuals in the original population, we can sample from it as though it were an infinite-sized population.

Biological populations are, of course, finite, but they are frequently so large that for purposes of sampling experiments we can consider them effectively infinite whether we replace sampled individuals or not. After all, even in this relatively small population of 10,000 students, the probability of sampling a second foreign graduate student (without replacement) is only minutely different from 0.03. For the rest of this section we shall consider sampling to be with replacement, so that the probability level of obtaining a foreign student does not change.

There is a second potential source of difficulty in this design. We have to assume not only that the probability of sampling a second foreign student is equal to that of the first, but also that it is *independent* of it. By *independence of events* we mean that *the probability that one event will occur is not affected by whether or not another event has or has not occurred*. In the case of the students, if we have sampled one foreign student, is it more or less likely that a second student sampled in the same manner will also be a foreign student? Independence of the events may depend on where we sample the students or on the method of sampling. If we have sampled students on campus, it is quite likely that the events are not independent; that is, if one foreign student has been sampled, the probability that the second student will be foreign is increased, since foreign students tend to congregate. Thus, at Matchless University the probability that a student walking with a foreign graduate student is also an FG will be greater than 0.03.

Events **D** and **E** in a sample space will be defined as independent whenever $P[D \cap E] = P[D]P[E]$. The probability values assigned to the sixteen points in the sample-space lattice of Figure 4.1 have been computed to satisfy the above condition. Thus, letting $P[D]$ equal the probability that the first student will be an AU, that is, $P[\{AU_1AU_2, AU_1AG_2, AU_1FU_2, AU_1FG_2\}]$, and letting $P[E]$ equal the probability that the second student will be an FG, that is, $P[\{AU_1FG_2, AG_1FG_2, FU_1FG_2, FG_1FG_2\}]$, we note that the intersection $D \cap E$ is $\{AU_1FG_2\}$. This has a value of 0.0210 in the probability space of Figure 4.1. We find that this value is the product $P[\{AU\}]P[\{FG\}] = 0.70 \times 0.03 = 0.0210$. These mutually independent relations have been deliberately imposed upon all points in the probability space. Therefore, if the sampling probabilities for the second student are independent of the type of student sampled first, we can compute the probabilities of the outcomes simply as the product of the independent probabilities. Thus the probability of obtaining two FG students is $P[\{FG\}]P[\{FG\}] = 0.03 \times 0.03 = 0.0009$.

The probability of obtaining one AU and one FG student in the sample should be the product 0.70×0.03. However, it is in fact twice that probability. It is easy to see why. There is only one way of obtaining two FG students, namely, by sampling first one FG and then again another FG. Similarly, there is only one way to sample two AU students. However, sampling one of each type of student can be done by sampling first an AU and then an FG or by sampling first an FG and then an AU. Thus the probability is $2P[\{AU\}]P[\{FG\}] = 2 \times 0.70 \times 0.03 = 0.0420$.

If we conducted such an experiment and obtain a sample of two FG students, we would be led to the following conclusions. Only 0.0009 of the samples ($\frac{9}{100}$ of 1% or 9 out of 10,000 cases) would be expected to consist of two foreign graduate students. It is quite improbable to obtain such a result by chance alone. Given $P[\{FG\}] = 0.03$ as a fact, we would therefore suspect that sampling was not random or that the events were not independent (or that both assumptions—random sampling and independence of events—were incorrect).

Random sampling is sometimes confused with randomness in nature. The former is the faithful representation in the sample of the distribution of the events in nature; the latter is the independence of the events in nature. The first of these generally is or should be under the control of the experimenter and is related to the strategy of good sampling. The second generally describes an innate property of the objects being sampled and thus is of greater biological interest. The confusion between random sampling and independence of events arises because lack of either can yield observed frequencies of events differing from expectation. We have already seen how lack of independence in samples of foreign students can be interpreted from both points of view in our illustrative example from Matchless University.

The above account of probability is adequate for our present purposes but far too sketchy to convey an understanding of the field. Readers interested in extending their knowledge of the subject are referred to Mosimann (1968) for a simple introduction.

4.2 The binomial distribution

For purposes of the discussion to follow we shall simplify our sample space to consist of only two elements, foreign and American students, and ignore whether the students are undergraduates or graduates; we shall represent the sample space by the set {F, A}. Let us symbolize the probability space by {p, q}, where $p = P[F]$, the probability that the student is foreign, and $q = P[A]$, the probability that the student is American. As before, we can compute the probability space of samples of two students as follows:

$$\{FF, FA, AA\}$$

$$\{ p^2, \; 2pq, \; q^2 \}$$

If we were to sample three students independently, the probability space of samples of three students would be as follows:

$$\{FFF, FFA, FAA, AAA\}$$

$$\{ p^3, \; 3p^2q, \; 3pq^2, \; q^3 \}$$

Samples of three foreign or three American students can again be obtained in only one way, and their probabilities are p^3 and q^3, respectively. However, in samples of three there are three ways of obtaining two students of one kind and one student of the other. As before, if A stands for American and F stands for foreign, then the sampling sequence can be AFF, FAF, FFA for two foreign students and one American. Thus the probability of this outcome will be $3p^2q$. Similarly, the probability for two Americans and one foreign student is $3pq^2$.

A convenient way to summarize these results is by means of the binomial expansion, which is applicable to samples of any size from populations in which objects occur independently in only two classes—students who may be foreign or American, or individuals who may be dead or alive, male or female, black or white, rough or smooth, and so forth. This is accomplished by expanding the binomial term $(p + q)^k$, where k equals sample size, p equals the probability of occurrence of the first class, and q equals the probability of occurrence of the second class. By definition, $p + q = 1$; hence q is a function of p: $q = 1 - p$. We shall expand the expression for samples of k from 1 to 3:

For samples of 1, $(p + q)^1 = p + q$
For samples of 2, $(p + q)^2 = p^2 + 2pq + q^2$
For samples of 3, $(p + q)^3 = p^3 + 3p^2q + 3pq^2 + q^3$

It will be seen that these expressions yield the same probability spaces discussed previously. The coefficients (the numbers before the powers of p and q) express the number of ways a particular outcome is obtained. An easy method for evaluating the coefficients of the expanded terms of the binomial expression is through the use of Pascal's triangle:

k

1					1	1				
2				1	2	1			
3			1	3	3	1		
4		1	4	6	4	1	
5	1	5	10	10	5	1

Pascal's triangle provides the coefficients of the binomial expression—that is, the number of possible outcomes of the various combinations of events. For $k = 1$ the coefficients are 1 and 1. For the second line ($k = 2$), write 1 at the left-hand margin of the line. The 2 in the middle of this line is the sum of the values to the left and right of it in the line above. The line is concluded with a 1. Similarly, the values at the beginning and end of the third line are 1, and the other numbers are sums of the values to their left and right in the line above; thus 3 is the sum of 1 and 2. This principle continues for every line. You can work out the coefficients for any size sample in this manner. The line for $k = 6$ would consist of the following coefficients: 1, 6, 15, 20, 15, 6, 1. The p and q values bear powers in a consistent pattern, which should be easy to imitate for any value of k. We give it here for $k = 4$:

$$p^4q^0 + p^3q^1 + p^2q^2 + p^1q^3 + p^0q^4$$

The power of p decreases from 4 to 0 (k to 0 in the general case) as the power of q increases from 0 to 4 (0 to k in the general case). Since any value to the power 0 is 1 and any term to the power 1 is simply itself, we can simplify this expression as shown below and at the same time provide it with the coefficients from Pascal's triangle for the case $k = 4$:

$$p^4 + 4p^3q + 6p^2q^2 + 4pq^3 + q^4$$

Thus we are able to write down almost by inspection the expansion of the binomial to any reasonable power. Let us now practice our newly learned ability to expand the binomial.

Suppose we have a population of insects, exactly 40% of which are infected with a given virus X. If we take samples of $k = 5$ insects each and examine each insect separately for presence of the virus, what distribution of samples could we expect if the probability of infection of each insect in a sample were independent of that of other insects in the sample? In this case $p = 0.4$, the proportion infected, and $q = 0.6$, the proportion not infected. It is assumed that the population is so large that the question of whether sampling is with or without replacement is irrelevant for practical purposes. The expected proportions would be the expansion of the binomial:

$$(p + q)^k = (0.4 + 0.6)^5$$

With the aid of Pascal's triangle this expansion is

$$\{p^5 + 5p^4q + 10p^3q^2 + 10p^2q^3 + 5pq^4 + q^5\}$$

or

$$(0.4)^5 + 5(0.4)^4(0.6) + 10(0.4)^3(0.6)^2 + 10(0.4)^2(0.6)^3 + 5(0.4)(0.6)^4 + (0.6)^5$$

representing the expected proportions of samples of five infected insects, four infected and one noninfected insects, three infected and two noninfected insects, and so on.

The reader has probably realized by now that the terms of the binomial expansion actually yield a type of frequency distribution for these different outcomes. Associated with each outcome, such as "five infected insects," there is a probability of occurrence—in this case $(0.4)^5 = 0.01024$. This is a theoretical frequency distribution or *probability distribution* of events that can occur in two classes. It describes the expected distribution of outcomes in random samples of five insects from a population in which 40% are infected. The probability distribution described here is known as the *binomial distribution*, and the binomial expansion yields the expected frequencies of the classes of the binomial distribution.

A convenient layout for presentation and computation of a binomial distribution is shown in Table 4.1. The first column lists the number of infected insects per sample, the second column shows decreasing powers of p from p^5 to p^0, and the third column shows increasing powers of q from q^0 to q^5. The binomial coefficients from Pascal's triangle are shown in column (4). The *relative*

TABLE **4.1**
Expected frequencies of infected insects in samples of 5 insects sampled from an infinitely large population with an assumed infection rate of 40%.

(1) Number of infected insects per sample Y	(2) Powers of $p = 0.4$	(3) Powers of $q = 0.6$	(4) Binomial coefficients	(5) Relative expected frequencies $\hat{f}_{rel}$	(6) Absolute expected frequencies $\hat{f}$	(7) Observed frequencies f
5	0.01024	1.00000	1	0.01024	24.8	29
4	0.02560	0.60000	5	0.07680	186.1	197
3	0.06400	0.36000	10	0.23040	558.3	535
2	0.16000	0.21600	10	0.34560	837.4	817
1	0.40000	0.12960	5	0.25920	628.0	643
0	1.00000	0.07776	1	0.07776	188.4	202
			$\sum \hat{f}$ or $\sum \hat{f} (= n)$	1.00000	2423.0	2423
			$\sum Y$	2.00000	4846.1	4815
		Mean		2.00000	2.00004	1.98721
		Standard deviation		1.09545	1.09543	1.11934

expected frequencies, which are the probabilities of the various outcomes, are shown in column (5). We label such expecred frequencies $\hat{f}_{rel}$. They are simply the product of columns (2), (3), and (4). Their sum is equal to 1.0, since the events listed in column (1) exhaust the possible outcomes. We see from column (5) in Table 4.1 that only about 1% of samples are expected to consist of 5 infected insects, and 25.9% are expected to contain 1 infected and 4 noninfected insects. We shall test whether these predictions hold in an actual experiment.

Experiment 4.1. Simulate the sampling of infected insects by using a table of random numbers such as Table I in Appendix A1. These are randomly chosen one-digit numbers in which each digit 0 through 9 has an equal probability of appearing. The numbers are grouped in blocks of 25 for convenience. Such numbers can also be obtained from random number keys on some pocket calculators and by means of pseudorandom number–generating algorithms in computer programs. (In fact, this entire experiment can be programmed and performed automatically—even on a small computer.) Since there is an equal probability for any one digit to appear, you can let any four digits (say, 0, 1, 2, 3) stand for the infected insects and the remaining digits (4, 5, 6, 7, 8, 9) stand for the noninfected insects. The probability that any one digit selected from the table will represent an infected insect (that is, will be a 0, 1, 2, or 3) is therefore 40%, or 0.4, since these are four of the ten possible digits. Also, successive digits are assumed to be independent of the values of previous digits. Thus the assumptions of the binomial distribution should be met in this experiment. Enter the table of random numbers at an arbitrary point (not always at the beginning!) and look at successive groups of five digits, noting in each group how many of the digits are 0, 1, 2, or 3. Take as many groups of five as you can find time to do, but no fewer than 100 groups.

Column (7) in Table 4.1 shows the results of one such experiment during one year by a biostatistics class. A total of 2423 samples of five numbers were obtained from the table of random numbers; the distribution of the four digits simulating the percentage of infection is shown in this column. The observed frequencies are labeled f. To calculate the expected frequencies for this actual example we multiplied the relative frequencies $\hat{f}_{rel}$ of column (5) times $n = 2423$, the number of samples taken. This results in *absolute expected frequencies*, labeled $\hat{f}$, shown in column (6). When we compare the observed frequencies in column (7) with the expected frequencies in column (6) we note general agreement between the two columns of figures. The two distributions are also illustrated in Figure 4.2. If the observed frequencies did not fit expected frequencies, we might believe that the lack of fit was due to chance alone. Or we might be led to reject one or more of the following hypotheses: (1) that the true proportion of digits 0, 1, 2, and 3 is 0.4 (rejection of this hypothesis would normally not be reasonable, for we may rely on the fact that the proportion of digits 0, 1, 2, and 3 in a table of random numbers *is* 0.4 or very close to it); (2) that sampling was at random; and (3) that events were independent.

These statements can be reinterpreted in terms of the original infection model with which we started this discussion. If, instead of a sampling experiment of digits by a biostatistics class, this had been a real sampling experiment of insects, we would conclude that the insects had indeed been randomly sampled

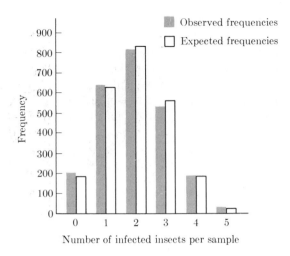

FIGURE 4.2
Bar diagram of observed and expected frequencies given in Table 4.1.

and that we had no evidence to reject the hypothesis that the proportion of infected insects was 40%. If the observed frequencies had not fitted the expected frequencies, the lack of fit might be attributed to chance, or to the conclusion that the true proportion of infection was not 0.4; or we would have had to reject one or both of the following assumptions: (1) that sampling was at random and (2) that the occurrence of infected insects in these samples was independent.

Experiment 4.1 was designed to yield random samples and independent events. How could we simulate a sampling procedure in which the occurrences of the digits 0, 1, 2, and 3 were not independent? We could, for example, instruct the sampler to sample as indicated previously, but, every time a 3 was found among the first four digits of a sample, to replace the following digit with another one of the four digits standing for infected individuals. Thus, once a 3 was found, the probability would be 1.0 that another one of the indicated digits would be included in the sample. After repeated samples, this would result in higher frequencies of samples of two or more indicated digits and in lower frequencies than expected (on the basis of the binomial distribution) of samples of one such digit. A variety of such different sampling schemes could be devised. It should be quite clear to the reader that the probability of the second event's occurring would be different from that of the first and dependent on it.

How would we interpret a large departure of the observed frequencies from expectation? We have not as yet learned techniques for testing whether observed frequencies differ from those expected by more than can be attributed to chance alone. This will be taken up in Chapter 13. Assume that such a test has been carried out and that it has shown us that our observed frequencies are significantly different from expectation. Two main types of departure from expectation can be characterized: (1) *clumping* and (2) *repulsion*, shown in fictitious

TABLE 4.2
Artificial distributions to illustrate clumping and repulsion. Expected frequencies from Table 4.1.

(1) Number of infected insects per sample Y	(2) Absolute expected frequencies $\hat{f}$	(3) Clumped (contagious) frequencies f	(4) Deviation from expectation	(5) Repulsed frequencies f	(6) Deviation from expectation
5	24.8	47	+	14	−
4	186.1	227	+	157	−
3	558.3	558	0	548	−
2	837.4	663	−	943	+
1	628.0	703	+	618	−
0	188.4	225	+	143	−
$\sum \hat{f}$ or n	2423.0	2423		2423.0	
$\sum Y$	4846.1	4846		4846	
Mean	2.00004	2.00000		2.00000	
Standard deviation	1.09543	1.20074		1.01435	

examples in Table 4.2. In actual examples we would have no a priori notions about the magnitude of p, the probability of one of the two possible outcomes. In such cases it is customary to obtain p from the observed sample and to calculate the expected frequencies, using the sample p. This would mean that the hypothesis that p is a given value cannot be tested, since by design the expected frequencies will have the same p value as the observed frequencies. Therefore, the hypotheses tested are whether the samples are random and the events independent.

The clumped frequencies in Table 4.2 have an excess of observations at the tails of the frequency distribution and consequently a shortage of observations at the center. Such a distribution is also said to be *contagious*. (Remember that the total number of items must be the same in both observed and expected frequencies in order to make them comparable.) In the repulsed frequency distribution there are more observations than expected at the center of the distribution and fewer at the tails. These discrepancies are most easily seen in columns (4) and (6) of Table 4.2, where the deviations of observed from expected frequencies are shown as plus or minus signs.

What do these phenomena imply? In the clumped frequencies, more samples were entirely infected (or largely infected), and similarly, more samples were entirely noninfected (or largely noninfected) than you would expect if probabilities of infection were independent. This could be due to poor sampling design. If, for example, the investigator in collecting samples of five insects always tended to pick out like ones—that is, infected ones or noninfected ones—then such a result would likely appear. But if the sampling design is sound, the results become more interesting. Clumping would then mean that the samples of five are in some way related, so that if one insect is infected, others in the same sample are more likely to be infected. This could be true if they come from adjacent locations in a situation in which neighbors are easily infected. Or they could be siblings jointly exposed to a source of infection. Or possibly the infection might spread among members of a sample between the time that the insects are sampled and the time they are examined.

The opposite phenomenon, repulsion, is more difficult to interpret biologically. There are fewer homogeneous groups and more mixed groups in such a distribution. This involves the idea of a compensatory phenomenon: if some of the insects in a sample are infected, the others in the sample are less likely to be. If the infected insects in the sample could in some way transmit immunity to their associates in the sample, such a situation could arise logically, but it is biologically improbable. A more reasonable interpretation of such a finding is that for each sampling unit, there were only a limited number of pathogens available; then once several of the insects have become infected, the others go free of infection, simply because there is no more infectious agent. This is an unlikely situation in microbial infections, but in situations in which a limited number of parasites enter the body of the host, repulsion may be more reasonable.

From the expected and observed frequencies in Table 4.1, we may calculate the mean and standard deviation of the number of infected insects per sample. These values are given at the bottom of columns (5), (6), and (7) in Table 4.1. We note that the means and standard deviations in columns (5) and (6) are almost identical and differ only trivially because of rounding errors. Column (7), being a sample from a population whose parameters are the same as those of the expected frequency distribution in column (5) or (6), differs somewhat. The mean is slightly smaller and the standard deviation is slightly greater than in the expected frequencies. If we wish to know the mean and standard deviation of expected binomial frequency distributions, we need not go through the computations shown in Table 4.1. The mean and standard deviation of a binomial frequency distribution are, respectively,

$$\mu = kp \qquad \sigma = \sqrt{kpq}$$

Substituting the values $k = 5$, $p = 0.4$, and $q = 0.6$ of the above example, we obtain $\mu = 2.0$ and $\sigma = 1.095,45$, which are identical to the values computed from column (5) in Table 4.1. Note that we use the Greek parametric notation here because μ and σ are parameters of an expected frequency distribution, not sample statistics, as are the mean and standard deviation in column (7). The proportions p and q are parametric values also, and strictly speaking, they should be distinguished from sample proportions. In fact, in later chapters we resort to $\hat{p}$ and $\hat{q}$ for parametric proportions (rather than π, which conventionally is used as the ratio of the circumference to the diameter of a circle). Here, however, we prefer to keep our notation simple. If we wish to express our variable as a proportion rather than as a count—that is, to indicate mean incidence of infection in the insects as 0.4, rather than as 2 per sample of 5—we can use other formulas for the mean and standard deviation in a binomial distribution:

$$\mu = p \qquad \sigma = \sqrt{\frac{pq}{k}}$$

It is interesting to look at the standard deviations of the clumped and replused frequency distributions of Table 4.2. We note that the clumped distribution has a standard deviation greater than expected, and that of the repulsed one is less than expected. Comparison of sample standard deviations with their expected values is a useful measure of dispersion in such instances.

We shall now employ the binomial distribution to solve a biological problem. On the basis of our knowledge of the cytology and biology of species A, we expect the sex ratio among its offspring to be 1:1. The study of a litter in nature reveals that of 17 offspring 14 were females and 3 were males. What conclusions can we draw from this evidence? Assuming that $p_\female$ (the probability of being a female offspring) $= 0.5$ and that this probability is independent among the members of the sample, the pertinent probability distribution is the binomial for sample size $k = 17$. Expanding the binomial to the power 17 is a formidable task, which, as we shall see, fortunately need not be done in its entirety. However, we must have the binomial coefficients, which can be obtained either from an expansion of Pascal's triangle (fairly tedious unless once obtained and stored for future use) or by working out the expected frequencies for any given class of Y from the general formula for any term of the binomial distribution

$$C(k, Y)p^Y q^{k-Y} \tag{4.1}$$

The expression $C(k, Y)$ stands for the number of combinations that can be formed from k items taken Y at a time. This can be evaluated as $k!/[Y!(k - Y)!]$, where ! means "factorial." In mathematics k factorial is the product of all the integers from 1 up to and including k. Thus, $5! = 1 \times 2 \times 3 \times 4 \times 5 = 120$. By convention, $0! = 1$. In working out fractions containing factorials, note that any factorial will always cancel against a higher factorial. Thus $5!/3! = (5 \times 4 \times 3!)/3! = 5 \times 4$. For example, the binomial coefficient for the expected frequency of samples of 5 items containing 2 infected insects is $C(5, 2) = 5!/2!3! = (5 \times 4)/2 = 10$.

The setup of the example is shown in Table 4.3. Decreasing powers of $p_\female$ from $p_\female^{17}$ down and increasing powers of $q_\male$ are computed (from power 0 to power 4). Since we require the probability of 14 females, we note that for the purposes of our problem, we need not proceed beyond the term for 13 females and 4 males. Calculating the relative expected frequencies in column (6), we note that the probability of 14 females and 3 males is 0.005,188,40, a very small value. If we add to this value all "worse" outcomes—that is, all outcomes that are even more unlikely than 14 females and 3 males on the assumption of a 1:1 hypothesis—we obtain a probability of 0.006,363,42, still a very small value. (In statistics, we often need to calculate the probability of observing a deviation as large as or larger than a given value.)

TABLE **4.3**

Some expected frequencies of males and females for samples of 17 offspring on the assumption that the sex ratio is 1:1 $[p_\female = 0.5, q_\male = 0.5; (p_\female + q_\male)^k = (0.5 + 0.5)^{17}]$.

(1) $\female\female$	(2) $\male\male$	(3) $p_\female$	(4) $q_\male$	(5) Binomial coefficients	(6) Relative expected frequencies $\hat{f}_{rel}$	
17	—	0.000,007,63	1	1	0.000,007,63	
16	1	0.000,015,26	0.5	17	0.000,129,71	
15	2	0.000,030,52	0.25	136	0.001,037,68	0.006,363,42
14	3	0.000,061,04	0.125	680	0.005,188,40	
13	4	0.000,122,07	0.0625	2380	0.018,157,91	

On the basis of these findings one or more of the following assumptions is unlikely: (1) that the true sex ratio in species A is 1:1, (2) that we have sampled at random in the sense of obtaining an unbiased sample, or (3) that the sexes of the offspring are independent of one another. Lack of independence of events may mean that although the average sex ratio is 1:1, the individual sibships, or litters, are largely unisexual, so that the offspring from a given mating would tend to be all (or largely) females or all (or largely) males. To confirm this hypothesis, we would need to have more samples and then examine the distribution of samples for clumping, which would indicate a tendency for unisexual sibships.

We must be very precise about the questions we ask of our data. There are really two questions we could ask about the sex ratio. First, are the sexes unequal in frequency so that females will appear more often than males? Second, are the sexes unequal in frequency? It may be that we know from past experience that in this particular group of organisms the males are never more frequent than females; in that case, we need be concerned only with the first of these two questions, and the reasoning followed above is appropriate. However, if we know very little about this group of organisms, and if our question is simply whether the sexes among the offspring are unequal in frequency, then we have to consider both tails of the binomial frequency distribution; departures from the 1:1 ratio could occur in either direction. We should then consider not only the probability of samples with 14 females and 3 males (and all worse cases) but also the probability of samples of 14 males and 3 females (and all worse cases in that direction). Since this probability distribution is symmetrical (because $p_\female = q_\male = 0.5$), we simply double the cumulative probability of 0.006,363,42 obtained previously, which results in 0.012,726,84. This new value is still very small, making it quite unlikely that the true sex ratio is 1:1.

This is your first experience with one of the most important applications of statistics—hypothesis testing. A formal introduction to this field will be deferred until Section 6.8. We may simply point out here that the two approaches followed above are known appropriately as *one-tailed tests* and *two-tailed tests*, respectively. Students sometimes have difficulty knowing which of the two tests to apply. In future examples we shall try to point out in each case why a one-tailed or a two-tailed test is being used.

We have said that a tendency for unisexual sibships would result in a clumped distribution of observed frequencies. An actual case of this nature is a classic in the literature, the sex ratio data obtained by Geissler (1889) from hospital records in Saxony. Table 4.4 reproduces sex ratios of 6115 sibships of 12 children each from the more extensive study by Geissler. All columns of the table should by now be familiar. The expected frequencies were not calculated on the basis of a 1:1 hypothesis, since it is known that in human populations the sex ratio at birth is not 1:1. As the sex ratio varies in different human populations, the best estimate of it for the population in Saxony was simply obtained using the mean proportion of males in these data. This can be obtained by calculating the average number of males per sibship ($\bar{Y} = 6.230,58$) for the 6115 sibships and converting this into a proportion. This value turns out to be 0.519,215. Consequently, the proportion of females = 0.480,785. In the deviations of the observed frequencies from the absolute expected frequencies shown in column (9) of Table 4.4, we notice considerable clumping. There are many more instances of families with all male or all female children (or nearly so) than independent probabilities would indicate. The genetic basis for this is not clear, but it is evident that there are some families which "run to girls" and similarly those which "run to boys." Evidence of clumping can also be seen from the fact that s^2 is much larger than we would expect on the basis of the binomial distribution ($\sigma^2 = kpq = 12(0.519,215)0.480,785 = 2.995,57$).

There is a distinct contrast between the data in Table 4.1 and those in Table 4.4. In the insect infection data of Table 4.1 we had a hypothetical proportion of infection based on outside knowledge. In the sex ratio data of Table 4.4 we had no such knowledge; we used an *empirical value of p obtained from the data*, rather than a *hypothetical value external to the data*. This is a distinction whose importance will become apparent later. In the sex ratio data of Table 4.3, as in much work in Mendelian genetics, a hypothetical value of p is used.

4.3 The Poisson distribution

In the typical application of the binomial we had relatively small samples (2 students, 5 insects, 17 offspring, 12 siblings) in which two alternative states occurred at varying frequencies (American and foreign, infected and noninfected, male and female). Quite frequently, however, we study cases in which sample size k is very large and one of the events (represented by probability q) is very much more frequent than the other (represented by probability p). We have seen that the expansion of the binomial $(p + q)^k$ is quite tiresome when k is large. Suppose you had to expand the expression $(0.001 + 0.999)^{1000}$. In such cases we are generally interested in one tail of the distribution only. This is the

TABLE 4.4
Sex ratios in 6115 sibships of twelve in Saxony.

(1) Y ♂♂	(2) ♀♀	(3) $p_\male$	(4) $q_\female$	(5) Binomial coefficients	(6) Relative expected frequencies f_{rel}	(7) Absolute expected frequencies $\hat f$	(8) Observed frequencies f	(9) Deviation from expectation $f - \hat f$
12	—	0.000,384	1	1	0.000,384	2.3	7	+
11	1	0.000,739	0.480,785	12	0.004,264	26.1	45	+
10	2	0.001,424	0.231,154	66	0.021,725	132.8	181	+
9	3	0.002,742	0.111,135	220	0.067,041	410.0	478	+
8	4	0.005,282	0.053,432	495	0.139,703	854.3	829	−
7	5	0.010,173	0.025,689	792	0.206,973	1265.6	1112	−
6	6	0.019,592	0.012,351	924	0.223,590	1367.3	1343	−
5	7	0.037,734	0.005,938	792	0.177,459	1085.2	1033	−
4	8	0.072,676	0.002,855	495	0.102,708	628.1	670	+
3	9	0.139,972	0.001,373	220	0.042,280	258.5	286	+
2	10	0.269,584	0.000,660	66	0.011,743	71.8	104	+
1	11	0.519,215	0.000,317	12	0.001,975	12.1	24	+
—	12	1	0.000,153	1	0.000,153	0.9	3	+
				Total	0.999,998	6115.0	6115	

$$\bar Y = 6.230,58 \qquad s^2 = 3.489,85$$

Source: Geissler (1889).

tail represented by the terms

$$p^0 q^k, \ C(k, 1)p^1 q^{k-1}, \ C(k, 2)p^2 q^{k-2}, \ C(k, 3)p^3 q^{k-3}, \ldots$$

The first term represents no rare events and k frequent events in a sample of k events. The second term represents one rare event and $k - 1$ frequent events. The third term represents two rare events and $k - 2$ frequent events, and so forth. The expressions of the form $C(k, i)$ are the binomial coefficients, represented by the combinatorial terms discussed in the previous section. Although the desired tail of the curve could be computed by this expression, as long as sufficient decimal accuracy is maintained, it is customary in such cases to compute another distribution, the Poisson distribution, which closely approximates the desired results. As a rule of thumb, we may use the Poisson distribution to approximate the binomial distribution when the probability of the rare event p is less than 0.1 and the product kp (sample size × probability) is less than 5.

The *Poisson distribution* is also a discrete frequency distribution of the number of times a rare event occurs. But, in contrast to the binomial distribution, the Poisson distribution applies to cases where the number of times that an event does not occur is infinitely large. For purposes of our treatment here, a Poisson variable will be studied in samples taken over space or time. An example of the first would be the number of moss plants in a sampling quadrat on a hillside or the number of parasites on an individual host; an example of a temporal sample is the number of mutations occurring in a genetic strain in the time interval of one month or the reported cases of influenza in one town during one week. The Poisson variable Y will be the number of events per sample. It can assume discrete values from 0 on up. To be distributed in Poisson fashion the variable must have two properties: (1) Its mean must be small relative to the maximum possible number of events per sampling unit. Thus the event should be "rare." But this means that our sampling unit of space or time must be large enough to accommodate a potentially substantial number of events. For example, a quadrat in which moss plants are counted must be large enough that a substantial number of moss plants could occur there physically if the biological conditions were such as to favor the development of numerous moss plants in the quadrat. A quadrat consisting of a 1-cm square would be far too small for mosses to be distributed in Poisson fashion. Similarly, a time span of 1 minute would be unrealistic for reporting new influenza cases in a town, but within 1 week a great many such cases could occur. (2) An occurrence of the event must be independent of prior occurrences within the sampling unit. Thus, the presence of one moss plant in a quadrat must not enhance or diminish the probability that other moss plants are developing in the quadrat. Similarly, the fact that one influenza case has been reported must not affect the probability of reporting subsequent influenza cases. Events that meet these conditions (*rare and random* events) should be distributed in Poisson fashion.

The purpose of fitting a Poisson distribution to numbers of rare events in nature is to test whether the events occur independently with respect to each other. If they do, they will follow the Poisson distribution. If the occurrence of one event enhances the probability of a second such event, we obtain a clumped, or contagious, distribution. If the occurrence of one event impedes that of a second such event in the sampling unit, we obtain a repulsed, or spatially or temporally uniform, distribution. The Poisson can be used as a test for randomness or independence of distribution not only spatially but also in time, as some examples below will show.

The Poisson distribution is named after the French mathematician Poisson, who described it in 1837. It is an infinite series whose terms add to 1 (as must be true for any probability distribution). The series can be represented as

$$\frac{1}{e^{\mu}}, \quad \frac{\mu}{1!e^{\mu}}, \quad \frac{\mu^2}{2!e^{\mu}}, \quad \frac{\mu^3}{3!e^{\mu}}, \quad \frac{\mu^4}{4!e^{\mu}}, \ldots, \quad \frac{\mu^r}{r!e^{\mu}}, \ldots \qquad (4.2)$$

where the terms are the relative expected frequencies corresponding to the following counts of the rare event Y:

$$0, \quad 1, \quad 2, \quad 3, \quad 4, \ldots, \quad r, \ldots$$

Thus, the first of these terms represents the relative expected frequency of samples containing no rare event; the second term, one rare event; the third term, two rare events; and so on. The denominator of each term contains e^{μ}, where e is the base of the natural, or Napierian, logarithms, a constant whose value, accurate to 5 decimal places, is 2.718,28. We recognize μ as the parametric mean of the distribution; it is a constant for any given problem. The exclamation mark after the coefficient in the denominator means "factorial," as explained in the previous section.

One way to learn more about the Poisson distribution is to apply it to an actual case. At the top of Box 4.1 is a well-known result from the early statistical literature based on the distribution of yeast cells in 400 squares of a hemacytometer, a counting chamber such as is used in making counts of blood cells and other microscopic objects suspended in liquid. Column (1) lists the number of yeast cells observed in each hemacytometer square, and column (2) gives the observed frequency—the number of squares containing a given number of yeast cells. We note that 75 squares contained no yeast cells, but that most squares held either 1 or 2 cells. Only 17 squares contained 5 or more yeast cells.

Why would we expect this frequency distribution to be distributed in Poisson fashion? We have here a relatively rare event, the frequency of yeast cells per hemacytometer square, the mean of which has been calculated and found to be 1.8. That is, on the average there are 1.8 cells per square. Relative to the amount of space provided in each square and the number of cells that could have come to rest in any one square, the actual number found is low indeed. We might also expect that the occurrence of individual yeast cells in a square is independent of the occurrence of other yeast cells. This is a commonly encountered class of application of the Poisson distribution.

The mean of the rare event is the only quantity that we need to know to calculate the relative expected frequencies of a Poisson distribution. Since we do

BOX 4.1
Calculation of expected Poisson frequencies.

Yeast cells in 400 squares of a hemacytometer: $\bar{Y} = 1.8$ cells per square; $n = 400$ squares sampled.

(1) Number of cells per square Y	(2) Observed frequencies f	(3) Absolute expected frequencies $\hat{f}$	(4) Deviation from expectation $f - \hat{f}$
0	75	66.1	+
1	103	119.0	−
2	121	107.1	+
3	54	64.3	−
4	30	28.9	+
5	13 ⎫	10.4 ⎫	+ ⎫
6	2 ⎪	3.1 ⎪	− ⎪
7	1 ⎬ 17	0.8 ⎬ 14.5	+ ⎬ +
8	0 ⎪	0.2 ⎪	− ⎪
9	1 ⎭	0.0 ⎭	+ ⎭
	$\overline{400}$	$\overline{399.9}$	

Source: "Student" (1907).

Computational steps

Flow of computation based on Expression (4.3) multiplied by n, since we wish to obtain absolute expected frequencies, $\hat{f}$.

1. Find $e^{\bar{Y}}$ in a table of exponentials or compute it using an exponential key:

$e^{\bar{Y}} = e^{1.8} = 6.0496$

2. $\hat{f}_0 = \dfrac{n}{e^{\bar{Y}}} = \dfrac{400}{6.0496} = 66.12$

3. $\hat{f}_1 = \hat{f}_0 \bar{Y} = 66.12(1.8) = 119.02$

4. $\hat{f}_2 = \hat{f}_1 \dfrac{\bar{Y}}{2} = 119.02\left(\dfrac{1.8}{2}\right) = 107.11$

5. $\hat{f}_3 = \hat{f}_2 \dfrac{\bar{Y}}{3} = 107.11\left(\dfrac{1.8}{3}\right) = 64.27$

6. $\hat{f}_4 = \hat{f}_3 \dfrac{\bar{Y}}{4} = 64.27\left(\dfrac{1.8}{4}\right) = 28.92$

7. $\hat{f}_5 = \hat{f}_4 \dfrac{\bar{Y}}{5} = 28.92\left(\dfrac{1.8}{5}\right) = 10.41$

8. $\hat{f}_6 = \hat{f}_5 \dfrac{\bar{Y}}{6} = 10.41\left(\dfrac{1.8}{6}\right) = 3.12$

BOX 4.1
Continued

9. $\hat{f}_7 = \hat{f}_6 \dfrac{\bar{Y}}{7} = 3.12\left(\dfrac{1.8}{7}\right) \quad = \quad 0.80$

10. $\hat{f}_8 = \hat{f}_7 \dfrac{\bar{Y}}{8} = 0.80\left(\dfrac{1.8}{8}\right) \quad = \quad 0.18$

    Total $\overline{399.95}$

    $\hat{f}_9$ and beyond   0.05

At step **3** enter $\bar{Y}$ as a constant multiplier. Then multiply it by n/e^Y (quantity **2**). At each subsequent step multiply the result of the previous step by $\bar{Y}$ and then divide by the appropriate integer.

---

not know the parametric mean of the yeast cells in this problem, we employ an estimate (the sample mean) and calculate expected frequencies of a Poisson distribution with μ equal to the mean of the observed frequency distribution of Box 4.1. It is convenient for the purpose of computation to rewrite Expression (4.2) as a recursion formula as follows:

$$\hat{f}_i = \hat{f}_{i-1}\left(\frac{\bar{Y}}{i}\right) \quad \text{for } i = 1, 2, \ldots, \quad \text{where } \hat{f}_0 = e^{-\bar{Y}} \qquad (4.3)$$

Note first of all that the parametric mean μ has been replaced by the sample mean $\bar{Y}$. Each term developed by this recursion formula is mathematically exactly the same as its corresponding term in Expression (4.2). It is important to make no computational error, since in such a chain multiplication the correctness of each term depends on the accuracy of the term before it. Expression (4.3) yields relative expected frequencies. If, as is more usual, absolute expected frequencies are desired, simply set the first term $\hat{f}_0$ to n/e^Y, where n is the number of samples, and then proceed with the computational steps as before. The actual computation is illustrated in Box 4.1, and the expected frequencies so obtained are listed in column (3) of the frequency distribution.

What have we learned from this computation? When we compare the observed with the expected frequencies, we notice quite a good fit of our observed frequencies to a Poisson distribution of mean 1.8, although we have not as yet learned a statistical test for goodness of fit (this will be covered in Chapter 13). No clear pattern of deviations from expectation is shown. We cannot test a hypothesis about the mean, because the mean of the expected distribution was taken from the sample mean of the observed variates. As in the binomial distribution, clumping or aggregation would indicate that the probability that a second yeast cell will be found in a square is not independent of the presence of the first one, but is higher than the probability for the first cell. This would result in a clumping of the items in the classes at the tails of the distribution so that there would be some squares with larger numbers of cells than expected, others with fewer numbers.

The biological interpretation of the dispersion pattern varies with the problem. The yeast cells seem to be randomly distributed in the counting chamber, indicating thorough mixing of the suspension. Red blood cells, on the other hand, will often stick together because of an electrical charge unless the proper suspension fluid is used. This so-called rouleaux effect would be indicated by clumping of the observed frequencies.

Note that in Box 4.1, as in the subsequent tables giving examples of the application of the Poisson distribution, we group the low frequencies at one tail of the curve, uniting them by means of a bracket. This tends to simplify the patterns of distribution somewhat. However, the main reason for this grouping is related to the G test for goodness of fit (of observed to expected frequencies), which is discussed in Section 13.2. For purposes of this test, no expected frequency $\hat{f}$ should be less than 5.

Before we turn to other examples, we need to learn a few more facts about the Poisson distribution. You probably noticed that in computing expected frequencies, we needed to know only one parameter—the mean of the distribution. By comparison, in the binomial distribution we needed two parameters, p and k. Thus, the mean completely defines the shape of a given Poisson distribution. From this it follows that the variance is some function of the mean. In a Poisson distribution, we have a very simple relationship between the two: $\mu = \sigma^2$, the variance being equal to the mean. The variance of the number of yeast cells per square based on the observed frequencies in Box 4.1 equals 1.965, not much larger than the mean of 1.8, indicating again that the yeast cells are distributed in Poisson fashion, hence randomly. This relationship between variance and mean suggests a rapid test of whether an observed frequency distribution is distributed in Poisson fashion even without fitting expected frequencies to the data. We simply compute a *coefficient of dispersion*

$$CD = \frac{s^2}{\bar{Y}}$$

This value will be near 1 in distributions that are essentially Poisson distributions, will be >1 in clumped samples, and will be <1 in cases of repulsion. In the yeast cell example, $CD = 1.092$.

The shapes of five Poisson distributions of different means are shown in Figure 4.3 as frequency polygons (a frequency polygon is formed by the line connecting successive midpoints in a bar diagram). We notice that for the low value of $\mu = 0.1$ the frequency polygon is extremely L-shaped, but with an increase in the value of μ the distributions become humped and eventually nearly symmetrical.

We conclude our study of the Poisson distribution with a consideration of two examples. The first example (Table 4.5) shows the distribution of a number

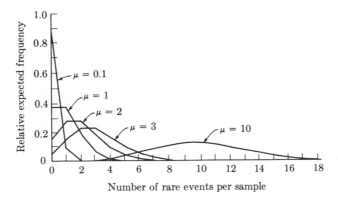

FIGURE 4.3
Frequency polygons of the Poisson distribution for various values of the mean.

of accidents per woman from an accident record of 647 women working in a munitions factory during a five-week period. The sampling unit is one woman during this period. The rare event is the number of accidents that happened to a woman in this period. The coefficient of dispersion is 1.488, and this is clearly reflected in the observed frequencies, which are greater than expected in the tails and less than expected in the center. This relationship is easily seen in the deviations in the last column (observed minus expected frequencies) and shows a characteristic clumped pattern. The model assumes, of course, that the accidents are not fatal or very serious and thus do not remove the individual from further exposure. The noticeable clumping in these data probably arises

TABLE 4.5
Accidents in 5 weeks to 647 women working on high-explosive shells.

(1) Number of accidents per woman	(2) Observed frequencies f	(3) Poisson expected frequencies $\hat{f}$	(4) Deviation from expectation $f - \hat{f}$
0	447	406.3	+
1	132	189.0	−
2	42	44.0	−
3	21 ⎫	6.8 ⎫	+ ⎫
4	3 ⎬ 26	0.8 ⎬ 7.7	+ ⎬ +
5+	2 ⎭	0.1 ⎭	+ ⎭
Total	647	647.0	

$$\bar{Y} = 0.4652 \qquad s^2 = 0.692 \qquad CD = 1.488$$

Source: Greenwood and Yule (1920).

TABLE 4.6

Azuki bean weevils (*Callosobruchus chinensis*) emerging from 112 Azuki beans (*Phaseolus radiatus*).

(1) Number of weevils emerging per bean Y	(2) Observed frequencies f	(3) Poisson expected frequencies f̂	(4) Deviation from expectation f − f̂
0	61	70.4	−
1	50	32.7	+
2	1 ⎫	7.6 ⎫	− ⎫
3	0 ⎬ 1	1.2 ⎬ 8.9	− ⎬ −
4	0 ⎭	0.1 ⎭	− ⎭
Total	112	112.0	

$$\bar{Y} = 0.4643 \qquad s^2 = 0.269 \qquad CD = 0.579$$

Source: Utida (1943).

either because some women are accident-prone or because some women have more dangerous jobs than others. Using only information on the distributions of accidents, one cannot distinguish between the two alternatives, which suggest very different changes that should be made to reduce the numbers of accidents.

The second example (Table 4.6) is extracted from an experimental study of the effects of different densities of the Azuki bean weevil. Larvae of these weevils enter the beans, feed and pupate inside them, and then emerge through an emergence hole. Thus the number of holes per bean is a good measure of the number of adults that have emerged. The rare event in this case is the presence of the weevil in the bean. We note that the distribution is strongly repulsed. There are many more beans containing one weevil than the Poisson distribution would predict. A statistical finding of this sort leads us to investigate the biology of the phenomenon. In this case it was found that the adult female weevils tended to deposit their eggs evenly rather than randomly over the available beans. This prevented the placing of too many eggs on any one bean and precluded heavy competition among the developing larvae on any one bean. A contributing factor was competition among remaining larvae feeding on the same bean, in which generally all but one were killed or driven out. Thus, it is easily understood how the above biological phenomena would give rise to a repulsed distribution.

Exercises

4.1 The two columns below give fertility of eggs of the CP strain of *Drosophila melanogaster* raised in 100 vials of 10 eggs each (data from R. R. Sokal). Find the expected frequencies on the assumption of independence of mortality for each egg in a vial. Use the observed mean. Calculate the expected variance and compare it with the observed variance. Interpret results, knowing that the eggs of each vial are siblings and that the different vials contain descendants from different parent pairs. ANS. $\sigma^2 = 2.417$, $s^2 = 6.636$. There is evidence that mortality rates are different for different vials.

Number of eggs hatched Y	Number of vials f
0	1
1	3
2	8
3	10
4	6
5	15
6	14
7	12
8	13
9	9
10	9

4.2 In human beings the sex ratio of newborn infants is about $100♀♀:105♂♂$. Were we to take 10,000 random samples of 6 newborn infants from the total population of such infants for one year, what would be the expected frequency of groups of 6 males, 5 males, 4 males, and so on?

4.3 The Army Medical Corps is concerned over the intestinal disease X. From previous experience it knows that soldiers suffering from the disease invariably harbor the pathogenic organism in their feces and that to all practical purposes every stool specimen from a diseased person contains the organism. However, the organisms are never abundant, and thus only 20% of all slides prepared by the standard procedure will contain some. (We assume that if an organism is present on a slide it will be seen.) How many slides should laboratory technicians be directed to prepare and examine per stool specimen, so that in case a specimen is positive, it will be erroneously diagnosed negative in fewer than 1% of the cases (on the average)? On the basis of your answer, would you recommend that the Corps attempt to improve its diagnostic methods? ANS. 21 slides.

4.4 Calculate Poisson expected frequencies for the frequency distribution given in Table 2.2 (number of plants of the sedge *Carex flacca* found in 500 quadrats).

4.5 A cross is made in a genetic experiment in *Drosophila* in which it is expected that $\frac{1}{4}$ of the progeny will have white eyes and $\frac{1}{2}$ will have the trait called "singed bristles." Assume that the two gene loci segregate independently. (a) What proportion of the progeny should exhibit both traits simultaneously? (b) If four flies are sampled at random, what is the probability that they will all be white-eyed? (c) What is the probability that none of the four flies will have either white eyes or "singed bristles?" (d) If two flies are sampled, what is the probability that at least one of the flies will have either white eyes or "singed bristles" or both traits? ANS. (a) $\frac{1}{8}$; (b) $(\frac{1}{4})^4$; (c) $[(1 - \frac{1}{4})(1 - \frac{1}{2})]^4$; (d) $1 - [(1 - \frac{1}{4})(1 - \frac{1}{2})]^2$.

4.6 Those readers who have had a semester or two of calculus may wish to try to prove that Expression (4.1) tends to Expression (4.2) as k becomes indefinitely large (and p becomes infinitesimal, so that $\mu = kp$ remains constant). HINT:

$$\left(1 - \frac{x}{n}\right)^n \rightarrow e^{-x} \quad \text{as} \quad n \rightarrow \infty$$

4.7   If the frequency of the gene A is p and the frequency of the gene a is q, what are the expected frequencies of the zygotes AA, Aa, and aa (assuming a diploid zygote represents a random sample of size 2)? What would the expected frequency be for an autotetraploid (for a locus close to the centromere a zygote can be thought of as a random sample of size 4)? ANS. $P\{AA\} = p^2$, $P\{Aa\} = 2pq$, $P\{aa\} = q^2$, for a diploid; and $P\{AAAA\} = p^4$, $P\{AAAa\} = 4p^3q$, $P\{AAaa\} = 6p^2q^2$, $P\{Aaaa\} = 4pq^3$, $P\{aaaa\} = q^4$, for a tetraploid.

4.8   Summarize and compare the assumptions and parameters on which the binomial and Poisson distributions are based.

4.9   A population consists of three types of individuals, A_1, A_2, and A_3, with relative frequencies of 0.5, 0.2, and 0.3, respectively. (a) What is the probability of obtaining only individuals of type A_1 in samples of size 1, 2, 3, ..., n? (b) What would be the probabilities of obtaining only individuals that were not of type A_1 or A_2 in a sample of size n? (c) What is the probability of obtaining a sample containing at least one representation of each type in samples of size 1, 2, 3, 4, 5, ..., n? ANS. (a) $\frac{1}{2}, \frac{1}{4}, \frac{1}{8}, \ldots, 1/2^n$. (b) $(0.3)^n$. (c) 0, 0, 0.18, 0.36, 0.507, ...,

$$\text{for } n: \quad \sum_{i=1}^{n-2} \sum_{j=1}^{n-i-1} \frac{n!}{i!j!(n-i-j)!} (0.5)^i (0.2)^j (0.3)^{n-i-j}$$

4.10   If the average number of weed seeds found in a $\frac{1}{4}$-ounce sample of grass seed is 1.1429, what would you expect the frequency distribution of weed seeds to be in ninety-eight $\frac{1}{4}$-ounce samples? (Assume there is random distribution of the weed seeds.)

CHAPTER 5

The Normal
Probability Distribution

The theoretical frequency distributions in Chapter 4 were discrete. Their variables assumed values that changed in integral steps (that is, they were meristic variables). Thus, the number of infected insects per sample could be 0 or 1 or 2 but never an intermediate value between these. Similarly, the number of yeast cells per hemacytometer square is a meristic variable and requires a discrete probability function to describe it. However, most variables encountered in biology either are continuous (such as the aphid femur lengths or the infant birth weights used as examples in Chapters 2 and 3) or can be treated as continuous variables for most practical purposes, even though they are inherently meristic (such as the neutrophil counts encountered in the same chapters). Chapter 5 will deal more extensively with the distributions of continuous variables.

Section 5.1 introduces frequency distributions of continuous variables. In Section 5.2 we show one way of deriving the most common such distribution, the normal probability distribution. Then we examine its properties in Section 5.3. A few applications of the normal distribution are illustrated in Section 5.4. A graphic technique for pointing out departures from normality and for estimating mean and standard deviation in approximately normal distributions is given in Section 5.5, as are some of the reasons for departure from normality in observed frequency distributions.

5.1 Frequency distributions of continuous variables

For continuous variables, the theoretical probability distribution, or *probability density function*, can be represented by a continuous curve, as shown in Figure 5.1. The ordinate of the curve gives the density for a given value of the variable shown along the abscissa. By *density* we mean the relative concentration of variates along the Y axis (as indicated in Figure 2.1). In order to compare the theoretical with the observed frequency distribution, it is necessary to divide the two into corresponding classes, as shown by the vertical lines in Figure 5.1. Probability density functions are defined so that the expected frequency of observations between two class limits (vertical lines) is given by the area between these limits under the curve. The total area under the curve is therefore equal to the sum of the expected frequencies (1.0 or n, depending on whether relative or absolute expected frequencies have been calculated).

When you form a frequency distribution of observations of a continuous variable, your choice of class limits is arbitrary, because all values of a variable are theoretically possible. In a continuous distribution, one cannot evaluate the probability that the variable will be exactly equal to a given value such as 3 or 3.5. One can only estimate the frequency of observations falling between two limits. This is so because the area of the curve corresponding to any point along the curve is an infinitesimal. Thus, to calculate expected frequencies for a continuous distribution, we have to calculate the area under the curve between the class limits. In Sections 5.3 and 5.4, we shall see how this is done for the normal frequency distribution.

Continuous frequency distributions may start and terminate at finite points along the Y axis, as shown in Figure 5.1, or one or both ends of the curve may extend indefinitely, as will be seen later in Figures 5.3 and 6.11. The idea of an area under a curve when one or both ends go to infinity may trouble those of you not acquainted with calculus. Fortunately, however, this is not a great conceptual stumbling block, since in all the cases that we shall encounter, the tail

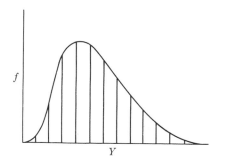

FIGURE 5.1
A probability distribution of a continuous variable.

of the curve will approach the Y axis rapidly enough that the portion of the area beyond a certain point will for all practical purposes be zero and the frequencies it represents will be infinitesimal.

We may fit continuous frequency distributions to some sets of meristic data (for example, the number of teeth in an organism). In such cases, we have reason to believe that underlying biological variables that cause differences in numbers of the structure are really continuous, even though expressed as a discrete variable.

We shall now proceed to discuss the most important probability density function in statistics, the normal frequency distribution.

5.2 Derivation of the normal distribution

There are several ways of deriving the normal frequency distribution from elementary assumptions. Most of these require more mathematics than we expect of our readers. We shall therefore use a largely intuitive approach, which we have found of heuristic value. Some inherently meristic variables, such as counts of blood cells, range into the thousands. Such variables can, for practical purposes, be treated as though they were continuous.

Let us consider a binomial distribution of the familiar form $(p + q)^k$ in which k becomes indefinitely large. What type of biological situation could give rise to such a binomial distribution? An example might be one in which many factors cooperate additively in producing a biological result. The following hypothetical case is possibly not too far removed from reality. The intensity of skin pigmentation in an animal will be due to the summation of many factors, some genetic, others environmental. As a simplifying assumption, let us state that every factor can occur in two states only: present or absent. When the factor is present, it contributes one unit of pigmentation to skin color, but it contributes nothing to pigmentation when it is absent. Each factor, regardless of its nature or origin, has the identical effect, and the effects are additive: if three out of five possible factors are present in an individual, the pigmentation intensity will be three units, or the sum of three contributions of one unit each. One final assumption: Each factor has an equal probability of being present or absent in a given individual. Thus, $p = P[F] = 0.5$, the probability that the factor is present; while $q = P[f] = 0.5$, the probability that the factor is absent.

With only one factor ($k = 1$), expansion of the binomial $(p + q)^1$ would yield two pigmentation classes among the animals, as follows:

$\{F, \quad f \ \}$ pigmentation classes (probability space)
$\{0.5, \quad 0.5\}$ expected frequency
$\{1, \quad 0 \ \}$ pigmentation intensity

Half the animals would have intensity 1, the other half 0. With $k = 2$ factors present in the population (the factors are assumed to occur independently of each other), the distribution of pigmentation intensities would be represented by

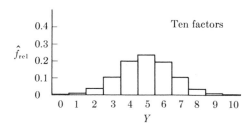

FIGURE 5.2
Histogram based on relative expected frequencies resulting from expansion of binomial $(0.5 + 0.5)^{10}$. The Y axis measures the number of pigmentation factors F.

the expansion of the binomial $(p + q)^2$:

$\{FF,$	$Ff,$	$ff\ \}$	pigmentation classes (probability space)
$\{0.25,$	$0.50,$	$0.25\}$	expected frequency
$\{2,$	$1,$	$0\ \}$	pigmentation intensity

One-fourth of the individuals would have pigmentation intensity 2; one-half, intensity 1; and the remaining fourth, intensity 0.

The number of classes in the binomial increases with the number of factors. The frequency distributions are symmetrical, and the expected frequencies at the tails become progressively less as k increases. The binomial distribution for $k = 10$ is graphed as a histogram in Figure 5.2 (rather than as a bar diagram, as it should be drawn). We note that the graph approaches the familiar bell-shaped outline of the normal frequency distribution (seen in Figures 5.3 and 5.4). Were we to expand the expression for $k = 20$, our histogram would be so close to a normal frequency distribution that we could not show the difference between the two on a graph the size of this page.

At the beginning of this procedure, we made a number of severe limiting assumptions for the sake of simplicity. What happens when these are removed? First, when $p \neq q$, the distribution also approaches normality as k approaches infinity. This is intuitively difficult to see, because when $p \neq q$, the histogram is at first asymmetrical. However, it can be shown that when k, p, and q are such that $kpq \geq 3$, the normal distribution will be closely approximated. Second, in a more realistic situation, factors would be permitted to occur in more than two states—one state making a large contribution, a second state a smaller contribution, and so forth. However, it can also be shown that the multinomial $(p + q + r + \cdots + z)^k$ approaches the normal frequency distribution as k approaches infinity. Third, different factors may be present in different frequencies and may have different quantitative effects. As long as these are additive and independent, normality is still approached as k approaches infinity.

Lifting these restrictions makes the assumptions leading to a normal distribution compatible with innumerable biological situations. It is therefore not surprising that so many biological variables are approximately normally distributed.

Let us summarize the conditions that tend to produce normal frequency distributions: (1) that there be many factors; (2) that these factors be independent in occurrence; (3) that the factors be independent in effect—that is, that their effects be additive; and (4) that they make equal contributions to the variance. The fourth condition we are not yet in a position to discuss; we mention it here only for completeness. It will be discussed in Chapter 7.

5.3 Properties of the normal distribution

Formally, the *normal probability density function* can be represented by the expression

$$Z = \frac{1}{\sigma\sqrt{2\pi}}\, e^{-\frac{1}{2}\left(\frac{Y-\mu}{\sigma}\right)^2} \tag{5.1}$$

Here Z indicates the height of the ordinate of the curve, which represents the density of the items. It is the dependent variable in the expression, being a function of the variable Y. There are two constants in the equation: π, well known to be approximately 3.141,59, making $1/\sqrt{2\pi}$ approximately 0.398,94, and e, the base of the natural logarithms, whose value approximates 2.718,28.

There are two parameters in a normal probability density function. These are the parametric mean μ and the parametric standard deviation σ, which determine the location and shape of the distribution. Thus, there is not just one normal distribution, as might appear to the uninitiated who keep encountering the same bell-shaped image in textbooks. Rather, there are an infinity of such curves, since these parameters can assume an infinity of values. This is illustrated by the three normal curves in Figure 5.3, representing the same total frequencies.

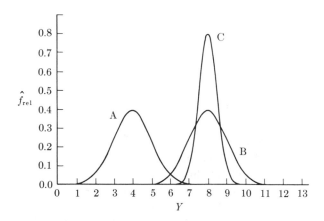

FIGURE 5.3
Illustration of how changes in the two parameters of the normal distribution affect the shape and location of the normal probability density function. (A) $\mu = 4$, $\sigma = 1$; (B) $\mu = 8$, $\sigma = 1$; (C) $\mu = 8$, $\sigma = 0.5$.

Curves A and B differ in their locations and hence represent populations with different means. Curves B and C represent populations that have identical means but different standard deviations. Since the standard deviation of curve C is only half that of curve B, it presents a much narrower appearance.

In theory, a normal frequency distribution extends from negative infinity to positive infinity along the axis of the variable (labeled Y, although it is frequently the abscissa). This means that a normally distributed variable can assume any value, however large or small, although values farther from the mean than plus or minus three standard deviations are quite rare, their relative expected frequencies being very small. This can be seen from Expression (5.1). When Y is very large or very small, the term $(Y - \mu)^2/2\sigma^2$ will necessarily become very large. Hence e raised to the negative power of that term will be very small, and Z will therefore be very small.

The curve is symmetrical around the mean. Therefore, the mean, median, and mode of the normal distribution are all at the same point. The following percentages of items in a normal frequency distribution lie within the indicated limits:

$\mu \pm \sigma$ contains 68.27% of the items
$\mu \pm 2\sigma$ contains 95.45% of the items
$\mu \pm 3\sigma$ contains 99.73% of the items

Conversely,

50% of the items fall in the range $\mu \pm 0.674\sigma$
95% of the items fall in the range $\mu \pm 1.960\sigma$
99% of the items fall in the range $\mu \pm 2.576\sigma$

These relations are shown in Figure 5.4.

How have these percentages been calculated? The direct calculation of any portion of the area under the normal curve requires an integration of the function shown as Expression (5.1). Fortunately, for those of you who do not know calculus (and even for those of you who do) the integration has already been carried out and is presented in an alternative form of the normal distribution: the *normal distribution function* (the theoretical *cumulative* distribution function of the normal probability density function), also shown in Figure 5.4. It gives the total frequency from negative infinity up to any point along the abscissa. We can therefore look up directly the probability that an observation will be less than a specified value of Y. For example, Figure 5.4 shows that the total frequency up to the mean is 50.00% and the frequency up to a point one standard deviation below the mean is 15.87%. These frequencies are found, graphically, by raising a vertical line from a point, such as $-\sigma$, until it intersects the cumulative distribution curve, and then reading the frequency (15.87%) off the ordinate. The probability that an observation will fall between two arbitrary points can be found by subtracting the probability that an observation will fall below the

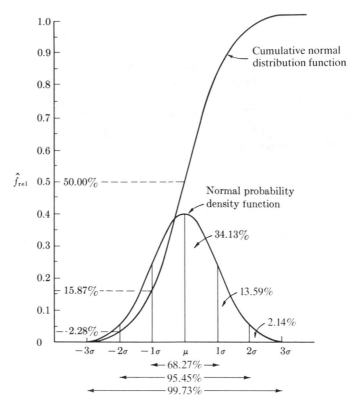

FIGURE 5.4
Areas under the normal probability density function and the cumulative normal distribution
function.

lower point from the probability that an observation will fall below the upper point. For example, we can see from Figure 5.4 that the probability that an observation will fall between the mean and a point one standard deviation below the mean is $0.5000 - 0.1587 = 0.3413$.

The normal distribution function is tabulated in Table **II** in Appendix A2, "Areas of the normal curve," where, for convenience in later calculations, 0.5 has been subtracted from all of the entries. This table therefore lists the proportion of the area between the mean and any point a given number of standard deviations above it. Thus, for example, the area between the mean and the point 0.50 standard deviations above the mean is 0.1915 of the total area of the curve. Similarly, the area between the mean and the point 2.64 standard deviations above the mean is 0.4959 of the curve. A point 4.0 standard deviations from the mean includes 0.499,968 of the area between it and the mean. However, since the normal distribution extends from negative to positive infinity, one needs to go an infinite distance from the mean to reach an area of 0.5. The use of the table of areas of the normal curve will be illustrated in the next section.

A sampling experiment will give you a "feel" for the distribution of items sampled from a normal distribution.

Experiment 5.1. You are asked to sample from two populations. The first one is an approximately normal frequency distribution of 100 wing lengths of houseflies. The second population deviates strongly from normality. It is a frequency distribution of the total annual milk yield of 100 Jersey cows. Both populations are shown in Table 5.1. You are asked to sample from them repeatedly in order to simulate sampling from an infinite population. Obtain samples of 35 items from each of the two populations. This can be done by obtaining two sets of 35 two-digit random numbers from the table of random numbers (Table I), with which you became familiar in Experiment 4.1. Write down the random numbers in blocks of five, and copy next to them the value of Y (for either wing length or milk yield) corresponding to the random number. An example of such a block of five numbers and the computations required for it are shown in the

TABLE 5.1
Populations of wing lengths and milk yields. *Column 1.* Rank number. *Column 2.* Lengths (in mm $\times 10^{-1}$) of 100 wings of houseflies arrayed in order of magnitude; $\mu = 45.5$, $\sigma^2 = 15.21$, $\sigma = 3.90$; distribution approximately normal. *Column 3.* Total annual milk yield (in hundreds of pounds) of 100 two-year-old registered Jersey cows arrayed in order of magnitude; $\mu = 66.61$, $\sigma^2 = 124.4779$, $\sigma = 11.1597$; distribution departs strongly from normality.

(1)	(2)	(3)	(1)	(2)	(3)	(1)	(2)	(3)	(1)	(2)	(3)	(1)	(2)	(3)
01	36	51	21	42	58	41	45	61	61	47	67	81	49	76
02	37	51	22	42	58	42	45	61	62	47	67	82	49	76
03	38	51	23	42	58	43	45	61	63	47	68	83	49	79
04	38	53	24	43	58	44	45	61	64	47	68	84	49	80
05	39	53	25	43	58	45	45	61	65	47	69	85	50	80
06	39	53	26	43	58	46	45	62	66	47	69	86	50	81
07	40	54	27	43	58	47	45	62	67	47	69	87	50	82
08	40	55	28	43	58	48	45	62	68	47	69	88	50	82
09	40	55	29	43	58	49	45	62	69	47	69	89	50	82
10	40	56	30	43	58	50	45	63	70	48	69	90	50	82
11	41	56	31	43	58	51	46	63	71	48	70	91	51	83
12	41	56	32	44	59	52	46	63	72	48	72	92	51	85
13	41	57	33	44	59	53	46	64	73	48	73	93	51	87
14	41	57	34	44	59	54	46	65	74	48	73	94	51	88
15	41	57	35	44	60	55	46	65	75	48	74	95	52	88
16	41	57	36	44	60	56	46	65	76	48	74	96	52	89
17	42	57	37	44	60	57	46	65	77	48	74	97	53	93
18	42	57	38	44	60	58	46	65	78	49	74	98	53	94
19	42	57	39	44	60	59	46	67	79	49	75	99	54	96
20	42	57	40	44	61	60	46	67	80	49	76	00	55	98

Source: Column 2—Data adapted from Sokal and Hunter (1955). Column 3—Data from Canadian government records.

following listing, using the housefly wing lengths as an example:

Random number	Wing length Y
16	41
59	46
99	54
36	44
21	42
$\sum Y =$	227
$\sum Y^2 = 10{,}413$	
$\bar{Y} =$	45.4

Those with ready access to a computer may prefer to program this exercise and take many more samples. These samples and the computations carried out for each sample will be used in subsequent chapters. Therefore, preserve your data carefully!

In this experiment, consider the 35 variates for each variable as a single sample, rather than breaking them down into groups of five. Since the true mean and standard deviation (μ and σ) of the two distributions are known, you can calculate the expression $(Y_i - \mu)/\sigma$ for each variate Y_i. Thus, for the first housefly wing length sampled above, you compute

$$\frac{41 - 45.5}{3.90} = -1.1538$$

This means that the first wing length is 1.1538 standard deviations below the true mean of the population. The deviation from the mean measured in standard deviation units is called a *standardized deviate* or *standard deviate*. The arguments of Table II, expressing distance from the mean in units of σ, are called *standard normal deviates*. Group all 35 variates in a frequency distribution; then do the same for milk yields. Since you know the parametric mean and standard deviation, you need not compute each deviate separately, but can simply write down class limits in terms of the actual variable as well as in standard deviation form. The class limits for such a frequency distribution are shown in Table 5.2. Combine the results of your sampling with those of your classmates and study the percentage of the items in the distribution one, two, and three standard deviations to each side of the mean. Note the marked differences in distribution between the housefly wing lengths and the milk yields.

5.4 Applications of the normal distribution

The normal frequency distribution is the most widely used distribution in statistics, and time and again we shall have recourse to it in a variety of situations. For the moment, we may subdivide its applications as follows.

TABLE 5.2
Table for recording frequency distributions of standard deviates $(Y_i - \mu)/\sigma$ for samples of Experiment 5.1.

	Wing lengths			Milk yields	
	Variates falling between these limits	f		Variates falling between these limits	f
$-\infty$			$-\infty$		
-3σ	_____	_____	-3σ	_____	_____
$-2\frac{1}{2}\sigma$	_____	_____	$-2\frac{1}{2}\sigma$	_____	_____
-2σ	36, 37	_____	-2σ	_____	_____
$-1\frac{1}{2}\sigma$	38, 39	_____	$-1\frac{1}{2}\sigma$	_____	_____
$-\sigma$	40, 41	_____	$-\sigma$	51–55	_____
$-\frac{1}{2}\sigma$	42, 43	_____	$-\frac{1}{2}\sigma$	56–61	_____
$\mu = 45.5$	44, 45	_____	$\mu = 66.61$	62–66	_____
$\frac{1}{2}\sigma$	46, 47	_____	$\frac{1}{2}\sigma$	67–72	_____
σ	48, 49	_____	σ	73–77	_____
$1\frac{1}{2}\sigma$	50, 51	_____	$1\frac{1}{2}\sigma$	78–83	_____
2σ	52, 53	_____	2σ	84–88	_____
$2\frac{1}{2}\sigma$	54, 55	_____	$2\frac{1}{2}\sigma$	89–94	_____
3σ	_____	_____	3σ	95–98	_____
$+\infty$			$+\infty$		

1. We sometimes have to know whether a given sample is normally distributed before we can apply a certain test to it. To test whether a given sample is normally distributed, we have to calculate expected frequencies for a normal curve of the same mean and standard deviation using the table of areas of the normal curve. In this book we shall employ only approximate graphic methods for testing normality. These are featured in the next section.

2. Knowing whether a sample is normally distributed may confirm or reject certain underlying hypotheses about the nature of the factors affecting the phenomenon studied. This is related to the conditions making for normality in a frequency distribution, discussed in Section 5.2. Thus, if we find a given variable to be normally distributed, we have no reason for rejecting the hypothesis that the causal factors affecting the variable are additive and independent and of equal variance. On the other hand, when we find departure from normality, this may indicate certain forces, such as selection, affecting the variable under study. For instance, bimodality may indicate a mixture

of observations from two populations. Skewness of milk yield data may indicate that these are records of selected cows and substandard milk cows have not been included in the record.
3. If we assume a given distribution to be normal, we may make predictions and tests of given hypotheses based upon this assumption. (An example of such an application follows.)

You will recall the birth weights of male Chinese children, illustrated in Box 3.2. The mean of this sample of 9465 birth weights is 109.9 oz, and its standard deviation is 13.593 oz. If you sample at random from the birth records of this population, what is your chance of obtaining a birth weight of 151 oz or heavier? Such a birth weight is considerably above the mean of our sample, the difference being $151 - 109.9 = 41.1$ oz. However, we cannot consult the table of areas of the normal curve with a difference in ounces. We must express it in *standardized* units—that is, divide it by the standard deviation to convert it into a standard deviate. When we divide the difference by the standard deviation, we obtain $41.1/13.593 = 3.02$. This means that a birth weight of 151 oz is 3.02 standard deviation units greater than the mean. Assuming that the birth weights are normally distributed, we may consult the table of areas of the normal curve (Table II), where we find a value of 0.4987 for 3.02 standard deviations. This means that 49.87% of the area of the curve lies between the mean and a point 3.02 standard deviations from it. Conversely, 0.0013, or 0.13%, of the area lies beyond 3.02 standard deviation units above the mean. Thus, assuming a normal distribution of birth weights and a value of $\sigma = 13.593$, only 0.13%, or 13 out of 10,000, of the infants would have a birth weight of 151 oz *or farther* from the mean. It is quite improbable that a single sampled item from that population would deviate by so much from the mean, and if such a random sample of one weight were obtained from the records of an unspecified population, we might be justified in doubting whether the observation did in fact come from the population known to us.

The above probability was calculated from one tail of the distribution. We found the probability that an individual would be *greater* than the mean by 3.02 or more standard deviations. If we are not concerned whether the individual is either heavier or lighter than the mean but wish to know only how different the individual is from the population mean, an appropriate question would be: Assuming that the individual belongs to the population, what is the probability of observing a birth weight of an individual deviant by a certain amount from the mean in either direction? That probability must be computed by using both tails of the distribution. The previous probability can be simply doubled, since the normal curve is symmetrical. Thus, $2 \times 0.0013 = 0.0026$. This, too, is so small that we would conclude that a birth weight as deviant as 151 oz is unlikely to have come from the population represented by our sample of male Chinese children.

We can learn one more important point from this example. Our assumption has been that the birth weights are normally distributed. Inspection of the frequency distribution in Box 3.2, however, shows clearly that the distribution is asymmetrical, tapering to the right. Though there are eight classes above the mean class, there are only six classes below the mean class. In view of this asymmetry, conclusions about one tail of the distribution would not necessarily pertain to the second tail. We calculated that 0.13% of the items would be found beyond 3.02 standard deviations above the mean, which corresponds to 151 oz. In fact, our sample contains 20 items (14 + 5 + 1) beyond the 147.5-oz class, the upper limit of which is 151.5 oz, almost the same as the single birth weight. However, 20 items of the 9465 of the sample is approximately 0.21%, more than the 0.13% expected from the normal frequency distribution. Although it would still be improbable to find a single birth weight as heavy as 151 oz in the sample, conclusions based on the assumption of normality might be in error if the exact probability were critical for a given test. Our statistical conclusions are only as valid as our assumptions about the population from which the samples are drawn.

5.5 Departures from normality: Graphic methods

In many cases an observed frequency distribution will depart obviously from normality. We shall emphasize two types of departure from normality. One is *skewness*, which is another name for asymmetry; skewness means that one tail of the curve is drawn out more than the other. In such curves the mean and the median will not coincide. Curves are said to be skewed to the right or left, depending upon whether the right or left tail is drawn out.

The other type of departure from normality is *kurtosis*, or "peakedness" of a curve. A *leptokurtic* curve has more items near the mean and at the tails, with fewer items in the intermediate regions relative to a normal distribution with the same mean and variance. A *platykurtic* curve has fewer items at the mean and at the tails than the normal curve but has more items in intermediate regions. A bimodal distribution is an extreme platykurtic distribution.

Graphic methods have been developed that examine the shape of an observed distribution for departures from normality. These methods also permit estimates of the mean and standard deviation of the distribution without computation.

The graphic methods are based on a cumulative frequency distribution. In Figure 5.4 we saw that a normal frequency distribution graphed in cumulative fashion describes an S-shaped curve, called a sigmoid curve. In Figure 5.5 the ordinate of the sigmoid curve is given as relative frequencies expressed as percentages. The slope of the cumulative curve reflects changes in height of the frequency distribution on which it is based. Thus the steep middle segment of the cumulative normal curve corresponds to the relatively greater height of the normal curve around its mean.

The ordinate in Figures 5.4 and 5.5 is in linear scale, as is the abscissa in Figure 5.4. Another possible scale is the *normal probability scale* (often simply called *probability scale*), which can be generated by dropping perpendiculars

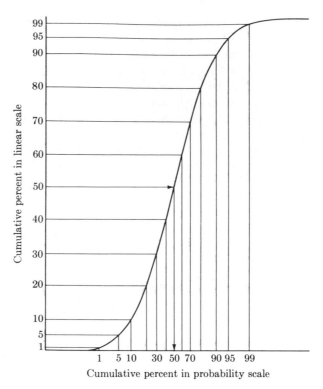

FIGURE 5.5
Transformation of cumulative percentages into normal probability scale.

from the cumulative normal curve, corresponding to given percentages on the ordinate, to the abscissa (as shown in Figure 5.5). The scale represented by the abscissa compensates for the nonlinearity of the cumulative normal curve. It contracts the scale around the median and expands it at the low and high cumulative percentages. This scale can be found on *arithmetic* or *normal probability graph paper* (or simply *probability graph paper*), which is generally available. Such paper usually has the long edge graduated in probability scale, while the short edge is in linear scale. Note that there are no 0% or 100% points on the ordinate. These points cannot be shown, since the normal frequency distribution extends from negative to positive infinity and thus however long we made our line we would never reach the limiting values of 0% and 100%.

If we graph a cumulative normal distribution with the ordinate in normal probability scale, it will lie exactly on a straight line. Figure 5.6A shows such a graph drawn on probability paper, while the other parts of Figure 5.6 show a series of frequency distributions variously departing from normality. These are graphed both as ordinary frequency distributions with density on a linear scale (ordinate not shown) and as cumulative distributions as they would appear on

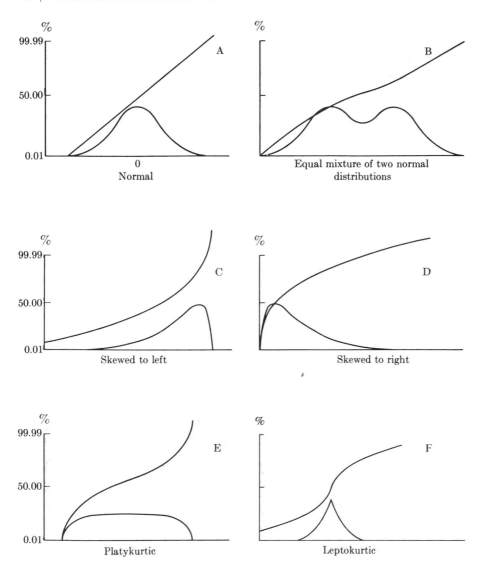

FIGURE 5.6
Examples of some frequency distributions with their cumulative distributions plotted with the ordinate in normal probability scale. (See Box 5.1 for explanation.)

probability paper. They are useful as guidelines for examining the distributions of data on probability paper.

Box 5.1 shows you how to use probability paper to examine a frequency distribution for normality and to obtain graphic estimates of its mean and standard deviation. The method works best for fairly large samples ($n > 50$). The method does not permit the plotting of the last cumulative frequency, 100%,

BOX 5.1
**Graphic test for normality of a frequency distribution and estimate of mean and
standard deviation. Use of arithmetic probability paper.**

Birth weights of male Chinese in ounces, from Box 3.2.

(1) Class mark Y	(2) Upper class limit	(3) f	(4) Cumulative frequencies F	(5) Percent cumulative frequencies
59.5	63.5	2	2	0.02
67.5	71.5	6	8	0.08
75.5	79.5	39	47	0.50
83.5	87.5	385	432	4.6
91.5	95.5	888	1320	13.9
99.5	103.5	1729	3049	32.2
107.5	111.5	2240	5289	55.9
115.5	119.5	2007	7296	77.1
123.5	127.5	1233	8529	90.1
131.5	135.5	641	9170	96.9
139.5	143.5	201	9371	99.0
147.5	151.5	74	9445	99.79
155.5	159.5	14	9459	99.94
163.5	167.5	5	9464	99.99
171.5	175.5	1	9465	100.0
		9465		

Computational steps

1. Prepare a frequency distribution as shown in columns (1), (2), and (3).

2. Form a cumulative frequency distribution as shown in column (4). It is obtained
   by successive summation of the frequency values. In column (5) express the
   cumulative frequencies as percentages of total sample size *n*, which is 9465 in
   this example. These percentages are 100 times the values of column (4) divided
   by 9465.

3. Graph the upper class limit of each class along the abscissa (in linear scale)
   against percent cumulative frequency along the ordinate (in probability scale)
   on normal probability paper (see Figure 5.7). A straight line is fitted to the points
   by eye, preferably using a transparent plastic ruler, which permits all the points
   to be seen as the line is drawn. In drawing the line, most weight should be
   given to the points between cumulative frequencies of 25% to 75%. This is
   because a difference of a single item may make appreciable changes in the
   percentages at the tails. We notice that the upper frequencies deviate to the right
   of the straight line. This is typical of data that are skewed to the right (see
   Figure 5.6D).

4. Such a graph permits the rapid estimation of the mean and standard deviation
   of a sample. The mean is approximated by a graphic estimation of the median.
   The more normal the distribution is, the closer the mean will be to the median.

BOX 5.1

Continued

The median is estimated by dropping a perpendicular from the intersection of the 50% point on the ordinate and the cumulative frequency curve to the abscissa (see Figure 5.7). The estimate of the mean of 110.7 oz is quite close to the computed mean of 109.9 oz.

5. The standard deviation can be estimated by dropping similar perpendiculars from the intersections of the 15.9% and the 84.1% points with the cumulative curve, respectively. These points enclose the portion of a normal curve represented by $\mu \pm \sigma$. By measuring the difference between these perpendiculars and dividing this by 2, we obtain an estimate of one standard deviation. In this instance the estimate is $s = 13.6$, since the difference is 27.2 oz divided by 2. This is a close approximation to the computed value of 13.59 oz.

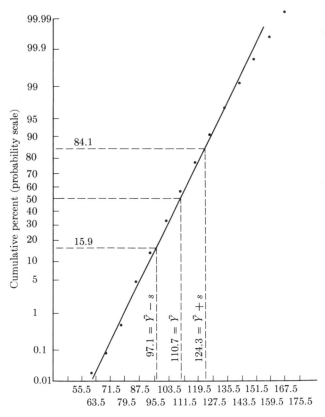

Birth weights of male Chinese (in oz.)

FIGURE 5.7
Graphic analysis of data from Box 5.1.

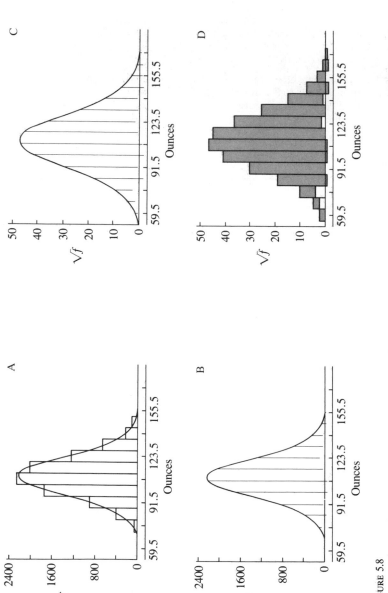

FIGURE 5.8

(A) Histogram of the observed frequency distribution of birth weights of male Chinese from Box 5.1 with expected normal curve superimposed. (B) The data in A displayed as a hanging histogram. Frequency bars are suspended from the expected normal curve. Bars that do not reach the abscissa indicate deficiencies from expectations. Bars extending below the abscissa indicate observed frequencies in excess of expectation. (C) The data in B shown as a hanging rootogram. Both observed and expected frequencies are given as square roots of the actual values. Departures from expectation in the tails of the distribution are accentuated. (D) Comparing observed and expected frequencies for the birth weight data in A. The histogram of the skyline indicates the expected frequencies; that of the "inverted skyline" indicates the observed frequencies. Both frequencies are given in square roots of actual values. Where the invered skyline does not reach the abscissa, there are fewer observed than expected frequencies. Wherever it reaches below the abscissa, there is an excess of observed frequencies over expected frequencies.

since that corresponds to an infinite distance from the mean. If you are interested in plotting all observations, you can plot, instead of cumulative frequencies F, the quantity $F - \frac{1}{2}$ expressed as a percentage of n.

Often it is desirable to compare observed frequency distributions with their expectations without resorting to cumulative frequency distributions. One method of doing so would be to superimpose a normal curve on the histogram of an observed frequency distribution. Fitting a normal distribution as a curve superimposed upon an observed frequency distribution in the form of a histogram is usually done only when graphic facilities (plotters) are available. Ordinates are computed by modifying Expression (5.1) to conform to a frequency distribution:

$$Z = \frac{ni}{s\sqrt{2\pi}} e^{-\frac{1}{2}\left(\frac{Y-\mu}{\sigma}\right)^2} \tag{5.2}$$

In this expression n is the sample size and i is the class interval of the frequency distribution. If this needs to be done without a computer program, a table of ordinates of the normal curve is useful. In Figure 5.8A we show the frequency distribution of birth weights of male Chinese from Box 5.1 with the ordinates of the normal curve superimposed. There is an excess of observed frequencies at the right tail due to the skewness of the distribution.

You will probably find it difficult to compare the heights of bars against the arch of a curve. For this reason, John Tukey has suggested that the bars of the histograms be suspended from the curve. Their departures from expectation can then be easily observed against the straight-line abscissa of the graph. Such a hanging histogram is shown in Figure 5.8B for the birth weight data. The departure from normality is now much clearer.

Because important departures are frequently noted in the tails of a curve, it has been suggested that square roots of expected frequencies should be compared with the square roots of observed frequencies. Such a "hanging rootogram" is shown in Figure 5.8C for the Chinese birth weight data. Note the accentuation of the departure from normality. Finally, one can also use an analogous technique for comparing expected with observed histograms. Figure 5.8D shows the same data plotted in this manner. Square roots of frequencies are again shown. The excess of observed over expected frequencies in the right tail of the distribution is quite evident.

Exercises

5.1   Using the information given in Box 3.2, what is the probability of obtaining an individual with a negative birth weight? What is this probability if we assume that birth weights are normally distributed? ANS. The empirical estimate is zero. If a normal distribution can be assumed, it is the probability that a standard normal deviate is less than $(0 - 109.9)/13.593 = -8.085$. This value is beyond the range of most tables, and the probability can be considered zero for practical purposes.

5.2 Carry out the operations listed in Exercise 5.1 on the transformed data generated in Exercise 2.6.

5.3 Assume you know that the petal length of a population of plants of species X is normally distributed with a mean of $\mu = 3.2$ cm and a standard deviation of $\sigma = 1.8$. What proportion of the population would be expected to have a petal length (a) greater than 4.5 cm? (b) Greater than 1.78 cm? (c) Between 2.9 and 3.6 cm? ANS. (a) $= 0.2353$, (b) $= 0.7845$, and (c) $= 0.154$.

5.4 Perform a graphic analysis of the butterfat data given in Exercise 3.3, using probability paper. In addition, plot the data on probability paper with the abscissa in logarithmic units. Compare the results of the two analyses.

5.5 Assume that traits A and B are independent and normally distributed with parameters $\mu_A = 28.6$, $\sigma_A = 4.8$, $\mu_B = 16.2$, and $\sigma_B = 4.1$. You sample two individuals at random (a) What is the probability of obtaining samples in which both individuals measure less than 20 for the two traits? (b) What is the probability that at least one of the individuals is greater than 30 for trait B? ANS. (a) $P\{A < 20\}P\{B < 20\} = (0.3654)(0.082,38) = 0.030$; (b) $1 - (P\{A < 30\}) \times (P\{B < 30\}) = 1 - (0.6147)(0.9960) = 0.3856$.

5.6 Perform the following operations on the data of Exercise 2.4. (a) If you have not already done so, make a frequency distribution from the data and graph the results in the form of a histogram. (b) Compute the expected frequencies for each of the classes based on a normal distribution with $\mu = \bar{Y}$ and $\sigma = s$. (c) Graph the expected frequencies in the form of a histogram and compare them with the observed frequencies. (d) Comment on the degree of agreement between observed and expected frequencies.

5.7 Let us approximate the observed frequencies in Exercise 2.9 with a normal frequency distribution. Compare the observed frequencies with those expected when a normal distribution is assumed. Compare the two distributions by forming and superimposing the observed and the expected histograms and by using a hanging histogram. ANS. The expected frequencies for the age classes are: 17.9, 48.2, 72.0, 51.4, 17.5, 3.0. This is clear evidence for skewness in the observed distribution.

5.8 Perform a graphic analysis on the following measurements. Are they consistent with what one would expect in sampling from a normal distribution?

11.44	12.88	11.06	7.02	10.25	6.26	7.92	12.53	6.74
15.81	9.46	21.27	9.72	6.37	5.40	3.21	6.50	3.40
5.60	14.20	6.60	10.42	8.18	11.09	8.74		

5.9 The following data are total lengths (in cm) of bass from a southern lake:

29.9	40.2	37.8	19.7	30.0	29.7	19.4	39.2	24.7	20.4
19.1	34.7	33.5	18.3	19.4	27.3	38.2	16.2	36.8	33.1
41.4	13.6	32.2	24.3	19.1	37.4	23.8	33.3	31.6	20.1
17.2	13.3	37.7	12.6	39.6	24.6	18.6	18.0	33.7	38.2

Compute the mean, the standard deviation, and the coefficient of variation. Make a histogram of the data. Do the data seem consistent with a normal distribution on the basis of a graphic analysis? If not, what type of departure is suggested? ANS. $\bar{Y} = 27.4475$, $s = 8.9035$, $V = 32.438$. There is a suggestion of bimodality.

CHAPTER 6

Estimation and
Hypothesis Testing

In this chapter we provide methods to answer two fundamental statistical questions that every biologist must ask repeatedly in the course of his or her work: (1) how reliable are the results I have obtained? and (2) how probable is it that the differences between observed results and those expected on the basis of a hypothesis have been produced by chance alone? The first question, about reliability, is answered through the setting of confidence limits to sample statistics. The second question leads into hypothesis testing. Both subjects belong to the field of statistical inference. The subject matter in this chapter is fundamental to an understanding of any of the subsequent chapters.

In Section 6.1 we consider the form of the distribution of means and their variance. In Section 6.2 we examine the distributions and variances of statistics other than the mean. This brings us to the general subject of standard errors, which are statistics measuring the reliability of an estimate. Confidence limits provide bounds to our estimates of population parameters. We develop the idea of a confidence limit in Section 6.3 and show its application to samples where the true standard deviation is known. However, one usually deals with small, more or less normally distributed samples with unknown standard deviations, in which case the t distribution must be used. We shall introduce the t distribution in Section 6.4. The application of t to the computation of confidence limits for statistics of small samples with unknown population standard deviations is shown in Section 6.5. Another important distribution, the chi-square distribution, is explained in Section 6.6. Then it is applied to setting confidence limits for the variance in Section 6.7. The theory of hypothesis testing is introduced in Section 6.8 and is applied in Section 6.9 to a variety of cases exhibiting the normal or t distributions. Finally, Section 6.10 illustrates hypothesis testing for variances by means of the chi-square distribution.

6.1 Distribution and variance of means

We commence our study of the distribution and variance of means with a sampling experiment.

Experiment 6.1 You were asked to retain from Experiment 5.1 the means of the seven samples of 5 housefly wing lengths and the seven similar means of milk yields. We can collect these means from every student in a class, possibly adding them to the sampling results of previous classes, and construct a frequency distribution of these means. For each variable we can also obtain the mean of the seven means, which is a mean of a sample 35 items. Here again we shall make a frequency distribution of these means, although it takes a considerable number of samplers to accumulate a sufficient number of samples of 35 items for a meaningful frequency distribution.

In Table 6.1 we show a frequency distribution of 1400 means of samples of 5 housefly wing lengths. Consider columns (1) and (3) for the time being. Actually, these samples were obtained not by biostatistics classes but by a digital computer, enabling us to collect these values with little effort. Their mean and standard deviation are given at the foot of the table. These values are plotted on probability paper in Figure 6.1. Note that the distribution appears quite normal, as does that of the means based on 200 samples of 35 wing lengths shown in the same figure. This illustrates an important theorem: *The means of samples from a normally distributed population are themselves normally distributed regardless of sample size n.* Thus, we note that the means of samples from the normally distributed housefly wing lengths are normally distributed whether they are based on 5 or 35 individual readings.

Similarly obtained distributions of means of the heavily skewed milk yields, as shown in Figure 6.2, appear to be close to normal distributions. However, the means based on five milk yields do not agree with the normal nearly as well as do the means of 35 items. This illustrates another theorem of fundamental importance in statistics: *As sample size increases, the means of samples drawn from a population of any distribution will approach the normal distribution.* This theorem, when rigorously stated (about sampling from populations with finite variances), is known as the *central limit theorem.* The importance of this theorem is that if n is large enough, it permits us to use the normal distri-

TABLE 6.1

Frequency distribution of means of 1400 random samples of 5 housefly wing lengths. (Data from Table 5.1.) Class marks chosen to give intervals of $\frac{1}{2}\sigma_{\bar{Y}}$ to each side of the parametric mean μ.

	(1) Class mark Y (in mm × 10^{-1})	(2) Class mark (in $\sigma_{\bar{Y}}$ units)	(3) f
	39.832	$-3\frac{1}{4}$	1
	40.704	$-2\frac{3}{4}$	11
	41.576	$-2\frac{1}{4}$	19
	42.448	$-1\frac{3}{4}$	64
	43.320	$-1\frac{1}{4}$	128
	44.192	$-\frac{3}{4}$	247
$\mu = 45.5 \rightarrow$	45.064	$-\frac{1}{4}$	226
	45.936	$\frac{1}{4}$	259
	46.808	$\frac{3}{4}$	231
	47.680	$1\frac{1}{4}$	121
	48.552	$1\frac{3}{4}$	61
	49.424	$2\frac{1}{4}$	23
	50.296	$2\frac{3}{4}$	6
	51.168	$3\frac{1}{4}$	3
			1400
	$\bar{Y} = 45.480$	$s = 1.778$	$\sigma_{\bar{Y}} = 1.744$

bution to make statistical inferences about means of populations in which the items are not at all normally distributed. The necessary size of n depends upon the distribution. (Skewed populations require larger sample sizes.)

The next fact of importance that we note is that the range of the means is considerably less than that of the original items. Thus, the wing-length means range from 39.4 to 51.6 in samples of 5 and from 43.9 to 47.4 in samples of 35, but the individual wing lengths range from 36 to 55. The milk-yield means range from 54.2 to 89.0 in samples of 5 and from 61.9 to 71.3 in samples of 35, but the individual milk yields range from 51 to 98. Not only do means show less scatter than the items upon which they are based (an easily understood phenomenon if you give some thought to it), but the range of the distribution of the means diminishes as the sample size upon which the means are based increases.

The differences in ranges are reflected in differences in the standard deviations of these distributions. If we calculate the standard deviations of the means

Samples of 5

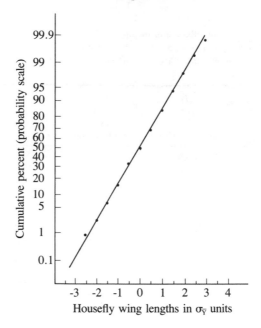

Samples of 35

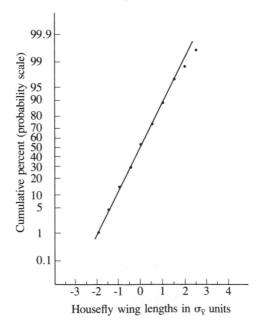

FIGURE 6.1
Graphic analysis of means of 1400 random samples of 5 housefly wing lengths (from Table 6.1) and of means of 200 random samples of 35 housefly wing lengths.

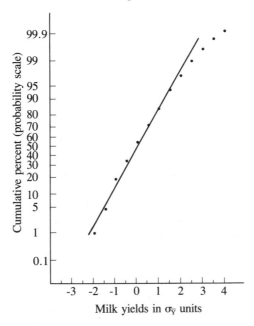

Samples of 5

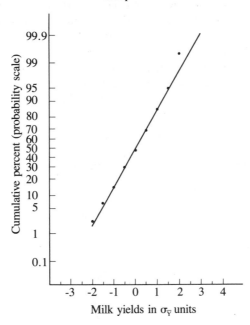

Samples of 35

FIGURE 6.2
Graphic analysis of means of 1400 random samples of 5 milk yields and of means of 200 random samples of 35 milk yields.

in the four distributions under consideration, we obtain the following values:

	Observed standard deviations of distributions of means	
	$n = 5$	$n = 35$
Wing lengths	1.778	0.584
Milk yields	5.040	1.799

Note that the standard deviations of the sample means based on 35 items are considerably less than those based on 5 items. This is also intuitively obvious. Means based on large samples should be close to the parametric mean, and means based on large samples will not vary as much as will means based on small samples. The variance of means is therefore partly a function of the sample size on which the means are based. It is also a function of the variance of the items in the samples. Thus, in the text table above, the means of milk yields have a much greater standard deviation than means of wing lengths based on comparable sample size simply because the standard deviation of the individual milk yields (11.1597) is considerably greater than that of individual wing lengths (3.90).

It is possible to work out the expected value of the variance of sample means. By *expected value* we mean *the average value to be obtained by infinitely repeated sampling.* Thus, if we were to take samples of a means of n items repeatedly and were to calculate the variance of these a means each time, the average of these variances would be the expected value. We can visualize the mean as a weighted average of the n independently sampled observations with each weight w_i equal to 1. From Expression (3.2) we obtain

$$\bar{Y}_w = \frac{\sum_{i}^{n} w_i Y_i}{\sum^{n} w_i}$$

for the weighted mean. We shall state without proof that the variance of the weighted sum of *independent* items $\sum^n w_i Y_i$ is

$$\text{Var}\left(\sum_{i}^{n} w_i Y_i\right) = \sum_{i}^{n} w_i^2 \sigma_i^2 \qquad (6.1)$$

where σ_i^2 is the variance of Y_i. It follows that

$$\sigma_{\bar{Y}_w}^2 = \frac{\sum_{i}^{n} w_i^2 \sigma_i^2}{\left(\sum^{n} w_i\right)^2}$$

Since the weights w_i in this case equal 1, $\sum^n w_i = n$, and we can rewrite the above expression as

$$\sigma_{\bar{Y}_w}^2 = \frac{\sum_{i}^{n} \sigma_i^2}{n^2}$$

If we assume that the variances σ_i^2 are all equal to σ^2, the expected variance of the mean is

$$\sigma_{\bar{Y}}^2 = \frac{n\sigma^2}{n^2} = \frac{\sigma^2}{n} \qquad (6.2)$$

and consequently, the expected standard deviation of means is

$$\sigma_{\bar{Y}} = \frac{\sigma}{\sqrt{n}} \qquad (6.2a)$$

From this formula it is clear that the standard deviation of means is a function of the standard deviation of items as well as of sample size of means. The greater the sample size, the smaller will be the standard deviation of means. In fact, as sample size increases to a very large number, the standard deviation of means becomes vanishingly small. This makes good sense. Very large sample sizes, averaging many observations, should yield estimates of means closer to the population mean and less variable than those based on a few items.

When working with samples from a population, we do not, of course, know its parametric standard deviation σ, and we can obtain only a sample estimate s of the latter. Also, we would be unlikely to have numerous samples of size n from which to compute the standard deviation of means directly. Customarily, we therefore have to estimate the standard deviation of means from a single sample by using Expression (6.2a), substituting s for σ:

$$s_{\bar{Y}} = \frac{s}{\sqrt{n}} \qquad (6.3)$$

Thus, from the standard deviation of a single sample, we obtain, an estimate of the standard deviation of means we would expect were we to obtain a collection of means based on equal-sized samples of n items from the same population. As we shall see, this estimate of the standard deviation of a mean is a very important and frequently used statistic.

Table 6.2 illustrates some estimates of the standard deviations of means that might be obtained from random samples of the two populations that we have been discussing. The means of 5 samples of wing lengths based on 5 individuals ranged from 43.6 to 46.8, their standard deviations from 1.095 to 4.827, and the estimate of standard deviation of the means from 0.490 to 2.159. Ranges for the other categories of samples in Table 6.2 similarly include the parametric values of these statistics. The estimates of the standard deviations of the means of the milk yields cluster around the expected value, since they are not dependent on normality of the variates. However, in a particular sample in which by chance the sample standard deviation is a poor estimate of the population standard deviation (as in the second sample of 5 milk yields), the estimate of the standard deviation of means is equally wide of the mark.

We should emphasize one point of difference between the standard deviation of items and the standard deviation of sample means. If we estimate a population standard deviation through the standard deviation of a sample, the magnitude of the estimate will not change as we increase our sample size. We may expect that the estimate will improve and will approach the true standard

TABLE **6.2**
Means, standard deviations, and standard deviations of means (standard errors) of five random samples of 5 and 35 housefly wing lengths and Jersey cow milk yields, respectively. (Data from Table 5.1.) Parametric values for the statistics are given in the sixth line of each category.

	(1) $\bar{Y}$	(2) s	(3) $s_{\bar{Y}}$
Wing lengths			
	45.8	1.095	0.490
	45.6	3.209	1.435
$n = 5$	43.6	4.827	2.159
	44.8	4.764	2.131
	46.8	1.095	0.490
	$\mu = 45.5$	$\sigma = 3.90$	$\sigma_{\bar{Y}} = 1.744$
	45.37	3.812	0.644
	45.00	3.850	0.651
$n = 35$	45.74	3.576	0.604
	45.29	4.198	0.710
	45.91	3.958	0.669
	$\mu = 45.5$	$\sigma = 3.90$	$\sigma_{\bar{Y}} = 0.659$
Milk yields			
	66.0	6.205	2.775
	61.6	4.278	1.913
$n = 5$	67.6	16.072	7.188
	65.0	14.195	6.348
	62.2	5.215	2.332
	$\mu = 66.61$	$\sigma = 11.160$	$\sigma_{\bar{Y}} = 4.991$
	65.429	11.003	1.860
	64.971	11.221	1.897
$n = 35$	66.543	9.978	1.687
	64.400	9.001	1.521
	68.914	12.415	2.099
	$\mu = 66.61$	$\sigma = 11.160$	$\sigma_{\bar{Y}} = 1.886$

deviation of the population. However, its order of magnitude will be the same, whether the sample is based on 3, 30, or 3000 individuals. This can be seen clearly in Table 6.2. The values of s are closer to σ in the samples based on $n = 35$ than in samples of $n = 5$. Yet the general magnitude is the same in both instances. The standard deviation of means, however, decreases as sample size increases, as is obvious from Expression (6.3). Thus, means based on 3000 items will have a standard deviation only one-tenth that of means based on 30 items. This is obvious from

$$\frac{s}{\sqrt{3000}} = \frac{s}{\sqrt{30} \times \sqrt{100}} = \frac{s}{\sqrt{30} \times 10}$$

6.2 Distribution and variance of other statistics

Just as we obtained a mean and a standard deviation from each sample of the wing lengths and milk yields, so we could also have obtained other statistics from each sample, such as a variance, a median, or a coefficient of variation. After repeated sampling and computation, we would have frequency distributions for these statistics and would be able to compute their standard deviations, just as we did for the frequency distribution of means. In many cases the statistics are normally distributed, as was true for the means. In other cases the statistics will be distributed normally only if they are based on samples from a normally distributed population, or if they are based on large samples, or if both these conditions hold. In some instances, as in variances, their distribution is never normal. An illustration is given in Figure 6.3, which shows a frequency distribution of the variances from the 1400 samples of 5 housefly wing lengths. We notice that the distribution is strongly skewed to the right, which is characteristic of the distribution of variances.

Standard deviations of various statistics are generally known as *standard errors*. Beginners sometimes get confused by an imagined distinction between standard deviations and standard errors. The standard error of a statistic such as the mean (or V) is the standard deviation of a distribution of means (or V's) for samples of a given sample size n. Thus, the terms "standard error" and "standard deviation" are used synonymously, with the following exception: it is not customary to use "standard error" as a synonym of "standard deviation" for items in a sample or population. Standard error or standard deviation has to be qualified by referring to a given statistic, such as the standard deviation

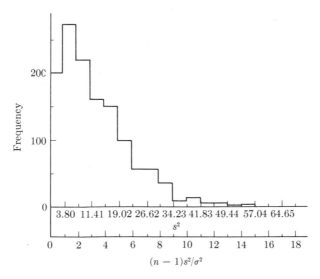

FIGURE 6.3
Histogram of variances based on 1400 samples of 5 housefly wing lengths from Table 5.1. Abscissa is given in terms of s^2 and $(n - 1)s^2/\sigma^2$.

of V, which is the same as the standard error of V. Used without any qualification, the term "standard error" conventionally implies the standard error of the mean. "Standard deviation" used without qualification generally means standard deviation of items in a sample or population. Thus, when you read that means, standard deviations, standard errors, and coefficients of variation are shown in a table, this signifies that arithmetic means, standard deviations of items in samples, standard deviations of their means ($=$ standard errors of means), and coefficients of variation are displayed. The following summary of terms may be helpful:

Standard deviation $= s = \sqrt{\Sigma y^2/(n-1)}$.
Standard deviation of a statistic $St =$ standard error of a statistic $St = s_{St}$.
Standard error $=$ standard error of a mean
$\qquad\qquad\quad = $ standard deviation of a mean $= s_{\bar{Y}}$.

Standard errors are usually not obtained from a frequency distribution by repeated sampling but are estimated from only a single sample and represent the expected standard deviation of the statistic in case a large number of such samples had been obtained. You will remember that we estimated the standard error of a distribution of means from a single sample in this manner in the previous section.

Box 6.1 lists the standard errors of four common statistics. Column (1) lists the statistic whose standard error is described; column (2) shows the formula

---

BOX 6.1
Standard errors for common statistics.

	(1) Statistic	(2) Estimate of standard error	(3) df	(4) Comments on applicability
1	$\bar{Y}$	$s_{\bar{Y}} = \dfrac{s}{\sqrt{n}} = \dfrac{s_Y}{\sqrt{n}} = \sqrt{\dfrac{s^2}{n}}$	$n-1$	True for any population with finite variance
2	Median	$s_{\text{med}} \approx (1.2533)s_{\bar{Y}}$	$n-1$	Large samples from normal populations
3	s	$s_s = (0.7071068)\dfrac{s}{\sqrt{n}}$	$n-1$	Samples from normal populations ($n > 15$)
4	V	$s_V = \dfrac{V}{\sqrt{2n}}\sqrt{1 + 2\left(\dfrac{V}{100}\right)^2}$	$n-1$	Samples from normal populations
		$s_V \approx \dfrac{V}{\sqrt{2n}}$	$n-1$	Used when $V < 15$

for the estimated standard error; column (3) gives the degrees of freedom on which the standard error is based (their use is explained in Section 6.5); and column (4) provides comments on the range of application of the standard error. The uses of these standard errors will be illustrated in subsequent sections.

6.3 Introduction to confidence limits

The various sample statistics we have been obtaining, such as means or standard deviations, are estimates of population parameters μ or σ, respectively. So far we have not discussed the reliability of these estimates. We first of all wish to know whether the sample statistics are *unbiased estimators* of the population parameters, as discussed in Section 3.7. But knowing, for example, that $\bar{Y}$ is an unbiased estimate of μ is not enough. We would like to find out how reliable a measure of μ it is. The true values of the parameters will almost always remain unknown, and we commonly estimate reliability of a sample statistic by setting confidence limits to it.

To begin our discussion of this topic, let us start with the unusual case of a population whose parametric mean and standard deviation are known to be μ and σ, respectively. The mean of a sample of n items is symbolized by $\bar{Y}$. The expected standard error of the mean is $\sigma/\sqrt{n}$. As we have seen, the sample means will be normally distributed. Therefore, from Section 5.3, the region from $1.96\sigma/\sqrt{n}$ below μ to $1.96\sigma/\sqrt{n}$ above μ includes 95% of the sample means of size n. Another way of stating this is to consider the ratio $(\bar{Y} - \mu)/(\sigma/\sqrt{n})$. This is the standard deviate of a sample mean from the parametric mean. Since they are normally distributed, 95% of such standard deviates will lie between -1.96 and $+1.96$. We can express this statement symbolically as follows:

$$P\left\{-1.96 \leq \frac{\bar{Y} - \mu}{\sigma/\sqrt{n}} \leq +1.96\right\} = 0.95$$

This means that the probability P that the sample means $\bar{Y}$ will differ by no more than 1.96 standard errors $\sigma/\sqrt{n}$ from the parametric mean μ equals 0.95. The expression between the brackets is an inequality, all terms of which can be multiplied by $\sigma/\sqrt{n}$ to yield

$$\left\{-1.96 \frac{\sigma}{\sqrt{n}} \leq (\bar{Y} - \mu) \leq +1.96 \frac{\sigma}{\sqrt{n}}\right\}$$

We can rewrite this expression as

$$\left\{-1.96 \frac{\sigma}{\sqrt{n}} \leq (\mu - \bar{Y}) \leq +1.96 \frac{\sigma}{\sqrt{n}}\right\}$$

because $-a \leq b \leq a$ implies $a \geq -b \geq -a$, which can be written as $-a \leq -b \leq a$. And finally, we can transfer $-\bar{Y}$ across the inequality signs, just as in an equation it could be transferred across the equal sign. This yields the final desired expression:

$$P\left\{\bar{Y} - \frac{1.96\sigma}{\sqrt{n}} \leq \mu \leq \bar{Y} + \frac{1.96\sigma}{\sqrt{n}}\right\} = 0.95 \qquad (6.4)$$

or

$$P\{\bar{Y} - 1.96\sigma_{\bar{Y}} \leq \mu \leq \bar{Y} + 1.96\sigma_{\bar{Y}}\} = 0.95 \qquad (6.4a)$$

This means that the probability P that the term $\bar{Y} - 1.96\sigma_{\bar{Y}}$ is less than or equal to the parametric mean μ and that the term $\bar{Y} + 1.96\sigma_{\bar{Y}}$ is greater than or equal to μ is 0.95. The two terms $\bar{Y} - 1.96\sigma_{\bar{Y}}$ and $\bar{Y} + 19.6\sigma_{\bar{Y}}$ we shall call L_1 and L_2, respectively, the lower and upper 95% *confidence limits* of the mean.

Another way of stating the relationship implied by Expression (6.4a) is that if we repeatedly obtained samples of size n from the population and constructed these limits for each, we could expect 95% of the intervals between these limits to contain the true mean, and only 5% of the intervals would miss μ. The interval from L_1 to L_2 is called a *confidence interval*.

If you were not satisfied to have the confidence interval contain the true mean only 95 times out of 100, you might employ 2.576 as a coefficient in place of 1.960. You may remember that 99% of the area of the normal curve lies in the range $\mu \pm 2.576\sigma$. Thus, to calculate 99% confidence limits, compute the two quantities $L_1 = \bar{Y} - 2.576\sigma/\sqrt{n}$ and $L_2 = \bar{Y} + 2.576\sigma/\sqrt{n}$ as lower and upper confidence limits, respectively. In this case 99 out of 100 confidence intervals obtained in repeated sampling would contain the true mean. The new confidence interval is wider than the 95% interval (since we have multiplied by a greater coefficient). If you were still not satisfied with the reliability of the confidence limit, you could increase it, multiplying the standard error of the mean by 3.291 to obtain 99.9% confidence limits. This value could be found by inverse interpolation in a more extensive table of areas of the normal curve or directly in a table of the inverse of the normal probability distribution. The new coefficient would widen the interval further. Notice that you can construct confidence intervals that will be expected to contain μ an increasingly greater percentage of the time. First you would expect to be right 95 times out of 100, then 99 times out of 100, finally 999 times out of 1000. But as your confidence increases, your statement becomes vaguer and vaguer, since the confidence interval lengthens. Let us examine this by way of an actual sample.

We obtain a sample of 35 housefly wing lengths from the population of Table 5.1 with known mean ($\mu = 45.5$) and standard deviation ($\sigma = 3.90$). Let us assume that the sample mean is 44.8. We can expect the standard deviation of means based on samples of 35 items to be $\sigma_{\bar{Y}} = \sigma/\sqrt{n} = 3.90/\sqrt{35} = 0.6592$. We compute confidence limits as follows:

The lower limit is $L_1 = 44.8 - (1.960)(0.6592) = 43.51$.
The upper limit is $L_2 = 44.8 + (1.960)(0.6592) = 46.09$.

Remember that this is an unusual case in which we happen to know the true mean of the population ($\mu = 45.5$) and hence we know that the confidence limits enclose the mean. We expect 95% of such confidence intervals obtained in repeated sampling to include the parametric mean. We could increase the reliability of these limits by going to 99% confidence intervals, replacing 1.960 in the above expression by 2.576 and obtaining $L_1 = 43.10$ and $L_2 = 46.50$. We could have greater confidence that our interval covers the mean, but we could be much less certain about the true value of the mean because of the wider limits. By increasing the degree of confidence still further, say, to 99.9%, we could be virtually certain that our confidence limits ($L_1 = 42.63$, $L_2 = 46.97$) contain the population mean, but the bounds enclosing the mean are now so wide as to make our prediction far less useful than previously.

Experiment 6.2. For the seven samples of 5 housefly wing lengths and the seven similar samples of milk yields last worked with in Experiment 6.1 (Section 6.1), compute 95% confidence limits to the parametric mean for each sample and for the total sample based on 35 items. Base the standard errors of the means on the parametric standard deviations of these populations (housefly wing lengths $\sigma = 3.90$, milk yields $\sigma = 11.1597$). Record how many in each of the four classes of confidence limits (wing lengths and milk yields, $n = 5$ and $n = 35$) are correct—that is, contain the parametric mean of the population. Pool your results with those of other class members.

We tried the experiment on a computer for the 200 samples of 35 wing lengths each, computing confidence limits of the parametric mean by employing the parametric standard error of the mean, $\sigma_{\bar{Y}} = 0.6592$. Of the 200 confidence intervals plotted parallel to the ordinate, 194 (97.0%) cross the parametric mean of the population.

To reduce the width of the confidence interval, we have to reduce the standard error of the mean. Since $\sigma_{\bar{Y}} = \sigma/\sqrt{n}$, this can be done only by reducing the standard deviation of the items or by increasing the sample size. The first of these alternatives is not always available. If we are sampling from a population in nature, we ordinarily have no way of reducing its standard deviation. However, in many experimental procedures we may be able to reduce the variance of the data. For example, if we are studying the effect of a drug on heart weight in rats and find that its variance is rather large, we might be able to reduce this variance by taking rats of only one age group, in which the variation of heart weight would be considerably less. Thus, by controlling one of the variables of the experiment, the variance of the response variable, heart weight, is reduced. Similarly, by keeping temperature or other environmental variables constant in a procedure, we can frequently reduce the variance of our response variable and hence obtain more precise estimates of population parameters.

A common way to reduce the standard error is to increase sample size. Obviously from Expression (6.2) as n increases, the standard error decreases; hence, as n approaches infinity, the standard error and the lengths of confidence intervals approach zero. This ties in with what we have learned: in samples whose size approaches infinity, the sample mean would approach the parametric mean.

We must guard against a common mistake in expressing the meaning of the confidence limits of a statistic. When we have set lower and upper limits (L_1 and L_2, respectively) to a statistic, we imply that the probability that this interval covers the mean is, for example, 0.95, or, expressed in another way, that on the average 95 out of 100 confidence intervals similarly obtained would cover the mean. We *cannot state* that there is a probability of 0.95 that the true mean is contained within a given pair of confidence limits, although this may seem to be saying the same thing. The latter statement is incorrect because the true mean is a parameter; hence it is a fixed value, and it is therefore either inside the interval or outside it. It cannot be inside the given interval 95% of the time. It is important, therefore, to learn the correct statement and meaning of confidence limits.

So far we have considered only means based on normally distributed samples with known parametric standard deviations. We can, however, extend the methods just learned to samples from populations where the standard deviation is unknown but where the distribution is known to be normal and the samples are large, say, $n \geq 100$. In such cases we use the sample standard deviation for computing the standard error of the mean.

However, when the samples are small ($n < 100$) and we lack knowledge of the parametric standard deviation, we must take into consideration the reliability of our sample standard deviation. To do so, we must make use of the so-called t or Student's distribution. We shall learn how to set confidence limits employing the t distribution in Section 6.5. Before that, however, we shall have to become familiar with this distribution in the next section.

6.4 Student's t distribution

The deviations $\bar{Y} - \mu$ of sample means from the parametric mean of a normal distribution are themselves normally distributed. If these deviations are divided by the parametric standard deviation, the resulting ratios, $(\bar{Y} - \mu)/\sigma_{\bar{Y}}$, are still normally distributed, with $\mu = 0$ and $\sigma = 1$. Subtracting the constant μ from every $\bar{Y}_i$ is simply an additive code (Section 3.8) and will not change the form of the distribution of sample means, which is normal (Section 6.1). Dividing each deviation by the constant $\sigma_{\bar{Y}}$ reduces the variance to unity, but proportionately so for the entire distribution, so that its shape is not altered and a previously normal distribution remains so.

If, on the other hand, we calculate the variance s_i^2 of each of the samples and calculate the deviation for each mean $\bar{Y}_i$ as $(\bar{Y}_i - \mu)/s_{\bar{Y}_i}$, where $s_{\bar{Y}_i}$ stands for the estimate of the standard error of the mean of the ith sample, we will find the distribution of the deviations wider and more peaked than the normal distribution. This is illustrated in Figure 6.4, which shows the ratio $(\bar{Y}_i - \mu)/s_{\bar{Y}_i}$ for the 1400 samples of five housefly wing lengths of Table 6.1. The new distribution ranges wider than the corresponding normal distribution, because the denominator is the sample standard error rather than the parametric standard error and will sometimes be smaller and sometimes greater than expected. This increased variation will be reflected in the greater variance of the ratio $(\bar{Y} - \mu)/s_{\bar{Y}}$. The

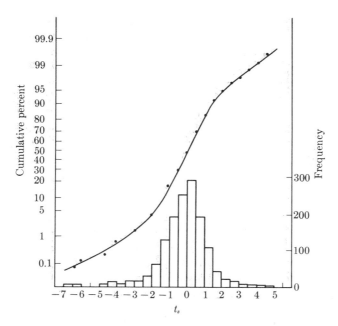

FIGURE 6.4
Distribution of quantity $t_s = (\bar{Y} - \mu)/s_{\bar{Y}}$ along abscissa computed for 1400 samples of 5 housefly wing lengths presented as a histogram and as a cumulative frequency distribution. Right-hand ordinate represents frequencies for the histogram; left-hand ordinate is cumulative frequency in probability scale.

expected distribution of this ratio is called the t distribution, also known as "*Student's*" *distribution,* named after W. S. Gossett, who first described it, publishing under the pseudonym "Student." The t distribution is a function with a complicated mathematical formula that need not be presented here.

The t distribution shares with the normal the properties of being symmetric and of extending from negative to positive infinity. However, it differs from the normal in that it assumes different shapes depending on the number of degrees of freedom. By "degrees of freedom" we mean the quantity $n - 1$, where n is the sample size upon which a variance has been based. It will be remembered that $n - 1$ is the divisor in obtaining an unbiased estimate of the variance from a sum of squares. The number of degrees of freedom pertinent to a given Student's distribution is the same as the number of degrees of freedom of the standard deviation in the ratio $(\bar{Y} - \mu)/s_{\bar{Y}}$. Degrees of freedom (abbreviated df or sometimes v) can range from 1 to infinity. A t distribution for $df = 1$ deviates most markedly from the normal. As the number of degrees of freedom increases, Student's distribution approaches the shape of the standard normal distribution ($\mu = 0$, $\sigma = 1$) ever more closely, and in a graph the size of this page a t distribution of $df = 30$ is essentially indistinguishable from a normal distribution. At

$df = \infty$, the t distribution *is* the normal distribution. Thus, we can think of the t distribution as the general case, considering the normal to be a special case of Student's distribution with $df = \infty$. Figure 6.5 shows t distributions for 1 and 2 degrees of freedom compared with a normal frequency distribution.

We were able to employ a single table for the areas of the normal curve by coding the argument in standard deviation units. However, since the t distributions differ in shape for differing degrees of freedom, it will be necessary to have a separate t table, corresponding in structure to the table of the areas of the normal curve, for each value of df. This would make for very cumbersome and elaborate sets of tables. Conventional t tables are therefore differently arranged. Table **III** shows degrees of freedom and probability as arguments and the corresponding values of t as functions. The probabilities indicate the percent of the area in both tails of the curve (to the right and left of the mean) beyond the indicated value of t. Thus, looking up the *critical value* of t at probability $P = 0.05$ and $df = 5$, we find $t = 2.571$ in Table **III**. Since this is a two-tailed table, the probability of 0.05 means that 0.025 of the area will fall to the left of a t value of -2.571 and 0.025 will fall to the right of $t = +2.571$. You will recall that the corresponding value for infinite degrees of freedom (for the normal curve) is 1.960. Only those probabilities generally used are shown in Table **III**.

You should become very familiar with looking up t values in this table. This is one of the most important tables to be consulted. A fairly conventional symbolism is $t_{\alpha[v]}$, meaning the tabled t value for v degrees of freedom and proportion α in both tails ($\alpha/2$ in each tail), which is equivalent to the t value for the cumulative probability of $1 - (\alpha/2)$. Try looking up some of these values to become familiar with the table. For example, convince yourself that $t_{0.05[7]}$, $t_{0.01[3]}$, $t_{0.02[10]}$, and $t_{0.05[\infty]}$ correspond to 2.365, 5.841, 2.764, and 1.960, respectively.

We shall now employ the t distribution for the setting of confidence limits to means of small samples.

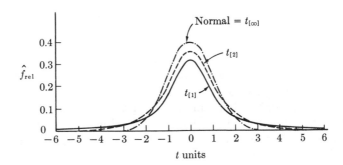

FIGURE 6.5
Frequency curves of t distributions for 1 and 2 degrees
of freedom compared with the normal distribution.

6.5 Confidence limits based on sample statistics

Armed with a knowledge of the t distribution, we are now able to set confidence limits to the means of samples from a normal frequency distribution whose parametric standard deviation is unknown. The limits are computed as $L_1 = \bar{Y} - t_{\alpha[n-1]}s_{\bar{Y}}$ and $L_2 = \bar{Y} + t_{\alpha[n-1]}s_{\bar{Y}}$ for confidence limits of probability $P = 1 - \alpha$. Thus, for 95% confidence limits we use values of $t_{0.05[n-1]}$. We can rewrite Expression (6.4a) as

$$P\{L_1 \leq \mu \leq L_2\} = P\{\bar{Y} - t_{\alpha[n-1]}s_{\bar{Y}} \leq \mu \leq \bar{Y} + t_{\alpha[n-1]}s_{\bar{Y}}\} = 1 - \alpha \quad (6.5)$$

An example of the application of this expression is shown in Box 6.2. We can

---

BOX 6.2

Confidence limits for μ.

Aphid stem mother femur lengths from Box 2.1: $\bar{Y} = 4.004$; $s = 0.366$; $n = 25$.

Values for $t_{\alpha[n-1]}$ from a two-tailed t table (Table III), where $1 - \alpha$ is the proportion expressing confidence and $n - 1$ are the degrees of freedom:

$$t_{0.05[24]} = 2.064 \qquad t_{0.01[24]} = 2.797$$

The 95% confidence limits for the population mean μ are given by the equations

$$L_1 \text{ (lower limit)} = \bar{Y} - t_{0.05[n-1]}\frac{s}{\sqrt{n}}$$

$$= 4.004 - \left(2.064\frac{0.366}{\sqrt{25}}\right) = 4.004 - 0.151$$

$$= 3.853$$

$$L_2 \text{ (upper limit)} = \bar{Y} + t_{0.05[n-1]}\frac{s}{\sqrt{n}}$$

$$= 4.004 + 0.151$$

$$= 4.155$$

The 99% confidence limits are

$$L_1 = \bar{Y} - t_{0.01[24]}\frac{s}{\sqrt{n}}$$

$$= 4.004 - \left(2.797\frac{0.366}{\sqrt{25}}\right) = 4.004 - 0.205$$

$$= 3.799$$

$$L_2 = \bar{Y} + t_{0.01[24]}\frac{s}{\sqrt{n}}$$

$$= 4.004 + 0.205$$

$$= 4.209$$

convince ourselves of the appropriateness of the t distribution for setting confidence limits to means of samples from a normally distributed population with unknown σ through a sampling experiment.

Experiment 6.3. Repeat the computations and procedures of Experiment 6.2 (Section 6.3), but base standard errors of the means on the standard deviations computed for each sample and use the appropriate t value in place of a standard normal deviate.

Figure 6.6 shows 95% confidence limits of 200 sampled means of 35 housefly wing lengths, computed with t and $s_{\bar{y}}$ rather than with the normal curve and $\sigma_{\bar{y}}$. We note that 191 (95.5%) of the 200 confidence intervals cross the parametric mean.

We can use the same technique for setting confidence limits to any given statistic as long as it follows the normal distribution. This will apply in an approximate way to all the statistics of Box 6.1. Thus, for example, we may set confidence limits to the coefficient of variation of the aphid femur lengths of Box 6.2. These are computed as

$$P\{V - t_{\alpha[n-1]}s_V \leq V_P \leq V + t_{\alpha[n-1]}s_V\} = 1 - \alpha$$

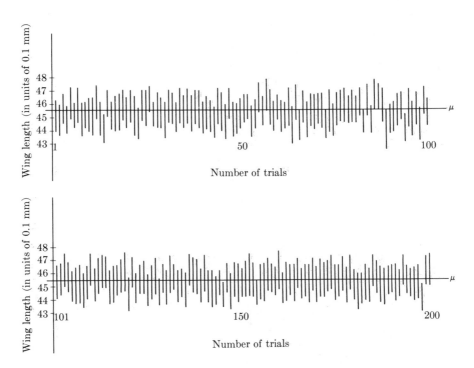

FIGURE 6.6
Ninety-five percent confidence intervals of means of 200 samples of 35 housefly wing lengths, based on sample standard errors $s_{\bar{y}}$. The heavy horizontal line is the parametric mean μ. The ordinate represents the variable.

where V_p stands for the parametric value of the coefficient of variation. Since the standard error of the coefficient of variation equals approximately $s_V = V/\sqrt{2n}$, we proceed as follows:

$$V = \frac{100s}{\bar{Y}} = \frac{100(0.3656)}{4.004} = 9.13$$

$$s_V = \frac{9.13}{\sqrt{2 \times 25}} = \frac{9.13}{7.0711} = 1.29$$

$$L_1 = V - t_{0.05[24]}s_V$$
$$= 9.13 - (2.064)(1.29)$$
$$= 9.13 - 2.66$$
$$= 6.47$$

$$L_2 = V + t_{0.05[24]}s_V$$
$$= 9.13 + 2.66$$
$$= 11.79$$

When sample size is very large or when σ is known, the distribution is effectively normal. However, rather than turn to the table of areas of the normal curve, it is convenient to simply use $t_{\alpha[\infty]}$, the t distribution with infinite degrees of freedom.

Although confidence limits are a useful measure of the reliability of a sample statistic, they are not commonly given in scientific publications, the statistic plus or minus its standard error being cited in their place. Thus, you will frequently see column headings such as "Mean $\pm$ S.E." This indicates that the reader is free to use the standard error to set confidence limits if so inclined.

It should be obvious to you from your study of the t distribution that you cannot set confidence limits to a statistic without knowing the sample size on which it is based, n being necessary to compute the correct degrees of freedom. Thus, the occasional citing of means and standard errors without also stating sample size n is to be strongly deplored.

It is important to state a statistic and its standard error to a sufficient number of decimal places. The following rule of thumb helps. Divide the standard error by 3, then note the decimal place of the first nonzero digit of the quotient; give the statistic significant to that decimal place and provide one further decimal for the standard error. This rule is quite simple, as an example will illustrate. If the mean and standard error of a sample are computed as 2.354 ± 0.363, we divide 0.363 by 3, which yields 0.121. Therefore the mean should be reported to one decimal place, and the standard error should be reported to two decimal places. Thus, we report this result as 2.4 ± 0.36. If, on the other hand, the same mean had a standard error of 0.243, dividing this standard error by 3 would have yielded 0.081, and the first nonzero digit would have been in the second decimal place. Thus the mean should have been reported as 2.35 ± 0.243.

6.6 The chi-square distribution

Another continuous distribution of great importance in statistics is the distribution of χ^2 (read *chi-square*). We need to learn it now in connection with the distribution and confidence limits of variances.

The chi-square distribution is a probability density function whose values range from zero to positive infinity. Thus, unlike the normal distribution or t, the function approaches the horizontal axis asymptotically only at the right-hand tail of the curve, not at both tails. The function describing the χ^2 distribution is complicated and will not be given here. As in t, there is not merely one χ^2 distribution, but there is one distribution for each number of degrees of freedom. Therefore, χ^2 is a function of v, the number of degrees of freedom. Figure 6.7 shows probability density functions for the χ^2 distributions for 1, 2, 3, and 6 degrees of freedom. Notice that the curves are strongly skewed to the right, ∟-shaped at first, but more or less approaching symmetry for higher degrees of freedom.

We can generate a χ^2 distribution from a population of standard normal deviates. You will recall that we standardize a variable Y_i by subjecting it to the operation $(Y_i - \mu)/\sigma$. Let us symbolize a standardized variable as $Y_i' = (Y_i - \mu)/\sigma$. Now imagine repeated samples of n variates Y_i from a normal population with mean μ and standard deviation σ. For each sample, we transform every variate Y_i to Y_i', as defined above. The quantities $\Sigma^n Y_i'^2$ computed for each sample will be distributed as a χ^2 distribution with n degrees of freedom.

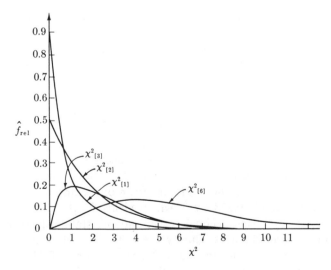

FIGURE 6.7
Frequency curves of χ^2 distribution for 1, 2, 3, and 6 degrees of freedom.

Using the definition of Y_i', we can rewrite $\Sigma^n\, Y_i'^2$ as

$$\sum_{}^{n} \frac{(Y_i - \mu)^2}{\sigma^2} = \frac{1}{\sigma^2} \sum_{}^{n} (Y_i - \mu)^2 \qquad (6.6)$$

When we change the parametric mean μ to a sample mean, this expression becomes

$$\frac{1}{\sigma^2} \sum_{}^{n} (Y_i - \bar{Y})^2 \qquad (6.7)$$

which is simply the sum of squares of the variable divided by a constant, the parametric variance. Another common way of stating this expression is

$$\frac{(n - 1)s^2}{\sigma^2} \qquad (6.8)$$

Here we have replaced the numerator of Expression (6.7) with $n - 1$ times the sample variance, which, of course, yields the sum of squares.

If we were to sample repeatedly n items from a normally distributed population, Expression (6.8) computed for each sample would yield a χ^2 distribution with $n - 1$ degrees of freedom. Notice that, although we have samples of n items, we have lost a degree of freedom because we are now employing a sample mean rather than the parametric mean. Figure 6.3, a sample distribution of variances, has a second scale along the abscissa, which is the first scale multiplied by the constant $(n - 1)/\sigma^2$. This scale converts the sample variances s^2 of the first scale into Expression (6.8). Since the second scale is proportional to s^2, the distribution of the sample variance will serve to illustrate a sample distribution approximating χ^2. The distribution is strongly skewed to the right, as would be expected in a χ^2 distribution.

Conventional χ^2 tables as shown in Table **IV** give the probability levels customarily required and degrees of freedom as arguments and list the χ^2 corresponding to the probability and the df as the functions. Each chi-square in Table **IV** is the value of χ^2 beyond which the area under the χ^2 distribution for v degrees of freedom represents the indicated probability. Just as we used subscripts to indicate the cumulative proportion of the area as well as the degrees of freedom represented by a given value of t, we shall subscript χ^2 as follows: $\chi^2_{\alpha[v]}$ indicates the χ^2 value to the right of which is found proportion α of the area under a χ^2 distribution for v degrees of freedom.

Let us learn how to use Table **IV**. Looking at the distribution of $\chi^2_{[2]}$, we note that 90% of all values of $\chi^2_{[2]}$ would be to the right of 0.211, but only 5% of all values of $\chi^2_{[2]}$ would be greater than 5.991. It can be shown that the expected value of $\chi^2_{[v]}$ (the mean of a χ^2 distribution) equals its degrees of freedom v. Thus the expected value of a $\chi^2_{[5]}$ distribution is 5. When we examine 50% values (the medians) in the χ^2 table, we notice that they are generally lower than the expected value (the means). Thus, for $\chi^2_{[5]}$ the 50% point is 4.351. This illustrates the asymmetry of the χ^2 distribution, the mean being to the right of the median.

Our first application of the χ^2 distribution will be in the next section. However, its most extensive use will be in connection with Chapter 13.

6.7 Confidence limits for variances

We saw in the last section that the ratio $(n - 1)s^2/\sigma^2$ is distributed as χ^2 with $n - 1$ degrees of freedom. We take advantage of this fact in setting confidence limits to variances.

First, we can make the following statement about the ratio $(n - 1)s^2/\sigma^2$:

$$P\left\{\chi^2_{(1-(\alpha/2))[n-1]} \leq \frac{(n-1)s^2}{\sigma^2} \leq \chi^2_{(\alpha/2)[n-1]}\right\} = 1 - \alpha$$

This expression is similar to those encountered in Section 6.3 and implies that the probability P that this ratio will be within the indicated boundary values of $\chi^2_{[n-1]}$ is $1 - \alpha$. Simple algebraic manipulation of the quantities in the inequality within brackets yields

$$P\left\{\frac{(n-1)s^2}{\chi^2_{(\alpha/2)[n-1]}} \leq \sigma^2 \leq \frac{(n-1)s^2}{\chi^2_{(1-(\alpha/2))[n-1]}}\right\} = 1 - \alpha \qquad (6.9)$$

Since $(n - 1)s^2 = \Sigma y^2$, we can simplify Expression (6.9) to

$$P\left\{\frac{\Sigma y^2}{\chi^2_{(\alpha/2)[n-1]}} \leq \sigma^2 \leq \frac{\Sigma y^2}{\chi^2_{(1-(\alpha/2))[n-1]}}\right\} = 1 - \alpha \qquad (6.10)$$

This still looks like a formidable expression, but it simply means that if we divide the sum of squares Σy^2 by the two values of $\chi^2_{[n-1]}$ that cut off tails each amounting to $\alpha/2$ of the area of the $\chi^2_{[n-1]}$-distribution, the two quotients will enclose the true value of the variance σ^2 with a probability of $P = 1 - \alpha$.

An actual numerical example will make this clear. Suppose we have a sample of 5 housefly wing lengths with a sample variance of $s^2 = 13.52$. If we wish to set 95% confidence limits to the parametric variance, we evaluate Expression (6.10) for the sample variance s^2. We first calculate the sum of squares for this sample: $4 \times 13.52 = 54.08$. Then we look up the values for $\chi^2_{0.025[4]}$ and $\chi^2_{0.975[4]}$. Since 95% confidence limits are required, α in this case is equal to 0.05. These χ^2 values span between them 95% of the area under the χ^2 curve. They correspond to 11.143 and 0.484, respectively, and the limits in Expression (6.10) then become

$$L_1 = \frac{54.08}{11.143} \quad \text{and} \quad L_2 = \frac{54.08}{0.484}$$

or

$$L_1 = 4.85 \quad \text{and} \quad L_2 = 111.74$$

This confidence interval is very wide, but we must not forget that the sample variance is, after all, based on only 5 individuals. Note also that the interval

BOX 6.3
Confidence limits for σ^2. Method of shortest unbiased confidence intervals.
Aphid stem mother femur lengths from Box 2.1: $n = 25$; $s^2 = 0.1337$.

The factors from Table **VII** for $v = n - 1 = 24$ df and confidence coefficient $(1 - \alpha) = 0.95$ are

$$f_1 = 0.5943 \qquad f_2 = 1.876$$

and for a confidence coefficient of 0.99 they are

$$f_1 = 0.5139 \qquad f_2 = 2.351$$

The 95% confidence limits for the population variance σ^2 are given by the equations

$$L_1 = \text{(lower limit)} = f_1 s^2 = 0.5943(0.1337) = 0.079,46$$

$$L_2 = \text{(upper limit)} = f_2 s^2 = 1.876(0.1337) = 0.2508$$

The 99% confidence limits are

$$L_1 = f_1 s^2 = 0.5139(0.1337) = 0.068,71$$

$$L_2 = f_2 s^2 = 2.351(0.1337) = 0.3143$$

is asymmetrical around 13.52, the sample variance. This is in contrast to the confidence intervals encountered earlier, which were symmetrical around the sample statistic.

The method described above is called the *equal-tails method*, because an equal amount of probability is placed in each tail (for example, $2\frac{1}{2}\%$). It can be shown that in view of the skewness of the distribution of variances, this method does not yield the shortest possible confidence intervals. One may wish the confidence interval to be "shortest" in the sense that the ratio L_2/L_1 be as small as possible. Box 6.3 shows how to obtain these shortest unbiased confidence intervals for σ^2 using Table **VII**, based on the method of Tate and Klett (1959). This table gives $(n - 1)/\chi^2_{p[n-1]}$, where p is an adjusted value of $\alpha/2$ or $1 - (\alpha/2)$ designed to yield the shortest unbiased confidence intervals. The computation is very simple.

6.8 Introduction to hypothesis testing

The most frequent application of statistics in biological research is to test some scientific hypothesis. Statistical methods are important in biology because results of experiments are usually not clear-cut and therefore need statistical tests to support decisions between alternative hypotheses. A statistical test examines a set of sample data and, on the basis of an expected distribution of the data, leads to a decision on whether to accept the hypothesis underlying the expected distribution or to reject that hypothesis and accept an alternative one. The nature of the tests varies with the data and the hypothesis, but the same general philosophy of hypothesis testing is common to all tests and will be discussed in this section. Study the material below very carefully, because it is fundamental to an understanding of every subsequent chapter in this book!

We would like to refresh your memory on the sample of 17 animals of species A, 14 of which were females and 3 of which were males. These data were examined for their fit to the binomial frequency distribution presented in Section 4.2, and their analysis was shown in Table 4.3. We concluded from Table 4.3 that if the sex ratio in the population was $1:1$ ($p_\female = q_\male = 0.5$), the probability of obtaining a sample with 14 males and 3 females would be 0.005,188, making it very unlikely that such a result could be obtained by chance alone. We learned that it is conventional to include all "worse" outcomes—that is, all those that deviate even more from the outcome expected on the hypothesis $p_\female = q_\male = 0.5$. Including all worse outcomes, the probability is 0.006,363, still a very small value. The above computation is based on the idea of a one-tailed test, in which we are interested only in departures from the $1:1$ sex ratio that show a preponderance of females. If we have no preconception about the direction of the departures from expectation, we must calculate the probability of obtaining a sample as deviant as 14 females and 3 males *in either direction* from expectation. This requires the probability either of obtaining a sample of 3 females and 14 males (and all worse samples) or of obtaining 14 females and 3 males (and all worse samples). Such a test is two-tailed, and since the distribution is symmetrical, we double the previously discussed probability to yield 0.012,726.

What does this probability mean? It is our hypothesis that $p_\female = q_\male = 0.5$. Let us call this hypothesis H_0, the *null hypothesis*, which is the hypothesis under test. It is called the null hypothesis because it assumes that there is no real difference between the true value of p in the population from which we sampled and the hypothesized value of $\hat{p} = 0.5$. Applied to the present example, the null hypothesis implies that the only reason our sample does not exhibit a $1:1$ sex ratio is because of sampling error. If the null hypothesis $p_\female = q_\male = 0.5$ is true, then approximately 13 samples out of 1000 will be as deviant as or more deviant than this one in either direction *by chance alone*. Thus, it is quite *possible* to have arrived at a sample of 14 females and 3 males by chance, but it is not very *probable*, since so deviant an event would occur only about 13 out of 1000 times, or 1.3% of the time. If we actually obtain such a sample, we may make one of two decisions. We may decide that the null hypothesis is in fact true (that is, the sex ratio is $1:1$) and that the sample obtained by us just happened to be one of those in the tail of the distribution, or we may decide that so deviant a sample is too improbable an event to justify acceptance of the null hypothesis. We may therefore decide that the hypothesis that the sex ratio is $1:1$ is not true. Either of these decisions may be correct, depending upon the truth of the matter. If in fact the $1:1$ hypothesis is correct, then the first decision (to accept the null hypothesis) will be correct. If we decide to reject the hypothesis under these circumstances, we commit an error. *The rejection of a true null hypothesis* is called a *type I error*. On the other hand, if in fact the true sex ratio of the population is other than 1:1, the first decision (to accept the 1:1 hypothesis) is an error, a so-called *type II error*, which is *the acceptance of a false null hypothesis*. Finally, if the 1:1 hypothesis is not true and we do decide to reject it, then we again make the correct decision. Thus, there are two kinds of correct decisions: accepting a true null hypothesis and rejecting a false null hypothesis, and there are two kinds of errors: type I, rejecting a true null hypothesis, and type II, accepting a false null hypothesis. These relationships between hypotheses and decisions can be summarized in the following table:

		Statistical decision	
		Null hypothesis	
Actual situation		Accepted	Rejected
Null hypothesis	True	Correct decision	Type I error
	False	Type II error	Correct decision

Before we carry out a test, we have to decide what magnitude of type I error (rejection of true hypothesis) we are going to allow. Even when we sample from a population of known parameters, there will always be some samples that by chance are very deviant. The most deviant of these are likely to mislead us into believing our hypothesis H_0 to be untrue. If we permit 5% of samples to lead us into a type I error, then we shall reject 5 out of 100 samples from the population, deciding that these are not samples from the given population. In the distribution under study, this means that we would reject all samples of 17 animals containing 13 of one sex plus 4 of the other sex. This can be seen by referring to column (3) of Table 6.3, where the expected frequencies of the various outcomes on the hypothesis $p_\female = q_\male = 0.5$ are shown. This table is an extension of the earlier Table 4.3, which showed only a tail of this distribution. Actually, you obtain a type I error slightly less than 5% if you sum relative expected frequencies for both tails starting with the class of 13 of one sex and 4 of the other. From Table 6.3 it can be seen that the relative expected frequency in the two tails will be $2 \times 0.024,520,9 = 0.049,041,8$. In a discrete frequency distribution, such as the binomial, we cannot calculate errors of exactly 5% as we can in a continuous frequency distribution, where we can measure off exactly 5% of the area. If we decide on an approximate 1% error, we will reject the hypothesis $p_\female = q_\male$ for all samples of 17 animals having 14 or more of one sex. (From Table 6.3 we find the $\hat{f}_{rel}$ in the tails equals $2 \times 0.006,362,9 = 0.012,725,8$.) Thus, the smaller the type I error we are prepared to accept, the more deviant a sample has to be for us to reject the null hypothesis H_0.

Your natural inclination might well be to have as little error as possible. You may decide to work with an extremely small type I error, such as 0.1% or even 0.01%, accepting the null hypothesis unless the sample is extremely deviant. The difficulty with such an approach is that, although guarding against a type I error, you might be falling into a type II error, accepting the null hypothesis

TABLE 6.3
Relative expected frequencies for samples of 17 animals under two hypotheses. Binomial distribution.

(1) ♀♀	(2) ♂♂	(3) H_0: $p_♀ = q_♂ = \frac{1}{2}$ $\hat{f}_{rel}$	(4) H_1: $p_♀ = 2q_♂ = \frac{2}{3}$ $\hat{f}_{rel}$
17	0	0.0000076	0.0010150
16	1	0.0001297	0.0086272
15	2	0.0010376	0.0345086
14	3	0.0051880	0.0862715
13	4	0.0181580	0.1509752
12	5	0.0472107	0.1962677
11	6	0.0944214	0.1962677
10	7	0.1483765	0.1542104
9	8	0.1854706	0.0963815
8	9	0.1854706	0.0481907
7	10	0.1483765	0.0192763
6	11	0.0944214	0.0061334
5	12	0.0472107	0.0015333
4	13	0.0181580	0.0002949
3	14	0.0051880	0.0000421
2	15	0.0010376	0.0000042
1	16	0.0001297	0.0000002
0	17	0.0000076	0.0000000
Total		1.0000002	0.9999999

when in fact it is not true and an alternative hypothesis H_1 is true. Presently, we shall show how this comes about.

First, let us learn some more terminology. Type I error is most frequently expressed as a probability and is symbolized by α. When a type I error is expressed as a percentage, it is also known as the *significance level.* Thus a type I error of $\alpha = 0.05$ corresponds to a significance level of 5% for a given test. When we cut off on a frequency distribution those areas proportional to α (the type I error), the portion of the abscissa under the area that has been cut off is called the *rejection region* or *critical region* of a test. The portion of the abscissa that would lead to acceptance of the null hypothesis is called the *acceptance region.* Figure 6.8A is a bar diagram showing the expected distribution of outcomes in the sex ratio example, given H_0. The dashed lines separate rejection regions from the 99% acceptance region.

Now let us take a closer look at the type II error. This is the probability of accepting the null hypothesis when in fact it is false. If you try to evaluate the probability of type II error, you immediately run into a problem. If the null hypothesis H_0 is false, some other hypothesis H_1 must be true. But unless you can specify H_1, you are not in a position to calculate type II error. An example will make this clear immediately. Suppose in our sex ratio case we have only two reasonable possibilities: (1) our old hypothesis H_0: $p_♀ = q_♂$, or (2) an alternative

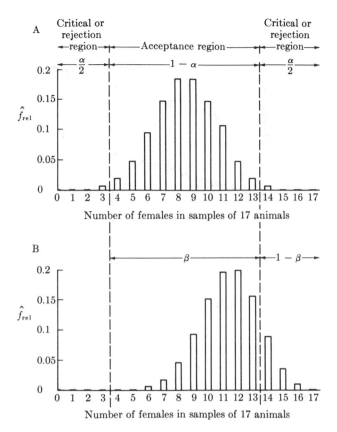

FIGURE 6.8
Expected distributions of outcomes when sampling 17 animals from two hypothetical populations. (A) $H_0: p_♀ = q_♂ = \frac{1}{2}$. (B) $H_1: p_♀ = 2q_♂ = \frac{2}{3}$. Dashed lines separate critical regions from acceptance region of the distribution of part A. Type I error α equals approximately 0.01.

hypothesis $H_1: p_♀ = 2q_♂$, which states that the sex ratio is 2:1 in favor of females so that $p_♀ = \frac{2}{3}$ and $q_♂ = \frac{1}{3}$. We now have to calculate expected frequencies for the binomial distribution $(p_♀ + q_♂)^k = (\frac{2}{3} + \frac{1}{3})^{17}$ to find the probabilities of the various outcomes under the alternative hypothesis. These are shown graphically in Figure 6.8B and are tabulated and compared with expected frequencies of the earlier distribution in Table 6.3.

Suppose we had decided on a type I error of $\alpha \approx 0.01$ ($\approx$ means "approximately equal to") as shown in Figure 6.8A. At this significance level we would accept the H_0 for all samples of 17 having 13 or fewer animals of one sex. Approximately 99% of all samples will fall into this category. However, what if H_0 is not true and H_1 is true? Clearly, from the population represented by hypothesis H_1 we could also obtain outcomes in which one sex was represented

13 or fewer times in samples of 17. We have to calculate what proportion of the curve representing hypothesis H_1 will overlap the acceptance region of the distribution representing hypothesis H_0. In this case we find that 0.8695 of the distribution representing H_1 overlaps the acceptance region of H_0 (see Figure 6.8B). Thus, if H_1 is really true (and H_0 correspondingly false), we would erroneously accept the null hypothesis 86.95% of the time. This percentage corresponds to the proportion of samples from H_1 that fall within the limits of the acceptance regions of H_0. This proportion is called β, the type II error expressed as a proportion. In this example β is quite large. Clearly, a sample of 17 animals is unsatisfactory to discriminate between the two hypotheses. Though 99% of the samples under H_0 would fall in the acceptance region, fully 87% would do so under H_1. A single sample that falls in the acceptance region would not enable us to reach a decision between the hypotheses with a high degree of reliability. If the sample had 14 or more females, we would conclude that H_1 was correct. If it had 3 or fewer females, we might conclude that neither H_0 nor H_1 was true. As H_1 approached H_0 (as in H_1: $p_\c{♀} = 0.55$, for example), the two distributions would overlap more and more and the magnitude of β would increase, making discrimination between the hypotheses even less likely. Conversely, if H_1 represented $p_\c{♀} = 0.9$, the distributions would be much farther apart and type II error β would be reduced. Clearly, then, the magnitude of β depends, among other things, on the parameters of the alternative hypothesis H_1 and cannot be specified without knowledge of the latter.

When the alternative hypothesis is fixed, as in the previous example (H_1: $p_{♀} = 2q_{♂}$), the magnitude of the type I error α we are prepared to tolerate will determine the magnitude of the type II error β. The smaller the rejection region α in the distribution under H_0, the greater will be the acceptance region $1 - \alpha$ in this distribution. The greater $1 - \alpha$, however, the greater will be its overlap with the distribution representing H_1, and hence the greater will be β. Convince yourself of this in Figure 6.8. By moving the dashed lines outward, we are reducing the critical regions representing type I error α in diagram A. But as the dashed lines move outward, more of the distribution of H_1 in diagram B will lie in the acceptance region of the null hypothesis. Thus, by decreasing α, we are increasing β and in a sense defeating our own purposes.

In most applications, scientists would wish to keep both of these errors small, since they do not wish to reject a null hypothesis when it is true, nor do they wish to accept it when another hypothesis is correct. We shall see in the following what steps can be taken to decrease β while holding α constant at a preset level.

Although significance levels α can be varied at will, investigators are frequently limited because, for many tests, cumulative probabilities of the appropriate distributions have not been tabulated and so published probability levels must be used. These are commonly 0.05, 0.01, and 0.001, although several others are occasionally encountered. When a null hypothesis has been rejected at a specified level of α, we say that the sample is *significantly different* from the parametric or hypothetical population at probability $P \leq \alpha$. Generally, values of α greater than 0.05 are not considered to be *statistically significant*. A significance level of 5% ($P = 0.05$) corresponds to one type I error in 20 trials, a level of 1% ($P = 0.01$) to one error in 100 trials. Significance levels of 1% or less ($P \leq 0.01$) are nearly always adjudged significant; those between 5% and 1% may be considered significant at the discretion of the investigator. Since statistical significance has a special technical meaning (H_0 rejected at $P \leq \alpha$), we shall use the adjective "significant" only in this sense; its use in scientific papers and reports, unless such a technical meaning is clearly implied, should be discouraged. For general descriptive purposes synonyms such as important, meaningful, marked, noticeable, and others can serve to underscore differences and effects.

A brief remark on null hypotheses represented by asymmetrical probability distributions is in order here. Suppose our null hypothesis in the sex ratio case had been $H_0: p_{\varphi} = \frac{2}{3}$, as discussed above. The distribution of samples of 17 offspring from such a population is shown in Figure 6.8B. It is clearly asymmetrical, and for this reason the critical regions have to be defined independently. For a given two-tailed test we can either double the probability P of a deviation in the direction of the closer tail and compare $2P$ with α, the conventional level of significance; or we can compare P with $\alpha/2$, half the conventional level of significance. In this latter case, 0.025 is the maximum value of P conventionally considered significant.

We shall review what we have learned by means of a second example, this time involving a continuous frequency distribution—the normally distributed housefly wing lengths—of parametric mean $\mu = 45.5$ and variance $\sigma^2 = 15.21$. Means based on 5 items sampled from these will also be normally distributed, as was demonstrated in Table 6.1 and Figure 6.1. Let us assume that someone presents you with a single sample of 5 housefly wing lengths and you wish to test whether they could belong to the specified population. Your null hypothesis will be $H_0: \mu = 45.5$ or $H_0: \mu = \mu_0$, where μ is the true mean of the population from which you have sampled and μ_0 stands for the hypothetical parametric mean of 45.5. We shall assume for the moment that we have no evidence that the variance of our sample is very much greater or smaller than the parametric variance of the housefly wing lengths. If it were, it would be unreasonable to assume that our sample comes from the specified population. There is a critical test of the assumption about the sample variance, which we shall take up later. The curve at the center of Figure 6.9 represents the expected distribution of means of samples of 5 housefly wing lengths from the specified population. Acceptance and rejection regions for a type I error $\alpha = 0.05$ are delimited along the abscissa. The boundaries of the critical regions are computed as follows (remember that $t_{[\infty]}$ is equivalent to the normal distribution):

$$L_1 = \mu_0 - t_{0.05[\infty]}\sigma_{\bar{Y}} = 45.5 - (1.96)(1.744) = 42.08$$

and

$$L_2 = \mu_0 + t_{0.05[\infty]}\sigma_Y = 45.5 + (1.96)(1.744) = 48.92$$

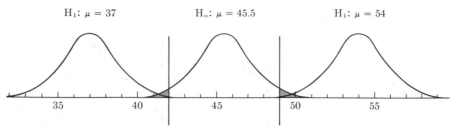

Wing length (in units of 0.1 mm)

FIGURE 6.9
Expected distribution of means of samples of 5 housefly wing lengths from normal populations specified by μ as shown above curves and $\sigma_{\bar{Y}} = 1.744$. Center curve represents null hypothesis, H_0: $\mu = 45.5$; curves at sides represent alternative hypotheses, $\mu = 37$ or $\mu = 54$. Vertical lines delimit 5% rejection regions for the null hypothesis ($2\frac{1}{2}$% in each tail, shaded).

Thus, we would consider it improbable for means less than 42.08 or greater than 48.92 to have been sampled from this population. For such sample means we would therefore reject the null hypothesis. The test we are proposing is two-tailed because we have no a priori assumption about the possible alternatives to our null hypothesis. If we could assume that the true mean of the population from which the sample was taken could only be equal to or greater than 45.5, the test would be one-tailed.

Now let us examine various alternative hypotheses. One alternative hypothesis might be that the true mean of the population from which our sample stems is 54.0, but that the variance is the same as before. We can express this assumption as H_1: $\mu = 54.0$ or H_1: $\mu = \mu_1$, where μ_1 stands for the alternative parametric mean 54.0. From Table **II** ("Areas of the normal curve") and our knowledge of the variance of the means, we can calculate the proportion of the distribution implied by H_1 that would overlap the acceptance region implied by H_0. We find that 54.0 is 5.08 measurement units from 48.92, the upper boundary of the acceptance region of H_0. This corresponds to $5.08/1.744 = 2.91\sigma_{\bar{Y}}$ units. From Table **II** we find that 0.0018 of the area will lie beyond 2.91σ at one tail of the curve. Thus, under this alternative hypothesis, 0.0018 of the distribution of H_1 will overlap the acceptance region of H_0. This is β, the type II error under this alternative hypothesis. Actually, this is not entirely correct. Since the left tail of the H_1 distribution goes all the way to negative infinity, it will leave the acceptance region and cross over into the left-hand rejection region of H_0. However, this represents only an infinitesimal amount of the area of H_1 (the lower critical boundary of H_0, 42.08, is $6.83\sigma_{\bar{Y}}$ units from $\mu_1 = 54.0$) and can be ignored.

Our alternative hypothesis H_1 specified that μ_1 is 8.5 units greater than μ_0. However, as said before, we may have no a priori reason to believe that the true mean of our sample is either greater or less than μ. Therefore, we may simply assume that it is 8.5 measurement units away from 45.5. In such a case we must similarly calculate β for the alternative hypothesis that $\mu_1 = \mu_0 - 8.5$. Thus the alternative hypothesis becomes H_1: $\mu = 54.0$ or 37.0, or H_1: $\mu = \mu_1$, where μ_1 represents either 54.0 or 37.0, the alternative parametric means. Since the distributions are symmetrical, β is the same for both alternative hypotheses. Type II error for hypothesis H_1 is therefore 0.0018, regardless of which of the two alternative hypotheses is correct. If H_1 is really true, 18 out of 10,000 samples will lead to an incorrect acceptance of H_0, a very low proportion of error. These relations are shown in Figure 6.9.

You may rightly ask what reason we have to believe that the alternative parametric value for the mean is 8.5 measurement units to either side of $\mu_0 = 45.5$. It would be quite unusual if we had any justification for such a belief. As a matter of fact, the true mean may just as well be 7.5 or 6.0 or any number of units to either side of μ_0. If we draw curves for H_1: $\mu = \mu_0 \pm 7.5$, we find that β has increased considerably, the curves for H_0 and H_1 now being closer together. Thus, the magnitude of β will depend on how far the alternative parametric mean is from the parametric mean of the null hypothesis. As the alternative mean approaches the parametric mean, β increases up to a maximum value of $1 - \alpha$, which is the area of the acceptance region under the null hypothesis. At this maximum, the two distributions would be superimposed upon each other. Figure 6.10 illustrates the increase in β as μ_1 approaches μ, starting with the test illustrated in Figure 6.9. To simplify the graph, the alternative distributions are shown for one tail only. Thus, we clearly see that β is not a fixed value but varies with the nature of the alternative hypothesis.

An important concept in connection with hypothesis testing is the *power* of a test. It is $1 - \beta$, the complement of β, and is the probability of rejecting the null hypothesis when in fact it is false and the alternative hypothesis is correct. Obviously, for any given test we would like the quantity $1 - \beta$ to be as large as possible and the quantity β as small as possible. Since we generally cannot specify a given alternative hypothesis, we have to describe β or $1 - \beta$ for a continuum of alternative values. When $1 - \beta$ is graphed in this manner, the result is called a *power curve* for the test under consideration. Figure 6.11 shows the power curve for the housefly wing length example just discussed. This figure can be compared with Figure 6.10, from which it is directly derived. Figure 6.10 emphasizes the type II error β, and Figure 6.11 graphs the complement of this value, $1 - \beta$. We note that the power of the test falls off sharply as the alternative hypothesis approaches the null hypothesis. Common sense confirms these conclusions: we can make clear and firm decisions about whether our sample comes from a population of mean 45.5 or 60.0. The power is essentially 1. But if the alternative hypothesis is that $\mu_1 = 45.6$, differing only by 0.1 from the value assumed under the null hypothesis, it will be difficult to decide which of these hypotheses is true, and the power will be very low.

To improve the power of a given test (or decrease β) while keeping α constant for a stated null hypothesis, we must increase sample size. If instead of sampling 5 wing lengths we had sampled 35, the distribution of means would be much narrower. Thus, rejection regions for the identical type I error would now commence at 44.21 and 46.79. Although the acceptance and rejection regions have

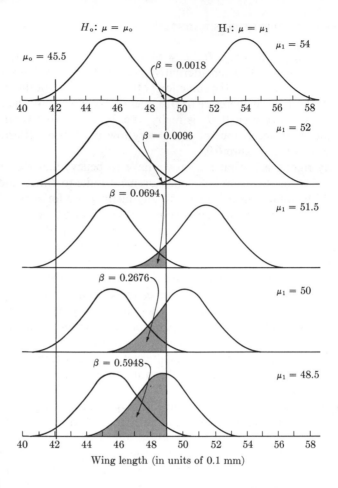

FIGURE 6.10
Diagram to illustrate increases in type II error β as alternative hypothesis H_1 approaches null hypothesis H_0—that is, μ_1 approaches μ. Shading represents β. Vertical lines mark off 5% critical regions ($2\frac{1}{2}\%$ in each tail) for the null hypothesis. To simplify the graph the alternative distributions are shown for one tail only. Data identical to those in Figure 6.9.

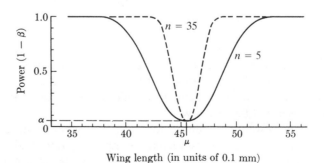

FIGURE 6.11
Power curves for testing H_0: $\mu = 45.5$, H_1: $\mu \neq 45.5$ for $n = 5$ (as in Figures 6.9 and 6.10) and for $n = 35$.

remained the same proportionately, the acceptance region has become much narrower in absolute value. Previously, we could not, with confidence, reject the null hypothesis for a sample mean of 48.0. Now, when based on 35 individuals, a mean as deviant as 48.0 would occur only 15 times out of 100,000 and the hypothesis would, therefore, be rejected.

What has happened to type II error? Since the distribution curves are not as wide as before, there is less overlap between them; if the alternative hypothesis H_1: $\mu = 54.0$ or 37.0 is true, the probability that the null hypothesis could be accepted by mistake (type II error) is infinitesimally small. If we let μ_1 approach μ_0, β will increase, of course, but it will always be smaller than the corresponding value for sample size $n = 5$. This comparison is shown in Figure 6.11, where the power for the test with $n = 35$ is much higher than that for $n = 5$. If we were to increase our sample size to 100 or 1000, the power would be still further increased. Thus, we reach an important conclusion: If a given test is not sensitive enough, we can increase its sensitivity (= power) by increasing sample size.

There is yet another way of increasing the power of a test. If we cannot increase sample size, the power may be raised by changing the nature of the test. Different statistical techniques testing roughly the same hypothesis may differ substantially both in the actual magnitude and in the slopes of their power curves. Tests that maintain higher power levels over substantial ranges of alternative hypotheses are clearly to be preferred. The popularity of the various nonparametric tests, mentioned in several places in this book, has grown not only because of their computational simplicity but also because their power curves are less affected by failure of assumptions than are those of the parametric methods. However, it is also true that nonparametric tests have lower overall power than parametric ones, when all the assumptions of the parametric test are met.

Let us briefly look at a one-tailed test. The null hypothesis is H_0: $\mu_0 = 45.5$, as before. However, the alternative hypothesis assumes that we have reason to believe that the parametric mean of the population from which our sample has been taken cannot possibly be less than $\mu_0 = 45.5$: if it is different from that value, it can only be greater than 45.5. We might have two grounds for such a hypothesis. First, we might have some biological reason for such a belief. Our parametric flies might be a dwarf population, so that any other population from which our sample could come must be bigger. A second reason might be that we are interested in only one direction of difference. For example, we may be testing the effect of a chemical in the larval food intended to increase the size of the flies in the sample. Therefore, we would expect that $\mu_1 \geq \mu_0$, and we would not be interested in testing for any μ_1 that is less than μ_0, because such an effect is the exact opposite of what we expect. Similarly, if we are investigating the effect of a certain drug as a cure for cancer, we might wish to compare the untreated population that has a mean fatality rate θ (from cancer) with the treated population, whose rate is θ_1. Our alternative hypotheses will be H_1: $\theta_1 < \theta$. That is, we are not interested in any θ_1 that is greater than θ, because if our drug will increase mortality from cancer, it certainly is not much of a prospect for a cure.

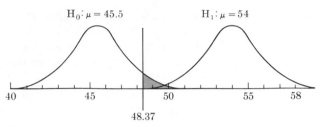

Wing length (in units of 0.1 mm)

FIGURE 6.12
One-tailed significance test for the distribution of Figure 6.9. Vertical line now cuts off 5% rejection region from one tail of the distribution (corresponding area of curve has been shaded).

When such a one-tailed test is performed, the rejection region along the abscissa is under only one tail of the curve representing the null hypothesis. Thus, for our housefly data (distribution of means of sample size $n = 5$), the rejection region will be in one tail of the curve only and for a 5% type I error will appear as shown in Figure 6.12. We compute the critical boundary as $45.5 + (1.645)(1.744) = 48.37$. The 1.645 is $t_{0.10[\infty]}$, which corresponds to the 5% value for a one-tailed test. Compare this rejection region, which rejects the null hypothesis for all means greater than 48.37, with the two rejection regions in Figure 6.10, which reject the null hypothesis for means lower than 42.08 and greater than 48.92. The alternative hypothesis is considered for one tail of the distribution only, and the power curve of the test is not symmetrical but is drawn out with respect to one side of the distribution only.

6.9 Tests of simple hypotheses employing the t distribution

We shall proceed to apply our newly won knowledge of hypothesis testing to a simple example involving the t distribution.

Government regulations prescribe that the standard dosage in a certain biological preparation should be 600 activity units per cubic centimeter. We prepare 10 samples of this preparation and test each for potency. We find that the mean number of activity units per sample is 592.5 units per cc and the standard deviation of the samples is 11.2. Does our sample conform to the government standard? Stated more precisely, our null hypothesis is $H_0: \mu = \mu_0$. The alternative hypothesis is that the dosage is not equal to 600, or $H_1: \mu \neq \mu_0$. We proceed to calculate the significance of the deviation $\bar{Y} - \mu_0$ expressed in standard deviation units. The appropriate standard deviation is that of means (the standard error of the mean), *not* the standard deviation of items, because the deviation is that of a sample mean around a parametric mean. We therefore calculate $s_{\bar{Y}} = s/\sqrt{n} = 11.2/\sqrt{10} = 3.542$. We next test the deviation $(\bar{Y} - \mu_0)/s_{\bar{Y}}$. We have seen earlier, in Section 6.4, that a deviation divided by an estimated standard deviation will be distributed according to the t distribution with $n - 1$ degrees of freedom. We therefore write

$$t_s = \frac{\bar{Y} - \mu_0}{s_{\bar{Y}}} \qquad (6.11)$$

This indicates that we would expect this deviation to be distributed as a t variate. Note that in Expression (6.11) we wrote t_s. In most textbooks you will find this ratio simply identified as t, but in fact the t distribution is a parametric and theoretical distribution that generally is only approached, but never equaled, by observed, sampled data. This may seem a minor distinction, but readers should be quite clear that in any hypothesis testing of samples we are only *assuming* that the distributions of the tested variables follow certain theoretical probability distributions. To conform with general statistical practice, the t distribution should really have a Greek letter (such as τ), with t serving as the sample statistic. Since this would violate long-standing practice, we prefer to use the subscript s to indicate the sample value.

The actual test is very simple. We calculate Expression (6.11),

$$t_s = \frac{592.5 - 600}{3.542} = \frac{-7.5}{3.542} = -2.12 \qquad df = n - 1 = 9$$

and compare it with the expected values for t at 9 degrees of freedom. Since the t distribution is symmetrical, we shall ignore the sign of t_s and always look up its positive value in Table **III**. The two values on either side of t_s are $t_{0.05[9]} = 2.26$ and $t_{0.10[9]} = 1.83$. These are t values for two-tailed tests, appropriate in this instance because the alternative hypothesis is that $\mu \neq 600$; that is, it can be smaller or greater. It appears that the significance level of our value of t_s is between 5% and 10%; if the null hypothesis is actually true, the probability of obtaining a deviation as great as or greater than 7.5 is somewhere between 0.05 and 0.10. By customary levels of significance, this is insufficient for declaring the sample mean significantly different from the standard. We consequently accept the null hypothesis. In conventional language, we would report the results of the statistical analysis as follows: "The sample mean is not significantly different from the accepted standard." Such a statement in a scientific report should always be backed up by a probability value, and the proper way of presenting this is to write "$0.10 > P > 0.05$." This means that the probability of such a deviation is between 0.05 and 0.10. Another way of saying this is that the value of t_s is *not significant* (frequently abbreviated as *ns*).

A convention often encountered is the use of asterisks after the computed value of the significance test, as in $t_s = 2.86^{**}$. The symbols generally represent the following probability ranges:

$$* = 0.05 \geq P > 0.01 \qquad ** = 0.01 \geq P > 0.001 \qquad *** = P \leq 0.001$$

However, since some authors occasionally imply other ranges by these asterisks, the meaning of the symbols has to be specified in each scientific report.

It might be argued that in a biological preparation the concern of the tester should not be whether the sample differs significantly from a standard, but whether it is significantly *below* the standard. This may be one of those biological preparations in which an excess of the active component is of no harm but a shortage would make the preparation ineffective at the conventional dosage. Then the test becomes one-tailed, performed in exactly the same manner except that the critical values of t for a one-tailed test are at half the probabilities of the two-tailed test. Thus 2.26, the former 0.05 value, becomes $t_{0.025[9]}$, and 1.83, the former 0.10 value, becomes $t_{0.05[9]}$, making our observed t_s value of 2.12 "significant at the 5% level" or, more precisely stated, significant at $0.05 > P > 0.025$. If we are prepared to accept a 5% significance level, we would consider the preparation significantly below the standard.

You may be surprised that the same example, employing the same data and significance tests, should lead to two different conclusions, and you may begin to wonder whether some of the things you hear about statistics and statisticians are not, after all, correct. The explanation lies in the fact that the two results are answers to different questions. If we test whether our sample is significantly different from the standard in either direction, we must conclude that it is not different enough for us to reject the null hypothesis. If, on the other hand, we exclude from consideration the fact that the true sample mean μ could be greater than the established standard μ_0, the difference as found by us is clearly significant. It is obvious from this example that in any statistical test one must clearly state whether a one-tailed or a two-tailed test has been performed if the nature of the example is such that there could be any doubt about the matter. We should also point out that such a difference in the outcome of the results is not necessarily typical. It is only because the outcome in this case is in a borderline area between clear significance and nonsignificance. Had the difference between sample and standard been 10.5 activity units, the sample would have been unquestionably significantly different from the standard by the one-tailed or the two-tailed test.

The promulgation of a standard mean is generally insufficient for the establishment of a rigid standard for a product. If the variance among the samples is sufficiently large, it will never be possible to establish a significant difference between the standard and the sample mean. This is an important point that should be quite clear to you. Remember that the standard error can be increased in two ways—by lowering sample size or by increasing the standard deviation of the replicates. Both of these are undesirable aspects of any experimental setup.

The test described above for the biological preparation leads us to a general test for the significance of any statistic—that is, for the significance of a deviation of any statistic from a parametric value, which is outlined in Box 6.4. Such a test applies whenever the statistics are expected to be normally distributed. When the standard error is estimated from the sample, the t distribution is used. However, since the normal distribution is just a special case $t_{[\infty]}$ of the t distribution, most statisticians uniformly apply the t distribution with the appro-

---

BOX 6.4

Testing the significance of a statistic—that is, the significance of a deviation of a sample statistic from a parametric value. For normally distributed statistics.

Computational steps

1. Compute t_s as the following ratio:

$$t_s = \frac{St - St_p}{s_{St}}$$

where St is a sample statistic, St_p is the parametric value against which the sample statistic is to be tested, and s_{St} is its estimated standard error, obtained from Box 6.1, or elsewhere in this book.

2. The pertinent hypotheses are

$$H_0: St = St_p \qquad H_1: St \neq St_p$$

for a two-tailed test, and

$$H_0: St = St_p \qquad H_1: St > St_p$$

or

$$H_0: St = St_p \qquad H_1: St < St_p$$

for a one-tailed test.

3. In the two-tailed test, look up the critical value of $t_{\alpha[v]}$, where α is the type I error agreed upon and v is the degrees of freedom pertinent to the standard error employed (see Box 6.1). In the one-tailed test look up the critical value of $t_{2\alpha[v]}$ for a significance level of α.

4. Accept or reject the appropriate hypothesis in **2** on the basis of the t_s value in **1** compared with critical values of t in **3**.

---

priate degrees of freedom from 1 to infinity. An example of such a test is the t test for the significance of a regression coefficient shown in step **2** of Box 11.4.

6.10 Testing the hypothesis $H_0: \sigma^2 = \sigma_0^2$

The method of Box 6.4 can be used only if the statistic is normally distributed. In the case of the variance, this is not so. As we have seen, in Section 6.6, sums of squares divided by σ^2 follow the χ^2 distribution. Therefore, for testing the hypothesis that a sample variance is different from a parametric variance, we must employ the χ^2 distribution.

Let us use the biological preparation of the last section as an example. We were told that the standard deviation was 11.2 based on 10 samples. Therefore, the variance must have been 125.44. Suppose the government postulates that the variance of samples from the preparation should be no greater than 100.0. Is our sample variance significantly above 100.0? Remembering from

Expression (6.8) that $(n - 1)s^2/\sigma^2$ is distributed as $\chi^2_{[n-1]}$, we proceed as follows. We first calculate

$$X^2 = \frac{(n - 1)s^2}{\sigma^2}$$

$$= \frac{(9)125.44}{100}$$

$$= 11.290$$

Note that we call the quantity X^2 rather than χ^2. This is done to emphasize that we are obtaining a sample statistic that we shall compare with the parametric distribution.

Following the general outline of Box 6.4, we next establish our null and alternative hypotheses, which are H_0: $\sigma^2 = \sigma_0^2$ and H_1: $\sigma^2 > \sigma_0^2$; that is, we are to perform a one-tailed test. The critical value of χ^2 is found next as $\chi^2_{\alpha[v]}$, where α is the proportion of the χ^2 distribution to the right of the critical value, as described in Section 6.6, and v is the pertinent degrees of freedom. You see now why we used the symbol α for that portion of the area. It corresponds to the probability of a type I error. For $v = 9$ degrees of freedom, we find in Table **IV** that

$$\chi^2_{0.05[9]} = 16.919 \qquad \chi^2_{0.10[9]} = 14.684 \qquad \chi^2_{0.50[9]} = 8.343$$

We notice that the probability of getting a χ^2 as large as 11.290 is therefore less than 0.50 but higher than 0.10, assuming that the null hypothesis is true. Thus X^2 is not significant at the 5% level, we have no basis for rejecting the null hypothesis, and we must conclude that the variance of the 10 samples of the biological preparation may be no greater than the standard permitted by the government. If we had decided to test whether the variance is different from the standard, permitting it to deviate in either direction, the hypotheses for this two-tailed test would have been H_0: $\sigma^2 = \sigma_0^2$ and H_1: $\sigma^2 \neq \sigma_0^2$, and a 5% type I error would have yielded the following critical values for the two-tailed test:

$$\chi^2_{0.975[9]} = 2.700 \qquad \chi^2_{0.025[9]} = 19.023$$

The values represent chi-squares at points cutting off $2\frac{1}{2}$% rejection regions at each tail of the χ^2 distribution. A value of $X^2 < 2.700$ or > 19.023 would have been evidence that the sample variance did not belong to this population. Our value of $X^2 = 11.290$ would again have led to an acceptance of the null hypothesis.

In the next chapter we shall see that there is another significance test available to test the hypotheses about variances of the present section. This is the mathematically equivalent F test, which is, however, a more general test, allowing us to test the hypothesis that two sample variances come from populations with equal variances.

Exercises

6.1 Since it is possible to test a statistical hypothesis with any size sample, why are larger sample sizes preferred? ANS. When the null hypothesis is false, the probability of a type II error decreases as n increases.

6.2 Differentiate between type I and type II errors. What do we mean by the power of a statistical test?

6.3 Set 99% confidence limits to the mean, median, coefficient of variation, and variance for the birth weight data given in Box 3.2. ANS. The lower limits are 109.540, 109.060, 12.136, and 178.698, respectively.

6.4 The 95% confidence limits for μ as obtained in a given sample were 4.91 and 5.67 g. Is it correct to say that 95 times out of 100 the population mean, μ, falls inside the interval from 4.91 to 5.67 g? If not, what would the correct statement be?

6.5 In a study of mating calls in the tree toad *Hyla ewingi*, Littlejohn (1965) found the note duration of the call in a sample of 39 observations from Tasmania to have a mean of 189 msec and a standard deviation of 32 msec. Set 95% confidence intervals to the mean and to the variance. ANS. The 95% confidence limits for the mean are from 178.6 to 199.4. The 95% shortest unbiased limits for the variance are from 679.5 to 1646.6.

6.6 Set 95% confidence limits to the means listed in Table 6.2. Are these limits all correct? (That is, do they contain μ?)

6.7 In Section 4.3 the coefficient of dispersion was given as an index of whether or not data agreed with a Poisson distribution. Since in a true Poisson distribution, the mean μ equals the parametric variance σ^2, the coefficient of dispersion is analogous to Expression (6.8). Using the mite data from Table 4.5, test the hypothesis that the true variance is equal to the sample mean—in other words, that we have sampled from a Poisson distribution (in which the coefficient of dispersion should equal unity). Note that in these examples the chi-square table is not adequate, so that approximate critical values must be computed using the method given with Table IV. In Section 7.3 an alternative significance test that avoids this problem will be presented. ANS. $X^2 = (n-1) \times CD = 1308.30$, $\chi^2_{0.05[588]} \approx 645.708$.

6.8 Using the method described in Exercise 6.7, test the agreement of the observed distribution with a Poisson distribution by testing the hypothesis that the true coefficient of dispersion equals unity for the data of Table 4.6.

6.9 In a study of bill measurements of the dusky flycatcher, Johnson (1966) found that the bill length for the males had a mean of 8.14 ± 0.021 and a coefficient of variation of 4.67%. On the basis of this information, infer how many specimens must have been used? ANS. Since $V = 100s/\bar{Y}$ and $s_{\bar{Y}} = s/\sqrt{n}$, $\sqrt{n} = Vs_{\bar{Y}}\bar{Y}/100$. Thus $n = 328$.

6.10 In direct klinokinetic behavior relating to temperature, animals turn more often in the warm end of a gradient and less often in the colder end, the direction of turning being at random, however. In a computer simulation of such behavior, the following results were found. The mean position along a temperature gradient was found to be -1.352. The standard deviation was 12.267, and n equaled 500 individuals. The gradient was marked off in units: zero corresponded to the middle of the gradient, the initial starting point of the animals; minus corresponded to the cold end; and plus corresponded to the warmer end. Test the hypothesis that direct klinokinetic behavior did not result in a tendency toward aggregation in either the warmer or colder end; that is, test the hypothesis that μ, the mean position along the gradient, was zero.

6.11   In an experiment comparing yields of three new varieties of corn, the following results were obtained.

Variety

	1	2	3
$\bar{Y}$	22.86	43.21	38.56
n	20	20	20

To compare the three varieties the investigator computed a weighted mean of the three means using the weights 2, -1, -1. Compute the weighted mean and its 95% confidence limits, assuming that the variance of each value for the weighted mean is zero. ANS. $\bar{Y}_w = -36.05$, $\sigma^2_{\bar{Y}_w} = 34.458$, the 95% confidence limits are -47.555 to -24.545, and the weighted mean is significantly different from zero even at the $P < 0.001$ level.

CHAPTER 7

Introduction to Analysis
of Variance

We now proceed to a study of the analysis of variance. This method, developed by R. A. Fisher, is fundamental to much of the application of statistics in biology and especially to experimental design. One use of the analysis of variance is to test whether two or more sample means have been obtained from populations with the same parametric mean. Where only two samples are involved, the t test can also be used. However, the analysis of variance is a more general test, which permits testing two samples as well as many, and we are therefore introducing it at this early stage in order to equip you with this powerful weapon for your statistical arsenal. We shall discuss the t test for two samples as a special case in Section 8.4.

In Section 7.1 we shall approach the subject on familiar ground, the sampling experiment of the housefly wing lengths. From these samples we shall obtain two independent estimates of the population variance. We digress in Section 7.2 to introduce yet another continuous distribution, the F distribution, needed for the significance test in analysis of variance. Section 7.3 is another digression; here we show how the F distribution can be used to test whether two samples may reasonably have been drawn from populations with the same variance. We are now ready for Section 7.4, in which we examine the effects of subjecting the samples to different treatments. In Section 7.5, we describe the partitioning of sums of squares and of degrees of freedom, the actual *analysis* of variance. The last two sections (7.6 and 7.7) take up in a more formal way the two scientific models for which the analysis of variance is appropriate, the so-called fixed treatment effects model (Model I) and the variance component model (Model II).

Except for Section 7.3, the entire chapter is largely theoretical. We shall postpone the practical details of computation to Chapter 8. However, a thorough understanding of the material in Chapter 7 is necessary for working out actual examples of analysis of variance in Chapter 8.

One final comment. We shall use J. W. Tukey's acronym "anova" interchangeably with "analysis of variance" throughout the text.

7.1 The variances of samples and their means

We shall approach analysis of variance through the familiar sampling experiment of housefly wing lengths (Experiment 5.1 and Table 5.1), in which we combined seven samples of 5 wing lengths to form samples of 35. We have reproduced one such sample in Table 7.1. The seven samples of 5, here called groups, are listed vertically in the upper half of the table.

Before we proceed to explain Table 7.1 further, we must become familiar with added terminology and symbolism for dealing with this kind of problem. We call our samples *groups*; they are sometimes called *classes* or are known by yet other terms we shall learn later. In any analysis of variance we shall have two or more such samples or groups, and we shall use the symbol a for the number of groups. Thus, in the present example $a = 7$. Each group or sample is based on n items, as before; in Table 7.1, $n = 5$. The total number of items in the table is a times n, which in this case equals 7×5 or 35.

The sums of the items in the respective groups are shown in the row underneath the horizontal dividing line. In an anova, summation signs can no longer be as simple as heretofore. We can sum either the items of one group only or the items of the entire table. We therefore have to use superscripts with the summation symbol. In line with our policy of using the simplest possible notation, whenever this is not likely to lead to misunderstanding, we shall use $\Sigma^n Y$ to indicate the sum of the items of a group and $\Sigma^{an} Y$ to indicate the sum of all the items in the table. The sum of the items of each group is shown in the first row under the horizontal line. The mean of each group, symbolized by $\bar{Y}$, is in the next row and is computed simply as $\Sigma^n Y / n$. The remaining two rows in that portion of Table 7.1 list $\Sigma^n Y^2$ and $\Sigma^n y^2$, separately for each group. These are the familiar quantities, the sum of the squared Y's and the sum of squares of Y.

From the sum of squares for each group we can obtain an estimate of the population variance of housefly wing length. Thus, in the first group $\Sigma^n y^2 = 29.2$. Therefore, our estimate of the population variance is

$$s^2 = \frac{\sum\limits^{n} y^2}{(n-1)} = \frac{29.2}{4} = 7.3$$

TABLE 7.1

Seven samples (groups) of 5 wing lengths of houseflies randomly selected. (Data from Experiment 5.1 and Table 5.1.) Parametric mean, $\mu = 45.5$; variance, $\sigma^2 = 15.21$.

	a groups (a = 7)							Computation of sum of squares of means	Computation of total sum of squares
	1	2	3	4	5	6	7		
n individuals per group (n = 5)	41	48	40	40	49	40	41		
	44	49	50	39	41	48	46		
	48	49	44	46	50	51	54		
	43	49	48	46	39	47	44		
	42	45	50	41	42	51	42		
$\sum\limits^{n} Y$	218	240	232	212	221	237	227	$\sum\limits^{a} \bar{Y} = 317.4$	$\sum\limits^{an} Y = 1587$
$\bar{Y}$	43.6	48.0	46.4	42.4	44.2	47.4	45.4	$\bar{\bar{Y}} = 45.34$	$\bar{\bar{Y}} = 45.34$
$\sum\limits^{n} Y^2$	9534	11,532	10,840	9034	9867	11,315	10,413	$\sum\limits^{a} \bar{Y}^2 = 14,417.24$	$\sum\limits^{an} Y^2 = 72,535$
$\sum\limits^{n} y^2$	29.2	12.0	75.2	45.2	98.8	81.2	107.2	$\sum\limits^{a} (\bar{Y} - \bar{\bar{Y}})^2 = 25.417$	$\sum\limits^{an} y^2 = 575.886$

a rather low estimate compared with those obtained in the other samples. Since we have a sum of squares for each group, we could obtain an estimate of the population variance from each of these. However, it stands to reason that we would get a better estimate if we averaged these separate variance estimates in some way. This is done by computing the weighted average of the variances by Expression (3.2) in Section 3.1. Actually, in this instance a simple average would suffice, since all estimates of the variance are based on samples of the same size. However, we prefer to give the general formula, which works equally well for this case as well as for instances of unequal sample sizes, where the weighted average is necessary. In this case each sample variance s_i^2 is weighted by its degrees of freedom, $w_i = n_i - 1$, resulting in a sum of squares (Σy_i^2), since $(n_i - 1)s_i^2 = \Sigma y_i^2$. Thus, the numerator of Expression (3.2) is the sum of the sums of squares. The denominator is $\Sigma^a (n_i - 1) = 7 \times 4$, the sum of the degrees of freedom of each group. The average variance, therefore, is

$$s^2 = \frac{29.2 + 12.0 + 75.2 + 45.2 + 98.8 + 81.2 + 107.2}{28} = \frac{448.8}{28} = 16.029$$

This quantity is an estimate of 15.21, the parametric variance of housefly wing lengths. This estimate, based on 7 independent estimates of variances of groups, is called the *average variance within groups* or simply *variance within groups*. Note that we use the expression *within* groups, although in previous chapters we used the term variance of groups. The reason we do this is that the variance estimates used for computing the average variance have so far all come from sums of squares measuring the variation within one column. As we shall see in what follows, one can also compute variances among groups, cutting across group boundaries.

To obtain a second estimate of the population variance, we treat the seven group means $\bar{Y}$ as though they were a sample of seven observations. The resulting statistics are shown in the lower right part of Table 7.1, headed "Computation of sum of squares of means." There are seven means in this example; in the general case there will be a means. We first compute $\Sigma^a \bar{Y}$, the sum of the means. Note that this is rather sloppy symbolism. To be entirely proper, we should identify this quantity as $\Sigma_{i=1}^{i=a} \bar{Y}_i$, summing the means of group 1 through group a. The next quantity computed is $\bar{\bar{Y}}$, the grand mean of the group means, computed as $\bar{\bar{Y}} = \Sigma^a \bar{Y}/a$. The sum of the seven means is $\Sigma^a \bar{Y} = 317.4$, and the grand mean is $\bar{\bar{Y}} = 45.34$, a fairly close approximation to the parametric mean $\mu = 45.5$. The sum of squares represents the deviations of the group means from the grand mean, $\Sigma^a (\bar{Y} - \bar{\bar{Y}})^2$. For this we first need the quantity $\Sigma^a \bar{Y}^2$, which equals 14,417.24. The customary computational formula for sum of squares applied to these means is $\Sigma^a \bar{Y}^2 - [(\Sigma^a \bar{Y})^2/a] = 25.417$. From the sum of squares of the means we obtain a *variance among the means* in the conventional way as follows: $\Sigma^a (\bar{Y} - \bar{\bar{Y}})^2/(a - 1)$. We divide by $a - 1$ rather than $n - 1$ because the sum of squares was based on a items (means). Thus, variance of the means $s_{\bar{Y}}^2 =$

25.417/6 = 4.2362. We learned in Chapter 6, Expression (6.1), that when we randomly sample from a single population,

$$\sigma_{\bar{Y}}^2 = \frac{\sigma^2}{n}$$

and hence

$$\sigma^2 = n\sigma_{\bar{Y}}^2$$

Thus, we can estimate a variance of items by multiplying the variance of means by the sample size on which the means are based (assuming we have sampled at random from a common population). When we do this for our present example, we obtain $s^2 = 5 \times 4.2362 = 21.181$. This is a second estimate of the parametric variance 15.21. It is not as close to the true value as the previous estimate based on the average variance within groups, but this is to be expected, since it is based on only 7 "observations." We need a name describing this variance to distinguish it from the variance of means from which it has been computed, as well as from the variance within groups with which it will be compared. We shall call it the *variance among groups*; it is n times the variance of means and is an independent estimate of the parametric variance σ^2 of the housefly wing lengths. It may not be clear at this stage why the two estimates of σ^2 that we have obtained, the variance within groups and the variance among groups, are independent. We ask you to take on faith that they are.

Let us review what we have done so far by expressing it in a more formal way. Table 7.2 represents a generalized table for data such as the samples of housefly wing lengths. Each individual wing length is represented by Y, subscripted to indicate the position of the quantity in the data table. The wing length of the jth fly from the ith sample or group is given by Y_{ij}. Thus, you will notice that the first subscript changes with each column representing a group in the

TABLE 7.2
Data arranged for simple analysis of variance, single classification, completely randomized.

				groups				
		1	*2*	*3* $\cdot\cdot$	*i*	$\cdots$	*a*	
n items	1	Y_{11}	Y_{21}	Y_{31} $\cdots$	Y_{i1}	$\cdots$	Y_{a1}	
	2	Y_{12}	Y_{22}	Y_{32} $\cdots$	Y_{i2}	$\cdots$	Y_{a2}	
	3	Y_{13}	Y_{23}	Y_{33} $\cdots$	Y_{i3}	$\cdots$	Y_{a3}	
	j	Y_{ij}	Y_{2j}	Y_{3j} $\cdots$	Y_{ij}	$\cdots$	Y_{aj}	
	n	Y_{1n}	Y_{2n}	Y_{3n} $\cdots$	Y_{in}	$\cdots$	Y_{an}	
Sums	$\sum\limits^{n} Y$	$\sum\limits^{n} Y_1$	$\sum\limits^{n} Y_2$	$\sum\limits^{n} Y_3$ $\cdots$	$\sum\limits^{n} Y_i$	$\cdots$	$\sum\limits^{n} Y_a$	
Means	$\bar{Y}$	$\bar{Y}_1$	$\bar{Y}_2$	$\bar{Y}_3$ $\cdots$	$\bar{Y}_i$	$\cdots$	$\bar{Y}_a$	

table, and the second subscript changes with each row representing an individual item. Using this notation, we can compute the variance of sample 1 as

$$\frac{1}{n-1} \sum_{j=1}^{j=n} (Y_{1j} - \bar{Y}_1)^2$$

The variance within groups, which is the average variance of the samples, is computed as

$$\frac{1}{a(n-1)} \sum_{i=1}^{i=a} \sum_{j=1}^{j=n} (Y_{ij} - \bar{Y}_i)^2$$

Note the double summation. It means that we start with the first group, setting $i = 1$ (i being the index of the outer Σ). We sum the squared deviations of all items from the mean of the first group, changing index j of the inner Σ from 1 to n in the process. We then return to the outer summation, set $i = 2$, and sum the squared deviations for group 2 from $j = 1$ to $j = n$. This process is continued until i, the index of the outer Σ, is set to a. In other words, we sum all the squared deviations within one group first and add this sum to similar sums from all the other groups.

The variance among groups is computed as

$$\frac{n}{a-1} \sum_{i=1}^{i=a} (\bar{Y}_i - \bar{\bar{Y}})^2$$

Now that we have two independent estimates of the population variance, what shall we do with them? We might wish to find out whether they do in fact estimate the same parameter. To test this hypothesis, we need a statistical test that will evaluate the probability that the two sample variances are from the same population. Such a test employs the F distribution, which is taken up next.

7.2 The F distribution

Let us devise yet another sampling experiment. This is quite a tedious one without the use of computers, so we will not ask you to carry it out. Assume that you are sampling at random from a normally distributed population, such as the housefly wing lengths with mean μ and variance σ^2. The sampling procedure consists of first sampling n_1 items and calculating their variance s_1^2, followed by sampling n_2 items and calculating their variance s_2^2. Sample sizes n_1 and n_2 may or may not be equal to each other, but are fixed for any one sampling experiment. Thus, for example, we might always sample 8 wing lengths for the first sample (n_1) and 6 wing lengths for the second sample (n_2). After each pair of values (s_1^2 and s_2^2) has been obtained, we calculate

$$F_s = \frac{s_1^2}{s_2^2}$$

This will be a ratio near 1, because these variances are estimates of the same quantity. Its actual value will depend on the relative magnitudes of variances s_1^2 and s_2^2. If we repeatedly take samples of sizes n_1 and n_2, calculating the ratios

F_s of their variances, the average of these ratios will in fact approach the quantity $(n_2 - 1)/(n_2 - 3)$, which is close to 1.0 when n_2 is large.

The distribution of this statistic is called the *F distribution*, in honor of R. A. Fisher. This is another distribution described by a complicated mathematical function that need not concern us here. Unlike the t and χ^2 distributions, the shape of the F distribution is determined by *two* values for degrees of freedom, v_1 and v_2 (corresponding to the degrees of freedom of the variance in the numerator and the variance in the denominator, respectively). Thus, for every possible combination of values v_1, v_2, each v ranging from 1 to infinity, there exists a separate F distribution. Remember that the F distribution is a theoretical probability distribution, like the t distribution and the χ^2 distribution. Variance ratios s_1^2/s_2^2, based on sample variances are sample statistics that may or may not follow the F distribution. We have therefore distinguished the sample variance ratio by calling it F_s, conforming to our convention of separate symbols for sample statistics as distinct from probability distributions (such as t_s and X^2 contrasted with t and χ^2).

We have discussed how to generate an F distribution by repeatedly taking two samples from the same normal distribution. We could also have generated it by sampling from two separate normal distributions differing in their mean but identical in their parametric variances; that is, with $\mu_1 \neq \mu_2$ but $\sigma_1^2 = \sigma_2^2$. Thus, we obtain an F distribution whether the samples come from the same normal population or from different ones, so long as their variances are identical.

Figure 7.1 shows several representative F distributions. For very low degrees of freedom the distribution is ∟-shaped, but it becomes humped and strongly skewed to the right as both degrees of freedom increase. Table **V** in Appendix

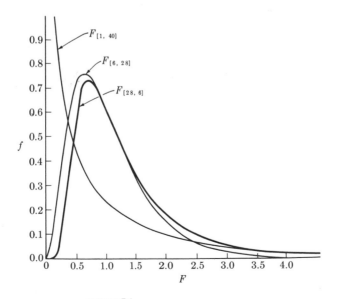

FIGURE 7.1
Three representative F distributions.

A2 shows the cumulative probability distribution of F for three selected probability values. The values in the table represent $F_{\alpha[v_1, v_2]}$, where α is the proportion of the F distribution to the right of the given F value (in one tail) and v_1, v_2 are the degrees of freedom pertaining to the variances in the numerator and the denominator of the ratio, respectively. The table is arranged so that across the top one reads v_1, the degrees of freedom pertaining to the upper (numerator) variance, and along the left margin one reads v_2, the degrees of freedom pertaining to the lower (denominator) variance. At each intersection of degree of freedom values we list three values of F decreasing in magnitude of α. For example, an F distribution with $v_1 = 6$, $v_2 = 24$ is 2.51 at $\alpha = 0.05$. By that we mean that 0.05 of the area under the curve lies to the right of $F = 2.51$. Figure 7.2 illustrates this. Only 0.01 of the area under the curve lies to the right of $F = 3.67$. Thus, if we have a null hypothesis H_0: $\sigma_1^2 = \sigma_2^2$, with the alternative hypothesis H_1: $\sigma_1^2 > \sigma_2^2$, we use a one-tailed F test, as illustrated by Figure 7.2.

We can now test the two variances obtained in the sampling experiment of Section 7.1 and Table 7.1. The variance among groups based on 7 means was 21.180, and the variance within 7 groups of 5 individuals was 16.029. Our null hypothesis is that the two variances estimate the same parametric variance; the alternative hypothesis in an anova is always that the parametric variance estimated by the variance among groups is greater than that estimated by the variance within groups. The reason for this restrictive alternative hypothesis, which leads to a one-tailed test, will be explained in Section 7.4. We calculate the variance ratio $F_s = s_1^2/s_2^2 = 21.181/16.029 = 1.32$. Before we can inspect the

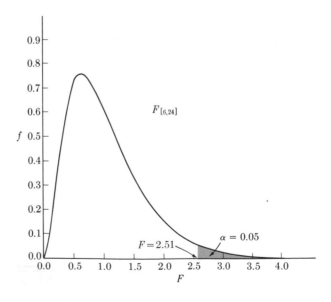

FIGURE 7.2
Frequency curve of the F distribution for 6 and 24 degrees of freedom, respectively. A one-tailed 5% rejection region is marked off at $F = 2.51$.

F table, we have to know the appropriate degrees of freedom for this variance ratio. We shall learn simple formulas for degrees of freedom in an anova later, but at the moment let us reason it out for ourselves. The upper variance (among groups) was based on the variance of 7 means; hence it should have $a - 1 = 6$ degrees of freedom. The lower variance was based on an average of 7 variances, each of them based on 5 individuals yielding 4 degrees of freedom per variance: $a(n - 1) = 7 \times 4 = 28$ degrees of freedom. Thus, the upper variance has 6, the lower variance 28 degrees of freedom. If we check Table V for $v_1 = 6$, $v_2 = 24$, the closest arguments in the table, we find that $F_{0.05[6,24]} = 2.51$. For $F = 1.32$, corresponding to the F_s value actually obtained, α is clearly > 0.05. Thus, we may expect more than 5% of all variance ratios of samples based on 6 and 28 degrees of freedom, respectively, to have F_s values greater than 1.32. We have no evidence to reject the null hypothesis and conclude that the two sample variances estimate the same parametric variance. This corresponds, of course, to what we knew anyway from our sampling experiment. Since the seven samples were taken from the same population, the estimate using the variance of their means is expected to yield another estimate of the parametric variance of housefly wing length.

Whenever the alternative hypothesis is that the two parametric variances are unequal (rather than the restrictive hypothesis $H_1: \sigma_1^2 > \sigma_2^2$), the sample variance s_1^2 can be smaller as well as greater than s_2^2. This leads to a two-tailed test, and in such cases a 5% type I error means that rejection regions of $2\frac{1}{2}\%$ will occur at each tail of the curve. In such a case it is necessary to obtain F values for $\alpha > 0.5$ (that is, in the left half of the F distribution). Since these values are rarely tabulated, they can be obtained by using the simple relationship

$$F_{\alpha[v_1, v_2]} = \frac{1}{F_{(1-\alpha)[v_2, v_1]}} \tag{7.1}$$

For example, $F_{0.05[5,24]} = 2.62$. If we wish to obtain $F_{0.95[5,24]}$ (the F value to the right of which lies 95% of the area of the F distribution with 5 and 24 degrees of freedom, respectively), we first have to find $F_{0.05[24,5]} = 4.53$. Then $F_{0.95[5,24]}$ is the reciprocal of 4.53, which equals 0.221. Thus 95% of an F distribution with 5 and 24 degrees of freedom lies to the right of 0.221.

There is an important relationship between the F distribution and the χ^2 distribution. You may remember that the ratio $X^2 = \sum y^2/\sigma^2$ was distributed as a χ^2 with $n - 1$ degrees of freedom. If you divide the numerator of this expression by $n - 1$, you obtain the ratio $F_s = s^2/\sigma^2$, which is a variance ratio with an expected distribution of $F_{[n-1, \infty]}$. The upper degrees of freedom are $n - 1$ (the degrees of freedom of the sum of squares or sample variance). The lower degrees of freedom are infinite, because only on the basis of an infinite number of items can we obtain the true, parametric variance of a population. Therefore, by dividing a value of X^2 by $n - 1$ degrees of freedom, we obtain an F_s value with $n - 1$ and ∞ df, respectively. In general, $\chi^2_{[v]}/v = F_{[v, \infty]}$. We can convince ourselves of this by inspecting the F and χ^2 tables. From the χ^2 table (Table IV) we find that $\chi^2_{0.05[10]} = 18.307$. Dividing this value by 10 df, we obtain 1.8307. From the F table (Table V) we find, for $v_1 = 10$, $v_2 = \infty$, that $F_{0.05[10,\infty]} = 1.83$.

Thus, the two statistics of significance are closely related and, lacking a χ^2 table, we could make do with an F table alone, using the values of $\nu F_{[\nu, \infty]}$ in place of $\chi^2_{[\nu]}$.

Before we return to analysis of variance, we shall first apply our newly won knowledge of the F distribution to testing a hypothesis about two sample variances.

BOX 7.1

Testing the significance of differences between two variances.

Survival in days of the cockroach *Blattella vaga* when kept without food or water.

Females	$n_1 = 10$	$\bar{Y}_1 = 8.5$ days	$s_1^2 = 3.6$
Males	$n_2 = 10$	$\bar{Y}_2 = 4.8$ days	$s_2^2 = 0.9$

$$H_0: \sigma_1^2 = \sigma_2^2 \qquad H_1: \sigma_1^2 \neq \sigma_2^2$$

Source: Data modified from Willis and Lewis (1957).

The alternative hypothesis is that the two variances are unequal. We have no reason to suppose that one sex should be more variable than the other. In view of the alternative hypothesis this is a two-tailed test. Since only the right tail of the F distribution is tabled extensively in Table **V** and in most other tables, we calculate F_s as the ratio of the greater variance over the lesser one:

$$F_s = \frac{s_1^2}{s_2^2} = \frac{3.6}{0.9} = 4.00$$

Because the test is two-tailed, we look up the critical value $F_{\alpha/2[\nu_1, \nu_2]}$, where α is the type I error accepted and $\nu_1 = n_1 - 1$ and $\nu_2 = n_2 - 1$ are the degrees of freedom for the upper and lower variance, respectively. Whether we look up $F_{\alpha/2[\nu_1, \nu_2]}$ or $F_{\alpha/2[\nu_2, \nu_1]}$ depends on whether sample 1 or sample 2 has the greater variance and has been placed in the numerator.

From Table **V** we find $F_{0.025[9,9]} = 4.03$ and $F_{0.05[9,9]} = 3.18$. Because this is a two-tailed test, we double these probabilities. Thus, the F value of 4.03 represents a probability of $\alpha = 0.05$, since the right-hand tail area of $\alpha = 0.025$ is matched by a similar left-hand area to the left of $F_{0.975[9,9]} = 1/F_{0.025[9,9]} = 0.248$. Therefore, assuming the null hypothesis is true, the probability of observing an F value greater than 4.00 and smaller than $1/4.00 = 0.25$ is $0.10 > P > 0.05$. Strictly speaking, the two sample variances are not significantly different—the two sexes are equally variable in their duration of survival. However, the outcome is close enough to the 5% significance level to make us suspicious that possibly the variances are in fact different. It would be desirable to repeat this experiment with larger sample sizes in the hope that more decisive results would emerge.

7.3 The hypothesis $H_0 : \sigma_1^2 = \sigma_2^2$

A test of the null hypothesis that two normal populations represented by two samples have the same variance is illustrated in Box 7.1. As will be seen later, some tests leading to a decision about whether two samples come from populations with the same mean assume that the population variances are equal. However, this test is of interest in its own right. We will repeatedly have to test whether two samples have the same variance. In genetics we may need to know whether an offspring generation is more variable for a character than the parent generation. In systematics we might like to find out whether two local populations are equally variable. In experimental biology we may wish to demonstrate under which of two experimental setups the readings will be more variable. In general, the less variable setup would be preferred; if both setups were equally variable, the experimenter would pursue the one that was simpler or less costly to undertake.

7.4 Heterogeneity among sample means

We shall now modify the data of Table 7.1, discussed in Section 7.1. Suppose the seven groups of houseflies did not represent random samples from the same population but resulted from the following experiment. Each sample was reared in a separate culture jar, and the medium in each of the culture jars was prepared in a different way. Some had more water added, others more sugar, yet others more solid matter. Let us assume that sample 7 represents the standard medium against which we propose to compare the other samples. The various changes in the medium affect the sizes of the flies that emerge from it; this in turn affects the wing lengths we have been measuring.

We shall assume the following effects resulting from treatment of the medium:

Medium 1—decreases average wing length of a sample by 5 units
       2—decreases average wing length of a sample by 2 units
       3—does not change average wing length of a sample
       4—increases average wing length of a sample by 1 unit
       5—increases average wing length of a sample by 1 unit
       6—increases average wing length of a sample by 5 units
       7—(control) does not change average wing length of a sample

The effect of treatment i is usually symbolized as α_i. (Please note that this use of α is not related to its use as a symbol for the probability of a type I error.) Thus α_i assumes the following values for the above treatment effects.

$$\alpha_1 = -5 \qquad \alpha_4 = 1$$
$$\alpha_2 = -2 \qquad \alpha_5 = 1$$
$$\alpha_3 = 0 \qquad \alpha_6 = 5$$
$$\alpha_7 = 0$$

TABLE 7.3
Data of Table 7.1 with fixed-treatment affects α_i or random effects A_i added to each sample.

	a Groups (a = 7)							Computation of sum of squares of means	Computation of total sum of squares
	1	2	3	4	5	6	7		
α_i or A_i	−5	−2	0	+1	+1	+5	0		
n individuals per group (n = 5)	36	46	40	41	50	45	41		
	39	47	50	40	42	53	46		
	43	47	44	47	51	56	54		
	38	47	48	47	40	52	44		
	37	43	50	42	43	56	42		
$\sum^n Y$	193	230	232	217	226	262	227	$\sum^a \bar Y = 317.4$	$\sum^{an} Y = 1587$
$\bar Y$	38.6	46.0	46.4	43.4	45.2	52.4	45.4	$\bar{\bar Y} = 45.34$	$\bar{\bar Y} = 45.34$
	$\left(\begin{smallmatrix}43.6\\-5\end{smallmatrix}\right)$	$\left(\begin{smallmatrix}48.0\\-2\end{smallmatrix}\right)$	$\left(\begin{smallmatrix}46.4\\+0\end{smallmatrix}\right)$	$\left(\begin{smallmatrix}42.4\\+1\end{smallmatrix}\right)$	$\left(\begin{smallmatrix}44.2\\+1\end{smallmatrix}\right)$	$\left(\begin{smallmatrix}47.4\\+5\end{smallmatrix}\right)$	$\left(\begin{smallmatrix}45.4\\+0\end{smallmatrix}\right)$	$\left(45.34 + \dfrac{-5-2+1+1+5}{7}\right)$	
$\sum^n Y^2$	7479	10,592	10,840	9463	10,314	13,810	10,413	$\sum^a \bar Y^2 = 14{,}492.44$	$\sum^{an} Y^2 = 72{,}911$
$\sum^n y^2$	29.2	12.0	75.2	45.2	98.8	81.2	107.2	$\sum^a (\bar Y - \bar{\bar Y})^2 = 100.617$	$\sum^{an} y^2 = 951.886$

Note that the α_i's have been defined so that $\Sigma^a\, \alpha_i = 0$; that is, the effects cancel out. This is a convenient property that is generally postulated, but it is unnecessary for our argument. We can now modify Table 7.1 by adding the appropriate values of α_i to each sample. In sample 1 the value of α_1 is -5; therefore, the first wing length, which was 41 (see Table 7.1), now becomes 36; the second wing length, formerly 44, becomes 39; and so on. For the second sample α_2 is -2, changing the first wing length from 48 to 46. Where α_i is 0, the wing lengths do not change; where α_i is positive, they are increased by the magnitude indicated. The changed values can be inspected in Table 7.3, which is arranged identically to Table 7.1.

We now repeat our previous computations. We first calculate the sum of squares of the first sample to find it to be 29.2. If you compare this value with the sum of squares of the first sample in Table 7.1, you find the two values to be identical. Similarly, all other values of $\Sigma^n\, y^2$, the sum of squares of each group, are identical to their previous values. Why is this so? The effect of adding α_i to each group is simply that of an additive code, since α_i is constant for any one group. From Appendix A1.2 we can see that additive codes do not affect sums of squares or variances. Therefore, not only is each separate sum of squares the same as before, but the average variance within groups is still 16.029. Now let us compute the variance of the means. It is $100.617/6 = 16.770$, which is a value much higher than the variance of means found before, 4.236. When we multiply by $n = 5$ to get an estimate of σ^2, we obtain the variance of groups, which now is 83.848 and is no longer even close to an estimate of σ^2. We repeat the F test with the new variances and find that $F_s = 83.848/16.029 = 5.23$, which is much greater than the closest critical value of $F_{0.05[6,24]} = 2.51$. In fact, the observed F_s is greater than $F_{0.01[6,24]} = 3.67$. Clearly, the upper variance, representing the variance among groups, has become significantly larger. The two variances are most unlikely to represent the same parametric variance.

What has happened? We can easily explain it by means of Table 7.4, which represents Table 7.3 symbolically in the manner that Table 7.2 represented Table 7.1. We note that each group has a constant α_i added and that this constant changes the sums of the groups by $n\alpha_i$ and the means of these groups by α_i. In Section 7.1 we computed the variance within groups as

$$\frac{1}{a(n-1)} \sum_{i=1}^{i=a} \sum_{j=1}^{j=n} (Y_{ij} - \bar{Y}_{ij})^2$$

When we try to repeat this, our formula becomes more complicated, because to each Y_{ij} and each $\bar{Y}_i$ there has now been added α_i. We therefore write

$$\frac{1}{a(n-1)} \sum_{i=1}^{i=a} \sum_{j=1}^{j=n} [(Y_{ij} + \alpha_i) - (\bar{Y}_i + \alpha_i)]^2$$

Then we open the parentheses inside the square brackets, so that the second α_i changes sign and the α_i's cancel out, leaving the expression exactly as before,

TABLE 7.4
Data of Table 7.3 arranged in the manner of Table 7.2.

		1	2	*a Groups* 3	$\cdots$	i	$\cdots$	a
	1	$Y_{11} + \alpha_1$	$Y_{21} + \alpha_2$	$Y_{31} + \alpha_3$	$\cdots$	$Y_{i1} + \alpha_i$	$\cdots$	$Y_{a1} + \alpha_a$
	2	$Y_{12} + \alpha_1$	$Y_{22} + \alpha_2$	$Y_{32} + \alpha_3$	$\cdots$	$Y_{i2} + \alpha_i$	$\cdots$	$Y_{a2} + \alpha_a$
n Items	3	$Y_{13} + \alpha_1$	$Y_{23} + \alpha_2$	$Y_{33} + \alpha_3$	$\cdots$	$Y_{i3} + \alpha_i$	$\cdots$	$Y_{a3} + \alpha_a$
	$\vdots$	$\vdots$	$\vdots$	$\vdots$		$\vdots$		$\vdots$
	j	$Y_{1j} + \alpha_1$	$Y_{2j} + \alpha_2$	$Y_{3j} + \alpha_3$	$\cdots$	$Y_{ij} + \alpha_i$	$\cdots$	$Y_{aj} + \alpha_a$
	$\vdots$	$\vdots$	$\vdots$	$\vdots$		$\vdots$		$\vdots$
	n	$Y_{1n} + \alpha_1$	$Y_{2n} + \alpha_2$	$Y_{3n} + \alpha_3$	$\cdots$	$Y_{in} + \alpha_i$	$\cdots$	$Y_{an} + \alpha_a$
Sums		$\sum\limits^{n} Y_1 + n\alpha_1$	$\sum\limits^{n} Y_2 + n\alpha_2$	$\sum\limits^{n} Y_3 + n\alpha_3$	$\cdots$	$\sum\limits^{n} Y_i + n\alpha_i$	$\cdots$	$\sum\limits^{n} Y_a + n\alpha_a$
Means		$\bar{Y}_1 + \alpha_1$	$\bar{Y}_1 + \alpha_2$	$\bar{Y}_3 + \alpha_3$	$\cdots$	$\bar{Y}_i + \alpha_i$	$\cdots$	$\bar{Y}_a + \alpha_a$

substantiating our earlier observation that the variance within groups does not change despite the treatment effects.

The variance of means was previously calculated by the formula

$$\frac{1}{a-1} \sum_{i=1}^{i=a} (\bar{Y}_i - \bar{\bar{Y}})^2$$

However, from Table 7.4 we see that the new grand mean equals

$$\frac{1}{a} \sum_{i=1}^{i=a} (\bar{Y}_i + \alpha_i) = \frac{1}{a} \sum_{i=1}^{i=a} \bar{Y}_i + \frac{1}{a} \sum_{i=1}^{i=a} \alpha_i = \bar{\bar{Y}} + \bar{\alpha}$$

When we substitute the new values for the group means and the grand mean, the formula appears as

$$\frac{1}{a-1} \sum_{i=1}^{i=a} [(\bar{Y}_i + \alpha_i) - (\bar{\bar{Y}} + \bar{\alpha})]^2$$

which in turn yields

$$\frac{1}{a-1} \sum_{i=1}^{i=a} [(\bar{Y}_i - \bar{\bar{Y}}) + (\alpha_i - \bar{\alpha})]^2$$

Squaring the expression in the square brackets, we obtain the terms

$$\frac{1}{a-1} \sum_{i=1}^{i=a} (\bar{Y}_i - \bar{\bar{Y}})^2 + \frac{1}{a-1} \sum_{i=1}^{i=a} (\alpha_i - \bar{\alpha})^2 + \frac{2}{a-1} \sum_{i=1}^{i=a} (\bar{Y}_i - \bar{\bar{Y}})(\alpha_i - \bar{\alpha})$$

The first of these terms we immediately recognize as the previous variance of the means, $s_{\bar{Y}}^2$. The second is a new quantity, but is familiar by general appearance; it clearly is a variance or at least a quantity akin to a variance. The third expression is a new type; it is a so-called covariance, which we have not yet encountered. We shall not be concerned with it at this stage except to say that in cases such as the present one, where the magnitude of the treatment effects α_i is assumed to be independent of the $\bar{Y}_i$ to which they are added, the expected value of this quantity is zero; hence it does not contribute to the new variance of means.

The independence of the treatments effects and the sample means is an important concept that we must understand clearly. If we had not applied different treatments to the medium jars, but simply treated all jars as controls, we would still have obtained differences among the wing length means. Those are the differences found in Table 7.1 with random sampling from the same population. By chance, some of these means are greater, some are smaller. In our planning of the experiment we had no way of predicting which sample means would be small and which would be large. Therefore, in planning our treatments, we had no way of matching up a large treatment effect, such as that of medium 6, with the mean that by chance would be the greatest, as that for sample 2. Also, the smallest sample mean (sample 4) is not associated with the smallest treatment effect. Only if the magnitude of the treatment effects were deliberately correlated with the sample means (this would be difficult to do in the experiment designed here) would the third term in the expression, the covariance, have an expected value other than zero.

The second term in the expression for the new variance of means is clearly added as a result of the treatment effects. It is analogous to a variance, but it cannot be called a variance, since it is not based on a random variable, but rather on deliberately chosen treatments largely under our control. By changing the magnitude and nature of the treatments, we can more or less alter the variancelike quantity at will. We shall therefore call it the *added component due to treatment effects*. Since the α_i's are arranged so that $\bar{\alpha} = 0$, we can rewrite the middle term as

$$\frac{1}{a-1} \sum_{i=1}^{i=a} (\alpha_i - \bar{\alpha})^2 = \frac{1}{a-1} \sum_{i=1}^{i=a} \alpha_i^2 = \frac{1}{a-1} \sum^{a} \alpha^2$$

In analysis of variance we multiply the variance of the means by n in order to estimate the parametric variance of the items. As you know, we call the quantity so obtained the variance of groups. When we do this for the case in which treatment effects are present, we obtain

$$n\left(s_{\bar{Y}}^2 + \frac{1}{a-1} \sum^{a} \alpha^2\right) = s^2 + \frac{n}{a-1} \sum^{a} \alpha^2$$

Thus we see that the estimate of the parametric variance of the population is increased by the quantity

$$\frac{n}{a-1} \sum^{a} \alpha^2$$

which is n times the added component due to treatment effects. We found the variance ratio F_s to be significantly greater than could be reconciled with the null hypothesis. It is now obvious why this is so. We were testing the variance ratio expecting to find F approximately equal to $\sigma^2/\sigma^2 = 1$. In fact, however, we have

$$F \approx \frac{\sigma^2 + \dfrac{n}{a-1}\sum^{a}\alpha^2}{\sigma^2}$$

It is clear from this formula (deliberately displayed in this lopsided manner) that the F test is sensitive to the presence of the added component due to treatment effects.

At this point, you have an additional insight into the analysis of variance. It permits us to test whether there are added treatment effects—that is, whether a group of means can simply be considered random samples from the same population, or whether treatments that have affected each group separately have resulted in shifting these means so much that they can no longer be considered samples from the same population. If the latter is so, an added component due to treatment effects will be present and may be detected by an F test in the significance test of the analysis of variance. In such a study, we are generally not interested in the magnitude of

$$\frac{n}{a-1}\sum^{a}\alpha^2$$

but we are interested in the magnitude of the separate values of α_i. In our example these are the effects of different formulations of the medium on wing length. If, instead of housefly wing length, we were measuring blood pressure in samples of rats and the different groups had been subjected to different drugs or different doses of the same drug, the quantities α_i would represent the effects of drugs on the blood pressure, which is clearly the issue of interest to the investigator. We may also be interested in studying differences of the type $\alpha_1 - \alpha_2$, leading us to the question of the significance of the differences between the effects of any two types of medium or any two drugs. But we are a little ahead of our story.

When analysis of variance involves treatment effects of the type just studied, we call it a *Model I anova*. Later in this chapter (Section 7.6), Model I will be defined precisely. There is another model, called a *Model II anova*, in which the added effects for each group are not fixed treatments but are random effects. By this we mean that we have not deliberately planned or fixed the treatment for any one group, but that the actual effects on each group are random and only partly under our control. Suppose that the seven samples of houseflies in Table 7.3 represented the offspring of seven randomly selected females from a population reared on a uniform medium. There would be genetic differences among these females, and their seven broods would reflect this. The exact nature of these differences is unclear and unpredictable. Before actually measuring them, we have no way of knowing whether brood 1 will have longer wings than brood 2, nor have we any way of controlling this experiment so that brood 1 will in fact grow longer wings. So far as we can ascertain, the genetic factors for wing length are distributed in an unknown manner in the population of houseflies (we might hope that they are normally distributed), and our sample of seven is a random sample of these factors.

In another example for a Model II anova, suppose that instead of making up our seven cultures from a single batch of medium, we have prepared seven batches separately, one right after the other, and are now analyzing the variation among the batches. We would not be interested in the exact differences from batch to batch. Even if these were measured, we would not be in a position to interpret them. Not having deliberately varied batch 3, we have no idea why, for example, it should produce longer wings than batch 2. We would, however, be interested in the magnitude of the variance of the added effects. Thus, if we used seven jars of medium derived from one batch, we could expect the variance of the jar means to be $\sigma^2/5$, since there were 5 flies per jar. But when based on different batches of medium, the variance could be expected to be greater, because all the imponderable accidents of formulation and environmental differences during medium preparation that make one batch of medium different from another would come into play. Interest would focus on the added variance component arising from differences among batches. Similarly, in the other example we would be interested in the added variance component arising from genetic differences among the females.

We shall now take a rapid look at the algebraic formulation of the anova in the case of Model II. In Table 7.3 the second row at the head of the data columns shows not only α_i but also A_i, which is the symbol we shall use for a random group effect. We use a capital letter to indicate that the effect is a variable. The algebra of calculating the two estimates of the population variance is the same as in Model I, except that in place of α_i we imagine A_i substituted in Table 7.4. The estimate of the variance among means now represents the quantity

$$\frac{1}{a-1} \sum_{i=1}^{i=a} (\bar{Y}_i - \bar{\bar{Y}})^2 + \frac{1}{a-1} \sum_{i=1}^{i=a} (A_i - \bar{A})^2 + \frac{2}{a-1} \sum_{i=1}^{i=a} (\bar{Y}_i - \bar{\bar{Y}})(A_i - \bar{A})$$

The first term is the variance of means $s_{\bar{Y}}^2$, as before, and the last term is the covariance between the group means and the random effects A_i, the expected value of which is zero (as before), because the random effects are independent of the magnitude of the means. The middle term is a true variance, since A_i is a random variable. We symbolize it by s_A^2 and call it the *added variance component among groups*. It would represent the added variance component among females or among medium batches, depending on which of the designs discussed above we were thinking of. The existence of this added variance component is demonstrated by the F test. If the groups are random samples, we may expect F to approximate $\sigma^2/\sigma^2 = 1$; but with an added variance component, the expected ratio, again displayed lopsidedly, is

$$F \approx \frac{\sigma^2 + n\sigma_A^2}{\sigma^2}$$

Note that σ_A^2, the parametric value of s_A^2, is multiplied by n, since we have to multiply the variance of means by n to obtain an independent estimate of the variance of the population. In a Model II anova we are interested not in the magnitude of any A_i or in differences such as $A_1 - A_2$, but in the magnitude of σ_A^2 and its relative magnitude with respect to σ^2, which is generally expressed as the percentage $100s_A^2/(s^2 + s_A^2)$. Since the variance among groups estimates $\sigma^2 + n\sigma_A^2$, we can calculate s_A^2 as

$\dfrac{1}{n}$ (variance among groups $-$ variance within groups)

$$= \frac{1}{n}[(s^2 + ns_A^2) - s^2] = \frac{1}{n}(ns_A^2) = s_A^2$$

For the present example, $s_A^2 = \frac{1}{5}(83.848 - 16.029) = 13.56$. This added variance component among groups is

$$\frac{100 \times 13.56}{16.029 + 13.56} = \frac{1356}{29.589} = 45.83\%$$

of the sum of the variances among and within groups. Model II will be formally discussed at the end of this chapter (Section 7.7); the methods of estimating variance components are treated in detail in the next chapter.

7.5 Partitioning the total sum of squares and degrees of freedom

So far we have ignored one other variance that can be computed from the data in Table 7.1. If we remove the classification into groups, we can consider the housefly data to be a single sample of $an = 35$ wing lengths and calculate the mean and variance of these items in the conventional manner. The various quantities necessary for this computation are shown in the last column at the right in Tables 7.1 and 7.3, headed "Computation of total sum of squares." We obtain a mean of $\bar{Y} = 45.34$ for the sample in Table 7.1, which is, of course, the same as the quantity $\bar{\bar{Y}}$ computed previously from the seven group means. The sum of squares of the 35 items is 575.886, which gives a variance of 16.938 when divided by 34 degrees of freedom. Repeating these computations for the data in Table 7.3, we obtain $\bar{Y} = 45.34$ (the same as in Table 7.1 because $\Sigma^a \alpha_i = 0$) and $s^2 = 27.997$, which is considerably greater than the corresponding variance from Table 7.1. The total variance computed from all an items is another estimate of σ^2. It is a good estimate in the first case, but in the second sample (Table 7.3), where added components due to treatment effects or added variance components are present, it is a poor estimate of the population variance.

However, the purpose of calculating the total variance in an anova is not for using it as yet another estimate of σ^2, but for introducing an important mathematical relationship between it and the other variances. This is best seen when we arrange our results in a conventional *analysis of variance table*, as

TABLE 7.5
Anova table for data in Table 7.1.

		(1) Source of variation	(2) df	(3) Sum of squares SS	(4) Mean square MS
$\bar{Y} - \bar{\bar{Y}}$		Among groups	6	127.086	21.181
$Y - \bar{Y}$		Within groups	28	448.800	16.029
$Y - \bar{\bar{Y}}$		Total	34	575.886	16.938

shown in Table 7.5. Such a table is divided into four columns. The first identifies the source of variation as among groups, within groups, and total (groups amalgamated to form a single sample). The column headed df gives the degrees of freedom by which the sums of squares pertinent to each source of variation must be divided in order to yield the corresponding variance. The degrees of freedom for variation among groups is $a - 1$, that for variation within groups is $a(n - 1)$, and that for the total variation is $an - 1$. The next two columns show sums of squares and variances, respectively. Notice that the sums of squares entered in the anova table are the sum of squares among groups, the sum of squares within groups, and the sum of squares of the total sample of an items. You will note that variances are not referred to by that term in anova, but are generally called *mean squares*, since, in a Model I anova, they do not estimate a population variance. These quantities are not true *mean* squares, because the sums of squares are divided by the degrees of freedom rather than sample size. The sum of squares and mean square are frequently abbreviated SS and MS, respectively.

The sums of squares and mean squares in Table 7.5 are the same as those obtained previously, except for minute rounding errors. Note, however, an important property of the sums of squares. They have been obtained independently of each other, but when we add the SS among groups to the SS within groups we obtain the total SS. The sums of squares are additive! Another way of saying this is that we can decompose the total sum of squares into a portion due to variation among groups and another portion due to variation within groups. Observe that the degrees of freedom are also additive and that the total of 34 df can be decomposed into 6 df among groups and 28 df within groups. Thus, if we know any two of the sums of squares (and their appropriate degrees of freedom), we can compute the third and complete our analysis of variance. Note that the mean squares are not additive. This is obvious, since generally $(a + b)/(c + d) \neq a/c + b/d$.

We shall use the computational formula for sum of squares (Expression (3.8)) to demonstrate why these sums of squares are additive. Although it is an algebraic derivation, it is placed here rather than in the Appendix because these formulas will also lead us to some common computational formulas for analysis of variance. Depending on computational equipment, the formulas we have used so far to obtain the sums of squares may not be the most rapid procedure.

The sum of squares of means in simplified notation is

$$SS_{means} = \sum^{a} (\bar{Y} - \bar{\bar{Y}})^2 = \sum^{a} \bar{Y}^2 - \frac{\left(\sum\limits^{a} \bar{Y}\right)^2}{a}$$

$$= \sum^{a} \left(\frac{1}{n} \sum^{n} Y\right)^2 - \frac{1}{a}\left[\sum^{a}\left(\frac{1}{n}\sum^{n} Y\right)\right]^2$$

$$= \frac{1}{n^2} \sum^{a} \left(\sum^{n} Y\right)^2 - \frac{1}{an^2}\left(\sum^{a}\sum^{n} Y\right)^2$$

Note that the deviation of means from the grand mean is first rearranged to fit the computational formula (Expression (3.8)), and then each mean is written in terms of its constituent variates. Collection of denominators outside the summation signs yields the final desired form. To obtain the sum of squares of groups, we multiply SS_{means} by n, as before. This yields

$$SS_{groups} = n \times SS_{means} = \frac{1}{n} \sum^{a} \left(\sum^{n} Y\right)^2 - \frac{1}{an}\left(\sum^{a}\sum^{n} Y\right)^2$$

Next we evaluate the sum of squares within groups:

$$SS_{within} = \sum^{a}\sum^{n} (Y - \bar{Y})^2 = \sum^{a}\left[\sum^{n} Y^2 - \frac{1}{n}\left(\sum^{n} Y\right)^2\right]$$

$$= \sum^{a}\sum^{n} Y^2 - \frac{1}{n}\sum^{a}\left(\sum^{n} Y\right)^2$$

The total sum of squares represents

$$SS_{total} = \sum^{a}\sum^{n} (Y - \bar{\bar{Y}})^2$$

$$= \sum^{a}\sum^{n} Y^2 - \frac{1}{an}\left(\sum^{a}\sum^{n} Y\right)^2$$

We now copy the formulas for these sums of squares, slightly rearranged as follows:

$$SS_{groups} = \frac{1}{n}\sum^{a}\left(\sum^{n} Y\right)^2 \qquad\qquad -\frac{1}{an}\left(\sum^{a}\sum^{n} Y\right)^2$$

$$SS_{within} = -\frac{1}{n}\sum^{a}\left(\sum^{n} Y\right)^2 + \sum^{a}\sum^{n} Y^2$$

$$SS_{total} = \qquad\qquad\qquad \sum^{a}\sum^{n} Y^2 - \frac{1}{an}\left(\sum^{a}\sum^{n} Y\right)^2$$

Adding the expression for SS_{groups} to that for SS_{within}, we obtain a quantity that is identical to the one we have just developed as SS_{total}. This demonstration explains why the sums of squares are additive.

We shall not go through any derivation, but simply state that the degrees of freedom pertaining to the sums of squares are also additive. The total degrees of freedom are split up into the degrees of freedom corresponding to variation among groups and those of variation of items within groups.

Before we continue, let us review the meaning of the three mean squares in the anova. The total MS is a statistic of dispersion of the 35 (an) items around their mean, the grand mean 45.34. It describes the variance in the entire sample due to all the sundry causes and estimates σ^2 when there are no added treatment effects or variance components among groups. The within-group MS, also known as the *individual* or *intragroup* or *error mean square*, gives the average dispersion of the 5 (n) items in each group around the group means. If the a groups are random samples from a common homogeneous population, the within-group MS should estimate σ^2. The MS among groups is based on the variance of group means, which describes the dispersion of the 7 (a) group means around the grand mean. If the groups are random samples from a homogeneous population, the expected variance of their mean will be σ^2/n. Therefore, in order to have all three variances of the same order of magnitude, we multiply the variance of means by n to obtain the variance among groups. If there are no added treatment effects or variance components, the MS among groups is an estimate of σ^2. Otherwise, it is an estimate of

$$\sigma^2 + \frac{n}{a-1} \sum^{a} \alpha^2 \qquad \text{or} \qquad \sigma^2 + n\sigma_A^2$$

depending on whether the anova at hand is Model I or II.

The additivity relations we have just learned are independent of the presence of added treatment or random effects. We could show this algebraically, but it is simpler to inspect Table 7.6, which summarizes the anova of Table 7.3 in which α_i or A_i is added to each sample. The additivity relation still holds, although the values for group SS and the total SS are different from those of Table 7.5.

TABLE 7.6
Anova table for data in Table 7.3.

	(1) Source of variation	(2) df	(3) Sum of squares SS	(4) Mean square MS
$\bar{Y} - \bar{\bar{Y}}$	Among groups	6	503.086	83.848
$Y - \bar{Y}$	Within groups	28	448.800	16.029
$Y - \bar{\bar{Y}}$	Total	34	951.886	27.997

Another way of looking at the partitioning of the variation is to study the deviation from means in a particular case. Referring to Table 7.1, we can look at the wing length of the first individual in the seventh group, which happens to be 41. Its deviation from its group mean is

$$Y_{71} - \bar{Y}_7 = 41 - 45.4 = -4.4$$

The deviation of the group mean from the grand mean is

$$\bar{Y}_7 - \bar{\bar{Y}} = 45.4 - 45.34 = 0.06$$

and the deviation of the individual wing length from the grand mean is

$$Y_{71} - \bar{\bar{Y}} = 41 - 45.34 = -4.34$$

Note that these deviations are additive. The deviation of the item from the group mean and that of the group mean from the grand mean add to the total deviation of the item from the grand mean. These deviations are stated algebraically as $(Y - \bar{Y}) + (\bar{Y} - \bar{\bar{Y}}) = (Y - \bar{\bar{Y}})$. Squaring and summing these deviations for an items will result in

$$\sum^{a}\sum^{n}(Y - \bar{Y})^2 + n\sum^{a}(\bar{Y} - \bar{\bar{Y}})^2 = \sum^{an}(Y - \bar{\bar{Y}})^2$$

Before squaring, the deviations were in the relationship $a + b = c$. After squaring, we would expect them to take the form $a^2 + b^2 + 2ab = c^2$. What happened to the cross-product term corresponding to $2ab$? This is

$$2\sum^{an}(Y - \bar{Y})(\bar{Y} - \bar{\bar{Y}}) = 2\sum^{a}[(\bar{Y} - \bar{\bar{Y}})\sum^{n}(Y - \bar{Y})]$$

a covariance-type term that is always zero, since $\sum^{n}(Y - \bar{Y}) = 0$ for each of the a groups (proof in Appendix A1.1).

We identify the deviations represented by each level of variation at the left margins of the tables giving the analysis of variance results (Tables 7.5 and 7.6). Note that the deviations add up correctly: the deviation among groups plus the deviation within groups equals the total deviation of items in the analysis of variance, $(\bar{Y} - \bar{\bar{Y}}) + (Y - \bar{Y}) = (Y - \bar{\bar{Y}})$.

7.6 Model I anova

An important point to remember is that the basic setup of data, as well as the actual computation and significance test, in most cases is the same for both models. The purposes of analysis of variance differ for the two models. So do some of the supplementary tests and computations following the initial significance test.

Let us now try to resolve the variation found in an analysis of variance case. This will not only lead us to a more formal interpretation of anova but will also give us a deeper understanding of the nature of variation itself. For purposes of discussion, we return to the housefly wing lengths of Table 7.3. We ask the question, What makes any given housefly wing length assume the value it does? The third wing length of the first sample of flies is recorded as 43 units. How can we explain such a reading?

If we knew nothing else about this individual housefly, our best guess of its wing length would be the grand mean of the population, which we know to be $\mu = 45.5$. However, we have additional information about this fly. It is a member of group 1, which has undergone a treatment shifting the mean of the group downward by 5 units. Therefore, $\alpha_1 = -5$, and we would expect our individual Y_{13} (the third individual of group 1) to measure $45.5 - 5 = 40.5$ units. In fact, however, it is 43 units, which is 2.5 units above this latest expectation. To what can we ascribe this deviation? It is individual variation of the flies within a group because of the variance of individuals in the population ($\sigma^2 = 15.21$). All the genetic and environmental effects that make one housefly different from another housefly come into play to produce this variance.

By means of carefully designed experiments, we might learn something about the causation of this variance and attribute it to certain specific genetic or environmental factors. We might also be able to eliminate some of the variance. For instance, by using only full sibs (brothers and sisters) in any one culture jar, we would decrease the genetic variation in individuals, and undoubtedly the variance within groups would be smaller. However, it is hopeless to try to eliminate all variance completely. Even if we could remove all genetic variance, there would still be environmental variance. And even in the most improbable case in which we could remove both types of variance, measurement error would remain, so that we would never obtain exactly the same reading even on the same individual fly. The within-groups MS always remains as a residual, greater or smaller from experiment to experiment—part of the nature of things. This is why the within-groups variance is also called the error variance or error mean square. It is not an error in the sense of our making a mistake, but in the sense of a measure of the variation you have to contend with when trying to estimate significant differences among the groups. The error variance is composed of individual deviations for each individual, symbolized by ϵ_{ij}, the random component of the jth individual variate in the ith group. In our case, $\epsilon_{13} = 2.5$, since the actual observed value is 2.5 units above its expectation of 40.5.

We shall now state this relationship more formally. In a Model I analysis of variance we assume that the differences among group means, if any, are due to the fixed treatment effects determined by the experimenter. The purpose of the analysis of variance is to estimate the true differences among the group means. Any single variate can be decomposed as follows:

$$Y_{ij} = \mu + \alpha_i + \epsilon_{ij} \tag{7.2}$$

where $i = 1, \ldots, a$, $j = 1, \ldots, n$; and ϵ_{ij} represents an independent, normally distributed variable with mean $\bar{\epsilon}_{ij} = 0$ and variance $\sigma_\epsilon^2 = \sigma^2$. Therefore, a given reading is composed of the grand mean μ of the population, a fixed deviation

α_i of the mean of group i from the grand mean μ, and a random deviation ϵ_{ij} of the jth individual of group i from its expectation, which is $(\mu + \alpha_i)$. Remember that both α_i and ϵ_{ij} can be positive as well as negative. The expected value (mean) of the ϵ_{ij}'s is zero, and their variance is the parametric variance of the population, σ^2. For all the assumptions of the analysis of variance to hold, the distribution of ϵ_{ij} must be normal.

In a Model I anova we test for differences of the type $\alpha_1 - \alpha_2$ among the group means by testing for the presence of an added component due to treatments. If we find that such a component is present, we reject the null hypothesis that the groups come from the same population and accept the alternative hypothesis that at least some of the group means are different from each other, which indicates that at least some of the α_i's are unequal in magnitude. Next, we generally wish to test which α_i's are different from each other. This is done by significance tests, with alternative hypotheses such as $H_1: \alpha_1 > \alpha_2$ or $H_1: \frac{1}{2}(\alpha_1 + \alpha_2) > \alpha_3$. In words, these test whether the mean of group 1 is greater than the mean of group 2, or whether the mean of group 3 is smaller than the average of the means of groups 1 and 2.

Some examples of Model I analyses of variance in various biological disciplines follow. An experiment in which we try the effects of different drugs on batches of animals results in a Model I anova. We are interested in the results of the treatments and the differences between them. The treatments are fixed and determined by the experimenter. This is true also when we test the effects of different doses of a given factor—a chemical or the amount of light to which a plant has been exposed or temperatures at which culture bottles of insects have been reared. The treatment does not have to be entirely understood and manipulated by the experimenter. So long as it is fixed and repeatable, Model I will apply.

If we wanted to compare the birth weights of the Chinese children in the hospital in Singapore with weights of Chinese children born in a hospital in China, our analysis would also be a Model I anova. The treatment effects then would be "China versus Singapore," which sums up a whole series of different factors, genetic and environmental—some known to us but most of them not understood. However, this is a definite treatment we can describe and also repeat: we can, if we wish, again sample birth weights of infants in Singapore as well as in China.

Another example of Model I anova would be a study of body weights for animals of several age groups. The treatments would be the ages, which are fixed. If we find that there are significant differences in weight among the ages, we might proceed with the question of whether there is a difference from age 2 to age 3 or only from age 1 to age 2.

To a very large extent, Model I anovas are the result of an experiment and of deliberate manipulation of factors by the experimenter. However, the study of differences such as the comparison of birth weights from two countries, while not an experiment proper, also falls into this category.

7.7 Model II anova

The structure of variation in a Model II anova is quite similar to that in
Model I:

$$Y_{ij} = \mu + A_i + \epsilon_{ij} \qquad (7.3)$$

where $i = 1, \ldots, a; j = 1, \ldots, n; \epsilon_{ij}$ represents an independent, normally distributed variable with mean $\bar{\epsilon}_{ij} = 0$ and variance $\sigma_\epsilon^2 = \sigma^2$; and A_i represents
a normally distributed variable, independent of all ϵ's, with mean $\bar{A}_i = 0$ and
variance σ_A^2. The main distinction is that in place of fixed-treatment effects α_i,
we now consider random effects A_i that differ from group to group. Since the
effects are random, it is uninteresting to estimate the magnitude of these random
effects on a group, or the differences from group to group. But we can estimate
their variance, the added variance component among groups σ_A^2. We test for its
presence and estimate its magnitude s_A^2, as well as its percentage contribution to
the variation in a Model II analysis of variance.

Some examples will illustrate the applications of Model II anova. Suppose
we wish to determine the DNA content of rat liver cells. We take five rats and
make three preparations from each of the five livers obtained. The assay readings will be for $a = 5$ groups with $n = 3$ readings per group. The five rats presumably are sampled at random from the colony available to the experimenter.
They must be different in various ways, genetically and environmentally, but we
have no definite information about the nature of the differences. Thus, if we learn
that rat 2 has slightly more DNA in its liver cells than rat 3, we can do little
with this information, because we are unlikely to have any basis for following
up this problem. We will, however, be interested in estimating the variance of
the three replicates within any one liver and the variance among the five rats;
that is, does variance σ_A^2 exist among rats in addition to the variance σ^2 expected
on the basis of the three replicates? The variance among the three preparations
presumably arises only from differences in technique and possibly from differences in DNA content in different parts of the liver (unlikely in a homogenate).
Added variance among rats, if it existed, might be due to differences in ploidy
or related phenomena. The relative amounts of variation among rats and
"within" rats (= among preparations) would guide us in designing further
studies of this sort. If there was little variance among the preparations and
relatively more variation among the rats, we would need fewer preparations and
more rats. On the other hand, if the variance among rats was proportionately
smaller, we would use fewer rats and more preparations per rat.

In a study of the amount of variation in skin pigment in human populations,
we might wish to study different families within a homogeneous ethnic or racial
group and brothers and sisters within each family. The variance within families
would be the error mean square, and we would test for an added variance
component among families. We would expect an added variance component
σ_A^2 because there are genetic differences among families that determine amount

of skin pigmentation. We would be especially interested in the relative proportions of the two variances σ^2 and σ_A^2, because they would provide us with
important genetic information. From our knowledge of genetic theory, we
would expect the variance among families to be greater than the variance among
brothers and sisters within a family.

The above examples illustrate the two types of problems involving Model
II analysis of variance that are most likely to arise in biological work. One is
concerned with the general problem of the design of an experiment and the
magnitude of the experimental error at different levels of replication, such as
error among replicates within rat livers and among rats, error among batches,
experiments, and so forth. The other relates to variation among and within
families, among and within females, among and within populations, and so
forth. Such problems are concerned with the general problem of the relation
between genetic and phenotypic variation.

Exercises

7.1   In a study comparing the chemical composition of the urine of chimpanzees
      and gorillas (Gartler, Firschein, and Dobzhansky, 1956), the following results
      were obtained. For 37 chimpanzees the variance for the amount of glutamic acid
      in milligrams per milligram of creatinine was 0.01069. A similar study based on
      six gorillas yielded a variance of 0.12442. Is there a significant difference be-
      tween the variability in chimpanzees and that in gorillas? ANS. $F_s = 11.639$,
      $F_{0.025[5,36]} \approx 2.90$.

7.2   The following data are from an experiment by Sewall Wright. He crossed Polish
      and Flemish giant rabbits and obtained 27 F_1 rabbits. These were inbred and
      112 F_2 rabbits were obtained. We have extracted the following data on femur
      length of these rabbits.

	n	$\bar{Y}$	s
F_1	27	83.39	1.65
F_2	112	80.5	3.81

Is there a significantly greater amount of variability in femur lengths among the
F_2 than among the F_1 rabbits? What well-known genetic phenomenon is illustrated by these data?

7.3   For the following data obtained by a physiologist, estimate σ^2 (the variance
      within groups), α_i (the fixed treatment effects), the variance among the groups,
      and the added component due to treatment $\Sigma \, \alpha^2/(a-1)$, and test the hypothesis
      that the last quantity is zero.

	Treatment			
	A	B	C	D
$\bar{Y}$	6.12	4.34	5.12	7.28
s^2	2.85	6.70	4.06	2.03
n	10	10	10	10

ANS. $s^2 = 3.91$, $\hat{\alpha}_1 = 0.405$, $\hat{\alpha}_2 = 1.375$, $\hat{\alpha}_3 = 0.595$, $\hat{\alpha}_4 = 1.565$, MS among groups $= 124.517$, and $F_s = 31.846$ (which is significant beyond the 0.01 level).

7.4 For the data in Table 7.3, make tables to represent partitioning of the value of each variate into its three components, $\bar{\bar{Y}}, (\bar{Y}_i - \bar{\bar{Y}}), (Y_{ij} - \bar{Y}_i)$. The first table would then consist of 35 values, all equal to the grand mean. In the second table all entries in a given column would be equal to the difference between the mean of that column and the grand mean. And the last table would consist of the deviations of the individual variates from their column means. These tables represent estimates of the individual components of Expression (7.3). Compute the mean and sum of squares for each table.

7.5 A geneticist recorded the following measurements taken on two-week-old mice of a particular strain. Is there evidence that the variance among mice in different litters is larger than one would expect on the basis of the variability found within each litter?

Litters

1	2	3	4	5	6	7
19.49	22.94	23.06	15.90	16.72	20.00	21.52
20.62	22.15	20.05	21.48	19.22	19.79	20.37
19.51	19.16	21.47	22.48	26.62	21.15	21.93
18.09	20.98	14.90	18.79	20.74	14.88	20.14
22.75	23.13	19.72	19.70	21.82	19.79	22.28

ANS. $s^2 = 5.987$, $MS_{\text{among}} = 4.416$, $s_A^2 = 0$, and $F_s = 0.7375$, which is clearly not significant at the 5% level.

7.6 Show that it is possible to represent the value of an individual variate as follows: $Y_{ij} = (\bar{\bar{Y}}) + (\bar{Y}_i - \bar{\bar{Y}}) + (Y_{ij} - \bar{Y}_i)$. What does each of the terms in parentheses estimate in a Model I anova and in a Model II anova?

CHAPTER 8

Single-Classification Analysis of Variance

We are now ready to study actual cases of analysis of variance in a variety of applications and designs. The present chapter deals with the simplest kind of anova, *single-classification analysis of variance.* By this we mean an analysis in which the groups (samples) are classified by only a single criterion. Either interpretations of the seven samples of housefly wing lengths (studied in the last chapter), different medium formulations (Model I), or progenies of different females (Model II) would represent a single criterion for classification. Other examples would be different temperatures at which groups of animals were raised or different soils in which samples of plants have been grown.

We shall start in Section 8.1 by stating the basic computational formulas for analysis of variance, based on the topics covered in the previous chapter. Section 8.2 gives an example of the common case with equal sample sizes. We shall illustrate this case by means of a Model I anova. Since the basic computations for the analysis of variance are the same in either model, it is not necessary to repeat the illustration with a Model II anova. The latter model is featured in Section 8.3, which shows the minor computational complications resulting from unequal sample sizes, since all groups in the anova need not necessarily have the same sample size. Some computations unique to a Model II anova are also shown; these estimate variance components. Formulas become especially simple for the two-sample case, as explained in Section 8.4. In Model I of this case, the mathematically equivalent t test can be applied as well.

When a Model I analysis of variance has been found to be significant, leading to the conclusion that the means are not from the same population, we will usually wish to test the means in a variety of ways to discover which pairs of means are different from each other and whether the means can be divided into groups that are significantly different from each other. To this end, Section 8.5 deals with so-called planned comparisons designed before the test is run; and Section 8.6, with unplanned multiple-comparison tests that suggest themselves to the experimenter as a result of the analysis.

8.1 Computational formulas

We saw in Section 7.5 that the total sum of squares and degrees of freedom can be additively partitioned into those pertaining to variation among groups and those to variation within groups. For the analysis of variance proper, we need only the sum of squares among groups and the sum of squares within groups. But when the computation is not carried out by computer, it is simpler to calculate the total sum of squares and the sum of squares among groups, leaving the sum of squares within groups to be obtained by the subtraction $SS_{\text{total}} - SS_{\text{groups}}$. However, it is a good idea to compute the individual variances so we can check for heterogeneity among them (see Section 10.1). This will also permit an independent computation of SS_{within} as a check. In Section 7.5 we arrived at the following computational formulas for the total and among-groups sums of squares:

$$SS_{\text{total}} = \sum^{a} \sum^{n} Y^2 - \frac{1}{an}\left(\sum^{a}\sum^{n} Y\right)^2$$

$$SS_{\text{groups}} = \frac{1}{n}\sum^{a}\left(\sum^{n} Y\right)^2 - \frac{1}{an}\left(\sum^{a}\sum^{n} Y\right)^2$$

These formulas assume equal sample size n for each group and will be modified in Section 8.3 for unequal sample sizes. However, they suffice in their present form to illustrate some general points about computational procedures in analysis of variance.

We note that the second, subtracted term is the same in both sums of squares. This term can be obtained by summing all the variates in the anova (this is the grand total), squaring the sum, and dividing the result by the total number of variates. It is comparable to the second term in the computational formula for the ordinary sum of squares (Expression (3.8)). This term is often called the *correction term* (abbreviated CT).

The first term for the total sum of squares is simple. It is the sum of all squared variates in the anova table. Thus the total sum of squares, which describes the variation of a single unstructured sample of an items, is simply the familiar sum-of-squares formula of Expression (3.8).

The first term of the sum of squares among groups is obtained by squaring the sum of the items of each group, dividing each square by its sample size, and summing the quotients from this operation for each group. Since the sample size of each group is equal in the above formulas, we can first sum all the squares of the group sums and then divide their sum by the constant n.

From the formula for the sum of squares among groups emerges an important computational rule of analysis of variance: *To find the sum of squares among any set of groups, square the sum of each group and divide by the sample size of the group; sum the quotients of these operations and subtract from the sum a correction term. To find this correction term, sum all the items in the set, square the sum, and divide it by the number of items on which this sum is based.*

8.2 Equal n

We shall illustrate a single-classification anova with equal sample sizes by a Model I example. The computation up to and including the first test of significance is identical for both models. Thus, the computation of Box 8.1 could also serve for a Model II anova with equal sample sizes.

The data are from an experiment in plant physiology. They are the lengths in coded units of pea sections grown in tissue culture with auxin present. The purpose of the experiment was to test the effects of the addition of various sugars on growth as measured by length. Four experimental groups, representing three different sugars and one mixture of sugars, were used, plus one control without sugar. Ten observations (replicates) were made for each treatment. The term "treatment" already implies a Model I anova. It is obvious that the five groups do not represent random samples from all possible experimental conditions but were deliberately designed to test the effects of certain sugars on the growth rate. We are interested in the effect of the sugars on length, and our null hypothesis will be that there is no added component due to treatment effects among the five groups; that is, the population means are all assumed to be equal.

The computation is illustrated in Box 8.1. After quantities **1** through **7** have been calculated, they are entered into an analysis-of-variance table, as shown in the box. General formulas for such a table are shown first; these are followed by a table filled in for the specific example. We note 4 degrees of freedom among groups, there being five treatments, and 45 df within groups, representing 5 times $(10 - 1)$ degrees of freedom. We find that the mean square among groups is considerably greater than the error mean square, giving rise to a suspicion that an added component due to treatment effects is present. If the MS_{groups} is equal to or less than the MS_{within}, we do not bother going on with the analysis, for we would not have evidence for the presence of an added component. You may wonder how it could be possible for the MS_{groups} to be less than the MS_{within}. You must remember that these two are independent estimates. If there is no added component due to treatment or variance component among groups, the estimate of the variance among groups is as likely to be less as it is to be greater than the variance within groups.

Expressions for the expected values of the mean squares are also shown in the first anova table of Box 8.1. They are the expressions you learned in the previous chapter for a Model I anova.

BOX 8.1

Single-classification anova with equal sample sizes.

The effect of the addition of different sugars on length, in ocular units ($\times 0.114 = $ mm), of pea sections grown in tissue culture with auxin present: $n = 10$ (replications per group). This is a Model I anova.

Observations, i.e., replications	Control	2% Glucose added	2% Fructose added	1% Glucose + 1% Fructose added	2% Sucrose added
			Treatments ($a = 5$)		
1	75	57	58	58	62
2	67	58	61	59	66
3	70	60	56	58	65
4	75	59	58	61	63
5	65	62	57	57	64
6	71	60	56	56	62
7	67	60	61	58	65
8	67	57	60	57	65
9	76	59	57	57	62
10	68	61	58	59	67
$\sum\limits^{n} Y$	701	593	582	580	641
$\bar{Y}$	70.1	59.3	58.2	58.0	64.1

Source: Data by W. Purves.

Preliminary computations

1. Grand total $= \sum\limits^{a}\sum\limits^{n} Y = 701 + 593 + \cdots + 641 = 3097$

2. Sum of the squared observations

$$= \sum\limits^{a}\sum\limits^{n} Y^2$$
$$= 75^2 + 67^2 + \cdots + 68^2 + 57^2 + \cdots + 67^2 = 193{,}151$$

3. Sum of the squared group totals divided by n

$$= \frac{1}{n}\sum\limits^{a}\left(\sum\limits^{n} Y\right)^2 = \tfrac{1}{10}(701^2 + 593^2 + \cdots + 641^2)$$
$$= \tfrac{1}{10}(1{,}929{,}055) = 192{,}905.50$$

4. Grand total squared and divided by total sample size $=$ correction term

$$CT = \frac{1}{an}\left(\sum\limits^{a}\sum\limits^{n} Y\right)^2 = \frac{(3097)^2}{5 \times 10} = \frac{9{,}591{,}409}{50} = 191{,}828.18$$

BOX 8.1
Continued

5. $SS_{\text{total}} = \sum^{a}\sum^{n} Y^2 - CT$
   $= \text{quantity } 2 - \text{quantity } 4 = 193{,}151 - 191{,}828.18 = 1322.82$

6. $SS_{\text{groups}} = \dfrac{1}{n}\sum^{a}\left(\sum^{n} Y\right)^2 - CT$
   $= \text{quantity } 3 - \text{quantity } 4 = 192{,}905.50 - 191{,}828.18 = 1077.32$

7. $SS_{\text{within}} = SS_{\text{total}} - SS_{\text{groups}}$
   $= \text{quantity } 5 - \text{quantity } 6 = 1322.82 - 1077.32 = 245.50$

The anova table is constructed as follows.

	Source of variation	df	SS	MS	F_s	Expected MS
$\bar{Y} - \bar{\bar{Y}}$	Among groups	$a-1$	6	$\dfrac{6}{(a-1)}$	$\dfrac{MS_{\text{groups}}}{MS_{\text{within}}}$	$\sigma^2 + \dfrac{n}{a-1}\sum^{a}\alpha^2$
$Y - \bar{Y}$	Within groups	$a(n-1)$	7	$\dfrac{7}{a(n-1)}$		σ^2
$Y - \bar{\bar{Y}}$	Total	$an-1$	5			

Substituting the computed values into the above table, we obtain the following:

Anova table

	Source of variation	df	SS	MS	F_s
$\bar{Y} - \bar{\bar{Y}}$	Among groups (among treatments)	4	1077.32	269.33	49.33**
$Y - \bar{Y}$	Within groups (error, replicates)	45	245.50	5.46	
$Y - \bar{\bar{Y}}$	Total	49	1322.82		

$$F_{0.05[4,45]} = 2.58 \qquad F_{0.01[4,45]} = 3.77$$

* $= 0.01 < P \le 0.05$.
** $= P \le 0.01$.

These conventions will be followed throughout the text and will no longer be explained in subsequent boxes and tables.

Conclusions. There is a highly significant ($P \ll 0.01$) added component due to treatment effects in the mean square among groups (treatments). The different sugar treatments clearly have a significant effect on growth of the pea sections.

See Sections 8.5 and 8.6 for the completion of a Model I analysis of variance: that is, the method for determining which means are significantly different from each other.

It may seem that we are carrying an unnecessary number of digits in the computations in Box 8.1. This is often necessary to ensure that the error sum of squares, quantity **7**, has sufficient accuracy.

Since v_2 is relatively large, the critical values of F have been computed by harmonic interpolation in Table V (see footnote to Table III for harmonic interpolation). The critical values have been given here only to present a complete record of the analysis. Ordinarily, when confronted with this example, you would not bother working out these values of F. Comparison of the observed variance ratio $F_s = 49.33$ with $F_{0.01[4,40]} = 3.83$, the conservative critical value (the next tabled F with fewer degrees of freedom), would convince you that the null hypothesis should be rejected. The probability that the five groups differ as much as they do by chance is almost infinitesimally small. Clearly, the sugars produce an added treatment effect, apparently inhibiting growth and consequently reducing the length of the pea sections.

At this stage we are not in a position to say whether each treatment is different from every other treatment, or whether the sugars are different from the control but not different from each other. Such tests are necessary to complete a Model I analysis, but we defer their discussion until Sections 8.5 and 8.6.

8.3 Unequal n

This time we shall use a Model II analysis of variance for an example. Remember that up to and including the F test for significance, the computations are exactly the same whether the anova is based on Model I or Model II. We shall point out the stage in the computations at which there would be a divergence of operations depending on the model.

The example is shown in Table 8.1. It concerns a series of morphological measurements of the width of the scutum (dorsal shield) of samples of tick larvae obtained from four different host individuals of the cottontail rabbit. These four hosts were obtained at random from one locality. We know nothing about their origins or their genetic constitution. They represent a random sample of the population of host individuals from the given locality. We would not be in a position to interpret differences between larvae from different hosts, since we know nothing of the origins of the individual rabbits. Population biologists are nevertheless interested in such analyses because they provide an answer to the following question: Are the variances of means of larval characters among hosts greater than expected on the basis of variances of the characters within hosts? We can calculate the average variance of width of larval scutum on a host. This will be our "error" term in the analysis of variance. We then test the observed mean square among groups and see if it contains an added component of variance. What would such an added component of variance represent? The mean square within host individuals (that is, of larvae on any one host) represents genetic differences among larvae and differences in environmental experiences of these larvae. Added variance among hosts demonstrates significant differentiation among the larvae possibly due to differences among the hosts affecting the larvae. It also may be due to genetic differences among

TABLE 8.1

Data and anova table for a single classification anova with unequal sample sizes. Width of scutum (dorsal shield) of larvae of the tick *Haemaphysalis leporispalustris* in samples from 4 cottontail rabbits. Measurements in microns. This is a Model II anova.

	Hosts ($a = 4$)			
	1	*2*	*3*	*4*
	380	350	354	376
	376	356	360	344
	360	358	362	342
	368	376	352	372
	372	338	366	374
	366	342	372	360
	374	366	362	
	382	350	344	
		344	342	
		364	358	
			351	
			348	
			348	
$\sum^{n_i} Y$	2978	3544	4619	2168
n_i	8	10	13	6
$\sum^{n_i} Y^2$	1,108,940	1,257,272	1,642,121	784,536
s^2	54.21	142.04	79.56	233.07

Source: Data by P. A. Thomas.

Anova table

	Source of variation	df	SS	MS	F_s
$\bar{Y} - \bar{\bar{Y}}$	Among groups (among hosts)	3	1808.7	602.6	5.26**
$Y - \bar{Y}$	Within groups (error; among larvae on a host)	33	3778.0	114.5	
$\bar{Y} - \bar{\bar{Y}}$	Total	36	5586.7		

$$F_{0.05[3,33]} = 2.89 \qquad F_{0.01[3,33]} = 4.44$$

Conclusion. There is a significant ($P < 0.01$) added variance component among hosts for width of scutum in larval ticks.

the larvae, should each host carry a family of ticks, or at least a population whose individuals are more related to each other than they are to tick larvae on other host individuals.

The emphasis in this example is on the magnitudes of the variances. In view of the random choice of hosts this is a clear case of a Model II anova. Because this is a Model II anova, the means for each host have been omitted from Table 8.1. We are not interested in the individual means or possible differences among them. A possible reason for looking at the means would be at the beginning of the analysis. One might wish to look at the group means to spot outliers, which might represent readings that for a variety of reasons could be in error.

The computation follows the outline furnished in Box 8.1, except that the symbol Σ^n now needs to be written Σ^{n_i}, since sample sizes differ for each group. Steps 1, 2, and 4 through 7 are carried out as before. Only step 3 needs to be modified appreciably. It is:

3. Sum of the squared group totals, each divided by its sample size,

$$= \sum^a \frac{\left(\sum^{n_i} Y\right)^2}{n_i} = \frac{(2978)^2}{8} + \frac{(3544)^2}{10} + \cdots + \frac{(2168)^2}{6} = 4{,}789{,}091$$

The critical 5% and 1% values of F are shown below the anova table in Table 8.1 (2.89 and 4.44, respectively). You should confirm them for yourself in Table V. Note that the argument $v_2 = 33$ is not given. You therefore have to interpolate between arguments representing 30 to 40 degrees of freedom, respectively. The values shown were computed using harmonic interpolation. However, again, it was not necessary to carry out such an interpolation. The conservative value of F, $F_{\alpha[3,30]}$, is 2.92 and 4.51, for $\alpha = 0.05$ and $\alpha = 0.01$, respectively. The observed value F_s is 5.26, considerably above the interpolated as well as the conservative value of $F_{0.01}$. We therefore reject the null hypothesis $(H_0: \sigma_A^2 = 0)$ that there is no added variance component among groups and that the two mean squares estimate the same variance, allowing a type I error of less than 1%. We accept, instead, the alternative hypothesis of the existence of an added variance component σ_A^2.

What is the biological meaning of this conclusion? For some reason, the ticks on different host individuals differ more from each other than do individual ticks on any one host. This may be due to some modifying influence of individual hosts on the ticks (biochemical differences in blood, differences in the skin, differences in the environment of the host individual—all of them rather unlikely in this case), or it may be due to genetic differences among the ticks. Possibly the ticks on each host represent a sibship (that is, are descendants of a single pair of parents) and the differences in the ticks among host individuals represent genetic differences among families; or perhaps selection has acted differently on the tick populations on each host, or the hosts have migrated to the collection locality from different geographic areas in which the ticks differ in width of scutum. Of these various possibilities, genetic differences among sibships seem most reasonable, in view of the biology of the organism.

The computations up to this point would have been identical in a Model I anova. If this had been Model I, the conclusion would have been that there is a significant treatment effect rather than an added variance component. Now, however, we must complete the computations appropriate to a Model II anova. These will include the estimation of the added variance component and the calculation of percentage variation at the two levels.

Since sample size n_i differs among groups in this example, we cannot write $\sigma^2 + n\sigma_A^2$ for the expected MS_{groups}. It is obvious that no single value of n would be appropriate in the formula. We therefore use an average n; this, however, is not simply $\bar{n}$, the arithmetic mean of the n_i's, but is

$$n_0 = \frac{1}{a-1}\left(\sum^a n_i - \frac{\sum^a n_i^2}{\sum^a n_i}\right) \tag{8.1}$$

which is an average usually close to but always less than $\bar{n}$, unless sample sizes are equal, in which case $n_0 = \bar{n}$. In this example,

$$n_0 = \frac{1}{4-1}\left[(8 + 10 + 13 + 6) - \frac{8^2 + 10^2 + 13^2 + 6^2}{8 + 10 + 13 + 6}\right] = 9.009$$

Since the Model II expected MS_{groups} is $\sigma^2 + n\sigma_A^2$ and the expected MS_{within} is σ^2, it is obvious how the variance component among groups σ_A^2 and the error variance σ^2 are obtained. Of course, the values that we obtain are sample estimates and therefore are written as s_A^2 and s^2. The added variance component s_A^2 is estimated as $(MS_{\text{groups}} - MS_{\text{within}})/n$. Whenever sample sizes are unequal, the denominator becomes n_0. In this example, $(602.7 - 114.5)/9.009 = 54.190$. We are frequently not so much interested in the actual values of these variance components as in their relative magnitudes. For this purpose we sum the components and express each as a percentage of the resulting sum. Thus $s^2 + s_A^2 = 114.5 + 54.190 = 168.690$, and s^2 and s_A^2 are 67.9% and 32.1% of this sum, respectively; relatively more variation occurs within groups (larvae on a host) than among groups (larvae on different hosts).

8.4 Two groups

A frequent test in statistics is to establish the *significance of the difference between two means*. This can easily be done by means of an *analysis of variance for two groups*. Box 8.2 shows this procedure for a Model I anova, the common case.

The example in Box 8.2 concerns the onset of reproductive maturity in water fleas, *Daphnia longispina*. This is measured as the average age (in days) at beginning of reproduction. Each variate in the table is in fact an average, and a possible flaw in the analysis might be that the averages are not based on equal sample sizes. However, we are not given this information and have to proceed on the assumption that each reading in the table is an equally reliable variate. The two series represent different genetic crosses, and the seven replicates in each series are clones derived from the same genetic cross. This example is clearly a Model I anova, since the question to be answered is whether series I differs from series II in average age at the beginning of reproduction. Inspection of the data shows that the mean age at beginning of reproduction

BOX 8.2

Testing the difference in means between two groups.

Average age (in days) at beginning of reproduction in *Daphnia longispina* (each variate is a mean based on approximately similar numbers of females). Two series derived from different genetic crosses and containing seven clones each are compared; $n = 7$ clones per series. This is a Model I anova.

	Series ($a = 2$)	
	I	*II*
	7.2	8.8
	7.1	7.5
	9.1	7.7
	7.2	7.6
	7.3	7.4
	7.2	6.7
	7.5	7.2
$\sum\limits^{n} Y$	52.6	52.9
$\bar{Y}$	7.5143	7.5571
$\sum\limits^{n} Y^2$	398.28	402.23
s^2	0.5047	0.4095

Source: Data by Ordway, from Banta (1939).

Single classification anova with two groups with equal sample sizes

Anova table

	Source of variation	df	SS	MS	F_s
$\bar{Y} - \bar{\bar{Y}}$	Between groups (series)	1	0.00643	0.00643	0.0141
$Y - \bar{Y}$	Within groups (error; clones within series)	12	5.48571	0.45714	
$Y - \bar{\bar{Y}}$	Total	13	5.49214		

$$F_{0.05[1,12]} = 4.75$$

Conclusions. Since $F_s \ll F_{0.05[1,12]}$, the null hypothesis is accepted. The means of the two series are not significantly different; that is, the two series do not differ in average age at beginning of reproduction.

A t test of the hypothesis that two sample means come from a population with equal μ; also confidence limits of the difference between two means

This test assumes that the variances in the populations from which the two samples were taken are identical. If in doubt about this hypothesis, test by method of Box 7.1, Section 7.3.

BOX 8.2
Continued

The appropriate formula for t_s is one of the following:

Expression (8.2), when sample sizes are unequal and n_1 or n_2 or both sample sizes are small (< 30): $df = n_1 + n_2 - 2$

Expression (8.3), when sample sizes are identical (regardless of size): $df = 2(n - 1)$

Expression (8.4), when n_1 and n_2 are unequal but both are large (> 30): $df = n_1 + n_2 - 2$

For the present data, since sample sizes are equal, we choose Expression (8.3):

$$t_s = \frac{(\bar{Y}_1 - \bar{Y}_2) - (\mu_1 - \mu_2)}{\sqrt{\frac{1}{n}(s_1^2 + s_2^2)}}$$

We are testing the null hypothesis that $\mu_1 - \mu_2 = 0$. Therefore we replace this quantity by zero in this example. Then

$$t_s = \frac{7.5143 - 7.5571}{\sqrt{(0.5047 + 0.4095)/7}} = \frac{-0.0428}{\sqrt{0.9142/7}} = \frac{-0.0428}{0.3614} = -0.1184$$

The degrees of freedom for this example are $2(n - 1) = 2 \times 6 = 12$. The critical value of $t_{0.05[12]} = 2.179$. Since the absolute value of our observed t_s is less than the critical t value, the means are found to be not significantly different, which is the same result as was obtained by the anova.

Confidence limits of the difference between two means

$$L_1 = (\bar{Y}_1 - \bar{Y}_2) - t_{\alpha[v]}s_{\bar{Y}_1 - \bar{Y}_2}$$

$$L_2 = (\bar{Y}_1 - \bar{Y}_2) + t_{\alpha[v]}s_{\bar{Y}_1 - \bar{Y}_2}$$

In this case $\bar{Y}_1 - \bar{Y}_2 = -0.0428$, $t_{0.05[12]} = 2.179$, and $s_{\bar{Y}_1 - \bar{Y}_2} = 0.3614$, as computed earlier for the denominator of the t test. Therefore

$$L_1 = -0.0428 - (2.179)(0.3614) = -0.8303$$

$$L_2 = -0.0428 + (2.179)(0.3614) = 0.7447$$

The 95% confidence limits contain the zero point (no difference), as was to be expected, since the difference $\bar{Y}_1 - \bar{Y}_2$ was found to be not significant.

is very similar for the two series. It would surprise us, therefore, to find that they are significantly different. However, we shall carry out a test anyway. As you realize by now, one cannot tell from the magnitude of a difference whether it is significant. This depends on the magnitude of the error mean square, representing the variance within series.

The computations for the analysis of variance are not shown. They would be the same as in Box 8.1. With equal sample sizes and only two groups, there is one further computational shortcut. Quantity **6**, SS_{groups}, can be directly computed by the following simple formula:

$$SS_{\text{groups}} = \frac{\left(\sum\limits^{n} Y_1 - \sum\limits^{n} Y_2\right)^2}{2n} = \frac{(52.6 - 52.9)^2}{14} = 0.00643$$

There is only 1 degree of freedom between the two groups. The critical value of $F_{0.05[1,12]}$ is given underneath the anova table, but it is really not necessary to consult it. Inspection of the mean squares in the anova shows that MS_{groups} is much smaller than MS_{within}; therefore the value of F_s is far below unity, and there cannot possibly be an added component due to treatment effects between the series. In cases where $MS_{\text{groups}} \leq MS_{\text{within}}$, we do not usually bother to calculate F_s, because the analysis of variance could not possibly be significant.

There is another method of solving a Model I two-sample analysis of variance. This is a *t test of the differences between two means*. This *t* test is the traditional method of solving such a problem; it may already be familiar to you from previous acquaintance with statistical work. It has no real advantage in either ease of computation or understanding, and as you will see, it is mathematically equivalent to the anova in Box 8.2. It is presented here mainly for the sake of completeness. It would seem too much of a break with tradition not to have *the t* test in a biostatistics text.

In Section 6.4 we learned about the *t* distribution and saw that a *t* distribution of $n - 1$ degree of freedom could be obtained from a distribution of the term $(\bar{Y}_i - \mu)/s_{\bar{Y}_i}$, where $s_{\bar{Y}_i}$ has $n - 1$ degrees of freedom and $\bar{Y}$ is normally distributed. The numerator of this term represents a deviation of a sample mean from a parametric mean, and the denominator represents a standard error for such a deviation. We now learn that the expression

$$t_s = \frac{(\bar{Y}_1 - \bar{Y}_2) - (\mu_1 - \mu_2)}{\sqrt{\left[\dfrac{(n_1 - 1)s_1^2 + (n_2 - 1)s_2^2}{n_1 + n_2 - 2}\right]\left(\dfrac{n_1 + n_2}{n_1 n_2}\right)}} \tag{8.2}$$

is also distributed as *t*. Expression (8.2) looks complicated, but it really has the same structure as the simpler term for *t*. The numerator is a deviation, this time, not between a single sample mean and the parametric mean, but between a single difference between two sample means, $\bar{Y}_1$ and $\bar{Y}_2$, and the true difference between the means of the populations represented by these means. In a test of this sort our null hypothesis is that the two samples come from the same population; that is, they must have the same parametric mean. Thus, the difference $\mu_1 - \mu_2$ is assumed to be zero. We therefore test the deviation of the difference $\bar{Y}_1 - \bar{Y}_2$ from zero. The denominator of Expression (8.2) is a standard error, the standard error of the difference between two means $s_{\bar{Y}_1 - \bar{Y}_2}$. The left portion of the expression, which is in square brackets, is a weighted average of the variances of the two samples, s_1^2 and s_2^2, computed in the manner of Section 7.1. The right term of the standard error is the computationally easier form of $(1/n_1) + (1/n_2)$, which is the factor by which the average variance within groups must be multiplied in order to convert it into a variance of the difference of means. The analogy with the multiplication of a sample variance s^2 by $1/n$ to transform it into a variance of a mean $s_{\bar{Y}}^2$ should be obvious.

The test as outlined here assumes equal variances in the two populations sampled. This is also an assumption of the analyses of variance carried out so far, although we have not stressed this. With only two variances, equality may be tested by the procedure in Box 7.1.

When sample sizes are equal in a two-sample test, Expression (8.2) simplifies to the expression

$$t_s = \frac{(\bar{Y}_1 - \bar{Y}_2) - (\mu_1 - \mu_2)}{\sqrt{\dfrac{1}{n}(s_1^2 + s_2^2)}} \tag{8.3}$$

which is what is applied in the present example in Box 8.2. When the sample sizes are unequal but rather large, so that the differences between n_i and $n_i - 1$ are relatively trivial, Expression (8.2) reduces to the simpler form

$$t_s = \frac{(\bar{Y}_1 - \bar{Y}_2) - (\mu_1 - \mu_2)}{\sqrt{\dfrac{s_1^2}{n_2} + \dfrac{s_2^2}{n_1}}} \tag{8.4}$$

The simplification of Expression (8.2) to Expressions (8.3) and (8.4) is shown in Appendix A1.3. The pertinent degrees of freedom for Expressions (8.2) and (8.4) are $n_1 + n_2 - 2$, and for Expression (8.3) df is $2(n - 1)$.

The test of significance for differences between means using the t test is shown in Box 8.2. This is a two-tailed test because our alternative hypothesis is $H_1: \mu_1 \neq \mu_2$. The results of this test are identical to those of the anova in the same box: the two means are not significantly different. We can demonstrate this mathematical equivalence by squaring the value for t_s. The result should be identical to the F_s value of the corresponding analysis of variance. Since $t_s = -0.1184$ in Box 8.2, $t_s^2 \doteq 0.0140$. Within rounding error, this is equal to the F_s obtained in the anova ($F_s = 0.0141$). Why is this so? We learned that $t_{[v]} = (\bar{Y} - \mu)/s_{\bar{Y}}$, where v is the degrees of freedom of the variance of the mean $s_{\bar{Y}}^2$; therefore $t_{[v]}^2 = (\bar{Y} - \mu)^2/s_{\bar{Y}}^2$. However, this expression can be regarded as a variance ratio. The denominator is clearly a variance with v degrees of freedom. The numerator is also a variance. It is a single deviation squared, which represents a sum of squares possessing 1 rather than zero degrees of freedom (since it is a deviation from the true mean μ rather than a sample mean). A sum of squares based on 1 degree of freedom is at the same time a variance. Thus, t^2 is a variance ratio, since $t_{[v]}^2 = F_{[1,v]}$, as we have seen. In Appendix A1.4 we demonstrate algebraically that the t_s^2 and the F_s value obtained in Box 8.2 are identical quantities. Since t approaches the normal distribution as its degrees of freedom approach infinity, $F_{\alpha[1,v]}$ approaches the distribution of the square of the normal deviate as $v \to \infty$. We also know (from Section 7.2) that $\chi^2_{[v_1]}/v_1 = F_{[v_1,\infty]}$. Therefore, when $v_1 = 1$ and $v_2 = \infty$, $\chi^2_{[1]} = F_{[1,\infty]} = t^2_{[\infty]}$ (this can be demonstrated from Tables **IV**, **V**, and **III**, respectively):

$$\chi^2_{0.05[1]} = 3.841$$

$$F_{0.05[1,\infty]} = 3.84$$

$$t_{0.05[\infty]} = 1.960 \qquad t^2_{0.05[\infty]} = 3.8416$$

The t test for differences between two means is useful when we wish to set confidence limits to such a difference. Box 8.2 shows how to calculate 95% confidence limits to the difference between the series means in the *Daphnia* example. The appropriate standard error and degrees of freedom depend on whether Expression (8.2), (8.3), or (8.4) is chosen for t_s. It does not surprise us to find that the confidence limits of the difference in this case enclose the value of zero, ranging from -0.8303 to $+0.7447$. This must be so when a difference is found to be not significantly different from zero. We can interpret this by saying that we cannot exclude zero as the true value of the difference between the means of the two series.

Another instance when you might prefer to compute the t test for differences between two means rather than use analysis of variance is when you are lacking the original variates and have only published means and standard errors available for the statistical test. Such an example is furnished in Exercise 8.4.

8.5 Comparisons among means: Planned comparisons

We have seen that after the initial significance test, a Model II analysis of variance is completed by estimation of the added variance components. We usually complete a Model I anova of more than two groups by examining the data in greater detail, testing which means are different from which other ones or which groups of means are different from other such groups or from single means. Let us look again at the Model I anovas treated so far in this chapter. We can dispose right away of the two-sample case in Box 8.2, the average age of water fleas at beginning of reproduction. As you will recall, there was no significant difference in age between the two genetic series. But even if there had been such a difference, no further tests are possible. However, the data on length of pea sections given in Box 8.1 show a significant difference among the five treatments (based on 4 degrees of freedom). Although we know that the means are not all equal, we do not know which ones differ from which other ones. This leads us to the subject of tests among pairs and groups of means. Thus, for example, we might test the control against the 4 experimental treatments representing added sugars. The question to be tested would be, Does the addition of sugars have an effect on length of pea sections? We might also test for differences among the sugar treatments. A reasonable test might be pure sugars (glucose, fructose, and sucrose) versus the mixed sugar treatment (1% glucose, 1% fructose).

An important point about such tests is that they are designed and chosen independently of the results of the experiment. They should be planned *before* the experiment has been carried out and the results obtained. Such comparisons are called *planned* or *a priori comparisons*. Such tests are applied regardless of the results of the preliminary overall anova. By contrast, after the experiment has been carried out, we might wish to compare certain means that we notice to be markedly different. For instance, sucrose, with a mean of 64.1, appears to have had less of a growth-inhibiting effect than fructose, with a mean of 58.2. We might therefore wish to test whether there is in fact a significant difference between the effects of fructose and sucrose. Such comparisons, which suggest themselves as a result of the completed experiment, are called *unplanned* or *a posteriori comparisons*. These tests are performed only if the preliminary overall anova is significant. They include tests of the comparisons between all possible pairs of means. When there are a means, there can, of course, be $a(a - 1)/2$ possible comparisons between pairs of means. The reason we make this distinction between a priori and a posteriori comparisons is that the tests of significance appropriate for the two comparisons are different. A simple example will show why this is so.

Let us assume we have sampled from an approximately normal population of heights on men. We have computed their mean and standard deviation. If we sample two men at a time from this population, we can predict the difference between them on the basis of ordinary statistical theory. Some men will be very similar, others relatively very different. Their differences will be distributed normally with a mean of 0 and an expected variance of $2\sigma^2$, for reasons that will be learned in Section 12.2. Thus, if we obtain a large difference between two randomly sampled men, it will have to be a sufficient number of standard deviations greater than zero for us to reject our null hypothesis that the two men come from the specified population. If, on the other hand, we were to look at the heights of the men before sampling them and then take pairs of men who seemed to be very different from each other, it is obvious that we would repeatedly obtain differences within pairs of men that were several standard deviations apart. Such differences would be outliers in the expected frequency distributon of differences, and time and again we would reject our null hypothesis when in fact it was true. The men would be sampled from the same population, but because they were not being sampled at random but being inspected before being sampled, the probability distribution on which our hypothesis testing rested would no longer be valid. It is obvious that the tails in a large sample from a normal distribution will be anywhere from 5 to 7 standard deviations apart. If we deliberately take individuals from each tail and compare them, they will appear to be highly significantly different from each other, according to the methods described in the present section, even though they belong to the same population.

When we compare means differing greatly from each other as the result of some treatment in the analysis of variance, we are doing exactly the same thing as taking the tallest and the shortest men from the frequency distribution of heights. If we wish to know whether these are significantly different from each other, we cannot use the ordinary probability distribution on which the analysis of variance rests, but we have to use special tests of significance. These unplanned tests will be discussed in the next section. The present section concerns itself with the carrying out of those comparisons planned before the execution of the experiment.

The general rule for making a planned comparison is extremely simple; it is related to the rule for obtaining the sum of squares for any set of groups (discussed at the end of Section 8.1). To compare k groups of any size n_i, take the sum of each group, square it, divide the result by the sample size n_i, and sum the k quotients so obtained. From the sum of these quotients, subtract a correction term, which you determine by taking the grand sum of all the groups in this comparison, squaring it, and dividing the result by the number of items in the grand sum. If the comparison includes all the groups in the anova, the correction term will be the main CT of the study. If, however, the comparison includes only some of the groups of the anova, the CT will be different, being restricted only to these groups.

These rules can best be learned by means of an example. Table 8.2 lists the means, group sums, and sample sizes of the experiment with the pea sections from Box 8.1. You will recall that there were highly significant differences among the groups. We now wish to test whether the mean of the control differs from that of the four treatments representing addition of sugar. There will thus be two groups, one the control group and the other the "sugars" groups, the latter with a sum of 2396 and a sample size of 40. We therefore compute

SS (control versus sugars)

$$= \frac{(701)^2}{10} + \frac{(593 + 582 + 580 + 641)^2}{40} - \frac{(701 + 593 + 582 + 580 + 641)^2}{50}$$

$$= \frac{(701)^2}{10} + \frac{(2396)^2}{40} - \frac{(3097)^2}{50} = 832.32$$

In this case the correction term is the same as for the anova, because it involves all the groups of the study. The result is a sum of squares for the comparison

TABLE 8.2

Means, group sums, and sample sizes from the data in Box 8.1. Length of pea sections grown in tissue culture (in ocular units).

	Control	2% glucose	2% fructose	1% glucose + 1% fructose	2% sucrose	Σ
$\bar{Y}$	70.1	59.3	58.2	58.0	64.1	$(61.94 = \bar{\bar{Y}})$
$\sum^n Y$	701	593	582	580	641	3097
n	10	10	10	10	10	50

between these two groups. Since a comparison between two groups has only 1 degree of freedom, the sum of squares is at the same time a mean square. This mean square is tested over the error mean square of the anova to give the following comparison:

$$F_s = \frac{MS \text{ (control versus sugars)}}{MS_{\text{within}}} = \frac{832.32}{5.46} = 152.44$$

$$F_{0.05[1,45]} = 4.05, \qquad F_{0.01[1,45]} = 7.23$$

This comparison is highly significant, showing that the additions of sugars have significantly retarded the growth of the pea sections.

Next we test whether the mixture of sugars is significantly different from the pure sugars. Using the same technique, we calculate

SS (mixed sugars versus pure sugars)

$$= \frac{(580)^2}{10} + \frac{(593 + 582 + 641)^2}{30} - \frac{(593 + 582 + 580 + 641)^2}{40}$$

$$= \frac{(580)^2}{10} + \frac{(1816)^2}{30} - \frac{(2396)^2}{40} = 48.13$$

Here the CT is different, since it is based on the sum of the sugars only. The appropriate test statistic is

$$F_s = \frac{MS \text{ (mixed sugars versus pure sugars)}}{MS_{\text{within}}} = \frac{48.13}{5.46} = 8.82$$

This is significant in view of the critical values of $F_{\alpha[1,45]}$ given in the preceding paragraph.

A final test is among the three sugars. This mean square has 2 degrees of freedom, since it is based on three means. Thus we compute

$$SS \text{ (among pure sugars)} = \frac{(593)^2}{10} + \frac{(582)^2}{10} + \frac{(641)^2}{10} - \frac{(1816)^2}{30} = 196.87$$

$$MS \text{ (among pure sugars)} = \frac{SS \text{ (among pure sugars)}}{df} = \frac{196.87}{2} = 98.433$$

$$F_s = \frac{MS \text{ (among pure sugars)}}{MS_{\text{within}}} = \frac{98.433}{5.46} = 18.03$$

This F_s is highly significant, since even $F_{0.01[2,40]} = 5.18$.

We conclude that the addition of the three sugars retards growth in the pea sections, that mixed sugars affect the sections differently from pure sugars, and that the pure sugars are significantly different among themselves, probably because the sucrose has a far higher mean. We cannot test the sucrose against the other two, because that would be an unplanned test, which suggests itself to us after we have looked at the results. To carry out such a test, we need the methods of the next section.

Our a priori tests might have been quite different, depending entirely on our initial hypotheses. Thus, we could have tested control versus sugars initially, followed by disaccharides (sucrose) versus monosaccharides (glucose, fructose, glucose + fructose), followed by mixed versus pure monosaccharides and finally by glucose versus fructose.

The pattern and number of planned tests are determined by one's hypotheses about the data. However, there are certain restrictions. It would clearly be a misuse of statistical methods to decide a priori that one wished to compare every mean against every other mean ($a(a - 1)/2$ comparisons). For a groups, the sum of the degrees of freedom of the separate planned tests should not exceed $a - 1$. In addition, it is desirable to structure the tests in such a way that each one tests an independent relationship among the means (as was done in the example above). For example, we would prefer not to test if means 1, 2, and 3 differed if we had already found that mean 1 differed from mean 3, since significance of the latter suggests significance of the former.

Since these tests are independent, the three sums of squares we have so far obtained, based on 1, 1, and 2 df, respectively, together add up to the sum of squares among treatments of the original analysis of variance based on 4 degrees of freedom. Thus:

$$\begin{array}{llr} & & df \\ SS \text{ (control versus sugars)} & = 832.32 & 1 \\ SS \text{ (mixed versus pure sugars)} = & 48.13 & 1 \\ \underline{SS \text{ (among pure sugars)}} & = \underline{196.87} & \underline{2} \\ SS \text{ (among treatments)} & = 1077.32 & 4 \end{array}$$

This again illustrates the elegance of analysis of variance. The treatment sums of squares can be decomposed into separate parts that are sums of squares in their own right, with degrees of freedom pertaining to them. One sum of squares measures the difference between the controls and the sugars, the second that between the mixed sugars and the pure sugars, and the third the remaining variation among the three sugars. We can present all of these results as an anova table, as shown in Table 8.3.

TABLE 8.3
Anova table from Box 8.1, with treatment sum of squares decomposed into planned comparisons.

Source of variation	df	SS	MS	F_s
Treatments	4	1077.32	269.33	49.33**
Control vs. sugars	1	832.32	832.32	152.44**
Mixed vs. pure sugars	1	48.13	48.13	8.82**
Among pure sugars	2	196.87	98.43	18.03**
Within	45	245.50	5.46	
Total	49	1322.82		

When the planned comparisons are not independent, and when the number of comparisons planned is less than the total number of comparisons possible between all pairs of means, which is $a(a - 1)/2$, we carry out the tests as just shown but we adjust the critical values of the type I error α. In comparisons that are not independent, if the outcome of a single comparison is significant, the outcomes of subsequent comparisons are more likely to be significant as well, so that decisions based on conventional levels of significance might be in doubt. For this reason, we employ a conservative approach, lowering the type I error of the statistic of significance for each comparison so that the probability of making any type I error at all in the entire series of tests does not exceed a predetermined value α. This value is called the *experimentwise error rate*. Assuming that the investigator plans a number of comparisons, adding up to k degrees of freedom, the appropriate critical values will be obtained if the probability α' is used for any one comparison, where

$$\alpha' = \frac{\alpha}{k}$$

The approach using this relation is called the *Bonferroni method*; it assures us of an experimentwise error rate $\leq \alpha$.

Applying this approach to the pea section data, as discussed above, let us assume that the investigator has good reason to test the following comparisons between and among treatments, given here in abbreviated form: (C) versus (G, F, S, G + F); (G, F, S) versus (G + F); and (G) versus (F) versus (S); as well as (G, F) versus (G + F). The 5 degrees of freedom in these tests require that each individual test be adjusted to a significance level of

$$\alpha' = \frac{\alpha}{k} = \frac{0.05}{5} = 0.01$$

for an experimentwise critical $\alpha = 0.05$. Thus, the critical value for the F_s ratios of these comparisons is $F_{0.01[1,45]}$ or $F_{0.01[2,45]}$, as appropriate. The first three tests are carried out as shown above. The last test is computed in a similar manner:

$$SS \begin{pmatrix} \text{average of glucose and} \\ \text{fructose vs. glucose} \\ \text{and fructose mixed} \end{pmatrix} = \frac{(593 + 582)^2}{20} + \frac{(580)^2}{10} - \frac{(593 + 582 + 580)^2}{30}$$

$$= \frac{(1175)^2}{20} + \frac{(580)^2}{10} - \frac{(1755)^2}{30} = 3.75$$

In spite of the change in critical value, the conclusions concerning the first three tests are unchanged. The last test, the average of glucose and fructose versus a mixture of the two, is not significant, since $F_s = \frac{3.75}{5.46} = 0.687$. Adjusting the critical value is a conservative procedure; individual comparisons using this approach are less likely to be significant.

The Bonferroni method generally will not employ the standard, tabled arguments of α for the F distribution. Thus, if we were to plan tests involving altogether 6 degrees of freedom, the value of α' would be 0.0083. Exact tables for Bonferroni critical values are available for the special case of single degree of freedom tests. Alternatively, we can compute the desired critical value by means of a computer program. A conservative alternative is to use the next smaller tabled value of α. For details, consult Sokal and Rohlf (1981), section 9.6.

The Bonferroni method (or a more recent refinement, the Dunn-Šidák method) should also be employed when you are reporting confidence limits for more than one group mean resulting from an analysis of variance. Thus, if you wanted to publish the means and $1 - \alpha$ confidence limits of all five treatments in the pea section example, you would not set confidence limits to each mean as though it were an independent sample, but you would employ $t_{\alpha'[\nu]}$, where ν is the degrees of freedom of the entire study and α' is the adjusted type I error explained earlier. Details of such a procedure can be learned in Sokal and Rohlf (1981), Section 14.10.

8.6 Comparisons among means: Unplanned comparisons

A single-classification anova is said to be significant if

$$\frac{MS_{\text{groups}}}{MS_{\text{within}}} \geq F_{\alpha[a-1, a(n-1)]} \tag{8.5}$$

Since $MS_{\text{groups}}/MS_{\text{within}} = SS_{\text{groups}}/[(a-1)MS_{\text{within}}]$, we can rewrite Expression (8.5) as

$$SS_{\text{groups}} \geq (a-1)MS_{\text{within}}F_{\alpha[a-1, a(n-1)]} \tag{8.6}$$

For example, in Box 8.1, where the anova is significant, $SS_{\text{groups}} = 1077.32$. Substituting into Expression (8.6), we obtain

$$1077.32 > (5-1)(5.46)(2.58) = 56.35 \qquad \text{for} \qquad \alpha = 0.05$$

It is therefore possible to compute a critical SS value for a test of significance of an anova. Thus, another way of calculating overall significance would be to see whether the SS_{groups} is greater than this critical SS. It is of interest to investigate why the SS_{groups} is as large as it is and to test for the significance of the various contributions made to this SS by differences among the sample means. This was discussed in the previous section, where separate sums of squares were computed based on comparisons among means planned before the data were examined. A comparison was called significant if its F_s ratio was $> F_{\alpha[k-1, a(n-1)]}$, where k is the number of means being compared. We can now also state this in terms of sums of squares: An SS is significant if it is greater than $(k-1)MS_{\text{within}}F_{\alpha[k-1, a(n-1)]}$.

The above tests were a priori comparisons. One procedure for testing a posteriori comparisons would be to set $k = a$ in this last formula, no matter how many means we compare; thus the critical value of the SS will be larger than in the previous method, making it more difficult to demonstrate the significance of a sample SS. Setting $k = a$ allows for the fact that we choose for testing those differences between group means that appear to be contributing substantially to the significance of the overall anova.

For an example, let us return to the effects of sugars on growth in pea sections (Box 8.1). We write down the means in ascending order of magnitude: 58.0 (glucose + fructose), 58.2 (fructose), 59.3 (glucose), 64.1 (sucrose), 70.1 (control). We notice that the first three treatments have quite similar means and suspect that they do not differ significantly among themselves and hence do not contribute substantially to the significance of the SS_{groups}.

To test this, we compute the SS among these three means by the usual formula:

$$SS = \frac{(593)^2 + (582)^2 + (580)^2}{10} - \frac{(593 + 582 + 580)^2}{3(10)}$$

$$= 102{,}677.3 - 102{,}667.5 = 9.8$$

The differences among these means are not significant, because this SS is less than the critical SS (56.35) calculated above.

The sucrose mean looks suspiciously different from the means of the other sugars. To test this we compute

$$SS = \frac{(641)^2}{10} + \frac{(593 + 582 + 580)^2}{30} - \frac{(641 + 593 + 582 + 580)^2}{10 + 30}$$

$$= 41{,}088.1 + 102{,}667.5 - 143{,}520.4 = 235.2$$

which is greater than the critical SS. We conclude, therefore, that sucrose retards growth significantly less than the other sugars tested. We may continue in this fashion, testing all the differences that look suspicious or even testing all possible sets of means, considering them 2, 3, 4, and 5 at a time. This latter approach may require a computer if there are more than 5 means to be compared, since there are very many possible tests that could be made. This procedure was proposed by Gabriel (1964), who called it a *sum of squares simultaneous test procedure* (*SS-STP*).

In the *SS-STP* and in the original anova, the chance of making any type I error at all is α, the probability selected for the critical F value from Table **V**. By "making any type I error at all" we mean making such an error in the overall test of significance of the anova and in any of the subsidiary comparisons among means or sets of means needed to complete the analysis of the experiment. This probability α therefore is an *experimentwise* error rate. Note that though the probability of any error at all is α, the probability of error for any particular test of some subset, such as a test of the difference among three or between two means, will always be less than α. Thus, for the test of each subset one is really using a significance level α', which may be much less than the experimentwise

α, and if there are many means in the anova, this actual error rate α' may be one-tenth, one one-hundredth, or even one one-thousandth of the experiment-wise α (Gabriel, 1964). For this reason, the unplanned tests discussed above and the overall anova are not very sensitive to differences between individual means or differences within small subsets. Obviously, not many differences are going to be considered significant if α' is minute. This is the price we pay for not planning our comparisons before we examine the data: if we were to make planned tests, the error rate of each would be greater, hence less conservative.

The *SS-STP* procedure is only one of numerous techniques for multiple unplanned comparisons. It is the most conservative, since it allows a large number of possible comparisons. Differences shown to be significant by this method can be reliably reported as significant differences. However, more sensitive and powerful comparisons exist when the number of possible comparisons is circumscribed by the user. This is a complex subject, to which a more complete introduction is given in Sokal and Rohlf (1981), Section 9.7.

Exercises

8.1 The following is an example with easy numbers to help you become familiar with the analysis of variance. A plant ecologist wishes to test the hypothesis that the height of plant species X depends on the type of soil it grows in. He has measured the height of three plants in each of four plots representing different soil types, all four plots being contained in an area of two miles square. His results are tabulated below. (Height is given in centimeters.) Does your analysis support this hypothesis? ANS. Yes, since $F_s = 6.951$ is larger than $F_{0.05[3,8]} = 4.07$.

Observation number	Localities			
	1	*2*	*3*	*4*
1	15	25	17	10
2	9	21	23	13
3	14	19	20	16

8.2 The following are measurements (in coded micrometer units) of the thorax length of the aphid *Pemphigus populitransversus*. The aphids were collected in 28 galls on the cottonwood *Populas deltoides*. Four alate (winged) aphids were randomly selected from each gall and measured. The alate aphids of each gall are isogenic (identical twins), being descended parthenogenetically from one stem mother. Thus, any variance within galls can be due to environment only. Variance between galls may be due to differences in genotype and also to environmental differences between galls. If this character, thorax length, is affected by genetic variation, significant intergall variance must be present. The converse is not necessarily true: significant variance between galls need not indicate genetic variation; it could as well be due to environmental differences between galls (data by Sokal, 1952). Analyze the variance of thorax length. Is there significant intergall variance present? Give estimates of the added component of intergall variance, if present. What percentage of the variance is controlled by intragall and what percentage by intergall factors? Discuss your results.

Gall no.					Gall no.				
1.	6.1,	6.0,	5.7,	6.0	15.	6.3,	6.5,	6.1,	6.3
2.	6.2,	5.1,	6.1,	5.3	16.	5.9,	6.1,	6.1,	6.0
3.	6.2,	6.2,	5.3,	6.3	17.	5.8,	6.0,	5.9,	5.7
4.	5.1,	6.0,	5.8,	5.9	18.	6.5,	6.3,	6.5,	7.0
5.	4.4,	4.9,	4.7,	4.8	19.	5.9,	5.2,	5.7,	5.7
6.	5.7,	5.1,	5.8,	5.5	20.	5.2,	5.3,	5.4,	5.3
7.	6.3,	6.6,	6.4,	6.3	21.	5.4,	5.5,	5.2,	6.3
8.	4.5,	4.5,	4.0,	3.7	22.	4.3,	4.7,	4.5,	4.4
9.	6.3,	6.2,	5.9,	6.2	23.	6.0,	5.8,	5.7,	5.9
10.	5.4,	5.3,	5.0,	5.3	24.	5.5,	6.1,	5.5,	6.1
11.	5.9,	5.8,	6.3,	5.7	25.	4.0,	4.2,	4.3,	4.4
12.	5.9,	5.9,	5.5,	5.5	26.	5.8,	5.6,	5.6,	6.1
13.	5.8,	5.9,	5.4,	5.5	27.	4.3,	4.0,	4.4,	4.6
14.	5.6,	6.4,	6.4,	6.1	28.	6.1,	6.0,	5.6,	6.5

8.3 Millis and Seng (1954) published a study on the relation of birth order to the birth weights of infants. The data below on first-born and eighth-born infants are extracted from a table of birth weights of male infants of Chinese third-class patients at the Kandang Kerbau Maternity Hospital in Singapore in 1950 and 1951.

Birth weight (lb:oz)	Birth order 1	8
3:0–3:7	—	—
3:8–3:15	2	—
4:0–4:7	3	—
4:8–4:15	7	4
5:0–5:7	111	5
5:8–5:15	267	19
6:0–6:7	457	52
6:8–6:15	485	55
7:0–7:7	363	61
7:8–7:15	162	48
8:0–8:7	64	39
8:8–8:15	6	19
9:0–9:7	5	4
9:8–9:15	—	—
10:0–10:7	—	1
10:8–10:15	—	—
	1932	307

Which birth order appears to be accompanied by heavier infants? Is this difference significant? Can you conclude that birth order causes differences in birth weight? (Computational note: The variable should be coded as simply as possible.) Reanalyze, using the t test, and verify that $t_s^2 = F_s$. ANS. $t_s = 11.016$ and $F_s = 121.352$

8.4 The following cytochrome oxidase assessments of male *Periplaneta* roaches in cubic millimeters per ten minutes per milligram were taken from a larger study by Brown and Brown (1956):

		n	$\bar{Y}$	$s_{\bar{Y}}$
24 hours after				
methoxychlor injection		5	24.8	0.9
Control		3	19.7	1.4

Are the two means significantly different?

8.5 P. E. Hunter (1959, detailed data unpublished) selected two strains of *D. melanogaster*, one for short larval period (SL) and one for long larval period (LL). A nonselected control strain (CS) was also maintained. At generation 42 these data were obtained for the larval period (measured in hours). Analyze and interpret.

	SL	Strain CS	LL	
n_i	80	69	33	
$\sum^{n_i} Y$	8070	7291	3640	$\sum^{3}\sum^{n_i} Y^2 = 1{,}994{,}650$

Note that part of the computation has already been performed for you. Perform unplanned tests among the three means (short vs. long larval periods and each against the control). Set 95% confidence limits to the observed differences of means for which these comparisons are made. ANS. $MS_{(SL\,vs.\,LL)} = 2076.6697$.

8.6 These data are measurements of five random samples of domestic pigeons collected during January, February, and March in Chicago in 1955. The variable is the length from the anterior end of the narial opening to the tip of the bony beak and is recorded in millimeters. Data from Olson and Miller (1958).

		Samples		
1	2	3	4	5
5.4	5.2	5.5	5.1	5.1
5.3	5.1	4.7	4.6	5.5
5.2	4.7	4.8	5.4	5.9
4.5	5.0	4.9	5.5	6.1
5.0	5.9	5.9	5.2	5.2
5.4	5.3	5.2	5.0	5.0
3.8	6.0	4.8	4.8	5.9
5.9	5.2	4.9	5.1	5.0
5.4	6.6	6.4	4.4	4.9
5.1	5.6	5.1	6.5	5.3
5.4	5.1	5.1	4.8	5.3
4.1	5.7	4.5	4.9	5.1
5.2	5.1	5.3	6.0	4.9
4.8	4.7	4.8	4.8	5.8
4.6	6.5	5.3	5.7	5.0
5.7	5.1	5.4	5.5	5.6
5.9	5.4	4.9	5.8	6.1
5.8	5.8	4.7	5.6	5.1
5.0	5.8	4.8	5.5	4.8
5.0	5.9	5.0	5.0	4.9

Are the five samples homogeneous?

8.7   The following data were taken from a study of blood protein variations in deer
(Cowan and Johnston, 1962). The variable is the mobility of serum protein fraction II expressed as 10^{-5} cm^2/volt-seconds.

	$\bar{Y}$	$s_{\bar{Y}}$
Sitka	2.8	0.07
California blacktail	2.5	0.05
Vancouver Island blacktail	2.9	0.05
Mule deer	2.5	0.05
Whitetail	2.8	0.07

$n = 12$ for each mean. Perform an analysis of variance and a multiple-comparison
test, using the sums of squares STP procedure. ANS. $MS_{\text{within}} = 0.0416$; maximal
nonsignificant sets (at $P = 0.05$) are samples 1, 3, 5 and 2, 4 (numbered in the
order given).

8.8   For the data from Exercise 7.3 use the Bonferroni method to test for differences
between the following 5 pairs of treatment means:

A, B
A, C
A, D
A, (B + C + D)/3
B, (C + D)/2

CHAPTER 9

Two-Way Analysis
of Variance

From the single-classification anova of Chapter 8 we progress to the two-way anova of the present chapter by a single logical step. Individual items may be grouped into classes representing the different possible combinations of two treatments or factors. Thus, the housefly wing lengths studied in earlier chapters, which yielded samples representing different medium formulations, might also be divided into males and females. Suppose we wanted to know not only whether medium 1 induced a different wing length than medium 2 but also whether male houseflies differed in wing length from females. Obviously, each combination of factors should be represented by a sample of flies. Thus, for seven media and two sexes we need at least $7 \times 2 = 14$ samples. Similarly, the experiment testing five sugar treatments on pea sections (Box 8.1) might have been carried out at three different temperatures. This would have resulted in a *two-way analysis of variance* of the effects of sugars as well as of temperatures.

It is the assumption of this two-way method of anova that a given temperature and a given sugar each contribute a certain amount to the growth of a pea section, and that these two contributions add their effects without influencing each other. In Section 9.1 we shall see how departures from the assumption are measured; we shall also consider the expression for decomposing variates in a two-way anova.

The two factors in the present design may represent either Model I or Model II effects or one of each, in which case we talk of a *mixed model*.

The computation of a two-way anova for replicated subclasses (more than one variate per subclass or factor combination) is shown in Section 9.1, which also contains a discussion of the meaning of interaction as used in statistics. Significance testing in a two-way anova is the subject of Section 9.2. This is followed by Section 9.3, on two-way anova without replication, or with only a single variate per subclass. The well-known method of paired comparisons is a special case of a two-way anova without replication.

We will now proceed to illustrate the computation of a two-way anova. You will obtain closer insight into the structure of this design as we explain the computations.

9.1 Two-way anova with replication

We illustrate the computation of a two-way anova in a study of oxygen consumption by two species of limpets at three concentrations of seawater. Eight replicate readings were obtained for each combination of species and seawater concentration. We have continued to call the number of columns a, and are calling the number of rows b. The sample size for each cell (row and column combination) of the table is n. The cells are also called subgroups or subclasses.

The data are featured in Box 9.1. The computational steps labeled *Preliminary computations* provide an efficient procedure for the analysis of variance, but we shall undertake several digressions to ensure that the concepts underlying this design are appreciated by the reader. We commence by considering the six subclasses as though they were six groups in a single-classification anova. Each subgroup or subclass represents eight oxygen consumption readings. If we had no further classification of these six subgroups by species or salinity, such an anova would test whether there was any variation among the six subgroups over and above the variance within the subgroups. But since we have the subdivision by species and salinity, our only purpose here is to compute some quantities necessary for the further analysis. Steps **1** through **3** in Box 9.1 correspond to the identical steps in Box 8.1, although the symbolism has changed slightly, since in place of a groups we now have ab subgroups. To complete the anova, we need a correction term, which is labeled step **6** in Box 9.1. From these quantities we obtain SS_{total}, and SS_{within} in steps **7**, **8**, and **12**, corresponding to steps **5**, **6**, and **7** in the layout of Box 8.1. The results of this preliminary anova are featured in Table 9.1.

The computation is continued by finding the sums of squares for rows and columns of the table. This is done by the general formula stated at the end of Section 8.1. Thus, for columns, we square the column sums, sum the resulting squares, and divide the result by 24, the number of items per row. This is step **4** in Box 9.1. A similar quantity is computed for rows (step **5**). From these

BOX 9.1

Two-way anova with replication.

Oxygen consumption rates of two species of limpets, *Acmaea scabra* and *A. digitalis*, at three concentrations of seawater. The variable measured is μl O_2/mg dry body weight/min at 22°C. There are eight replicates per combination of species and salinity ($n = 8$). This is a Model I anova.

Factor B: seawater concentrations ($b = 3$)	Factor A: Species ($a = 2$)				Σ
	Acmaea scabra		*Acmaea digitalis*		
100%	7.16	8.26	6.14	6.14	
	6.78	14.00	3.86	10.00	
	13.60	16.10	10.40	11.60	
	8.93	9.66	5.49	5.80	
	$\Sigma = 84.49$		$\Sigma = 59.43$		143.92
75%	5.20	13.20	4.47	4.95	
	5.20	8.39	9.90	6.49	
	7.18	10.40	5.75	5.44	
	6.37	7.18	11.80	9.90	
	$\Sigma = 63.12$		$\Sigma = 58.70$		121.82
50%	11.11	10.50	9.63	14.50	
	9.74	14.60	6.38	10.20	
	18.80	11.10	13.40	17.70	
	9.74	11.80	14.50	12.30	
	$\Sigma = 97.39$		$\Sigma = 98.61$		196.00
Σ	245.00		216.74		461.74

Source: Unpublished study by F. J. Rohlf.

BOX 9.1
Continued

Preliminary computations

1. Grand total $= \sum^{a}\sum^{b}\sum^{n} Y = 461.74$

2. Sum of the squared observations $= \sum^{a}\sum^{b}\sum^{n} Y^2 = (7.16)^2 + \cdots + (12.30)^2 = 5065.1530$

3. Sum of the squared subgroup (cell) totals, divided by the sample size of the subgroups

$$= \frac{\sum^{a}\sum^{b}\left(\sum^{n} Y\right)^2}{n} = \frac{(84.49)^2 + \cdots + (98.61)^2}{8} = 4663.6317$$

4. Sum of the squared column totals divided by the sample size of a column $= \dfrac{\sum^{a}\left(\sum^{b}\sum^{n} Y\right)^2}{bn} = \dfrac{(245.00)^2 + (216.74)^2}{(3 \times 8)} = 4458.3844$

5. Sum of the squared row totals divided by the sample size of a row $= \dfrac{\sum^{b}\left(\sum^{a}\sum^{n} Y\right)^2}{an} = \dfrac{(143.92)^2 + (121.82)^2 + (196.00)^2}{(2 \times 8)} = 4623.0674$

6. Grand total squared and divided by the total sample size = correction term CT $= \dfrac{\left(\sum^{a}\sum^{b}\sum^{n} Y\right)^2}{abn} = \dfrac{(\text{quantity } \mathbf{1})^2}{abn} = \dfrac{(461.74)^2}{(2 \times 3 \times 8)} = 4441.7464$

7. $SS_{\text{total}} = \sum\limits^{a}\sum\limits^{b}\sum\limits^{n} Y^2 - CT = $ quantity **2** – quantity **6** $= 5065.1530 - 4441.7464 = 623.4066$

8. $SS_{\text{subgr}} = \dfrac{\sum\limits^{a}\sum\limits^{b}\left(\sum\limits^{n} Y\right)^2}{n} - CT = $ quantity **3** – quantity **6** $= 4663.6317 - 4441.7464 = 221.8853$

9. SS_A (SS of columns) $= \dfrac{\sum\limits^{a}\left(\sum\limits^{b}\sum\limits^{n} Y\right)^2}{bn} - CT = $ quantity **4** – quantity **6** $= 4458.3844 - 4441.7464 = 16.6380$

10. SS_B (SS of rows) $= \dfrac{\sum\limits^{b}\left(\sum\limits^{a}\sum\limits^{n} Y\right)^2}{an} - CT = $ quantity **5** – quantity **6** $= 4623.0674 - 4441.7464 = 181.3210$

11. $SS_{A \times B}$ (interaction SS) $= SS_{\text{subgr}} - SS_A - SS_B = $ quantity **8** – quantity **9** – quantity **10**
$= 221.8853 - 16.6380 - 181.3210 = 23.9263$

12. SS_{within} (within subgroups; error SS) $= SS_{\text{total}} - SS_{\text{subgr}} = $ quantity **7** – quantity **8**
$= 623.4066 - 221.8853 = 401.5213$

As a check on your computations, ascertain that the following relations hold for some of the above quantities: $2 \geq 3 \geq 4 \geq 6$; $3 \geq 5 \geq 6$.

Explicit formulas for these sums of squares suitable for computer programs are as follows:

9a. $SS_A = nb \sum\limits^{a} (\bar{Y}_A - \bar{\bar{Y}})^2$

10a. $SS_B = na \sum\limits^{b} (\bar{Y}_B - \bar{\bar{Y}})^2$

11a. $SS_{AB} = n \sum\limits^{a}\sum\limits^{b} (\bar{Y} - \bar{Y}_A - \bar{Y}_B + \bar{\bar{Y}})^2$

12a. $SS_{\text{within}} = n \sum\limits^{a}\sum\limits^{b} (\bar{Y} - \bar{\bar{Y}})^2$

BOX 9.1
Continued

Such formulas may furnish more exact solutions in computer algorithms (Wilkinson and Dallal, 1977), although they are far more tedious to compute on a pocket or tabletop calculator that is not able to store the n data values.

Now fill in the anova table.

Source of variation		df	SS	MS	Expected MS (Model I)
$\bar{Y}_A - \bar{\bar{Y}}$	A (columns)	$a - 1$	**9**	$\dfrac{\mathbf{9}}{(a-1)}$	$\sigma^2 + \dfrac{nb}{a-1}\sum\limits^{a}\alpha^2$
$\bar{Y}_B - \bar{\bar{Y}}$	B (rows)	$b - 1$	**10**	$\dfrac{\mathbf{10}}{(b-1)}$	$\sigma^2 + \dfrac{na}{b-1}\sum\limits^{b}\beta^2$
$\bar{Y} - \bar{Y}_A - \bar{Y}_B + \bar{\bar{Y}}$	$A \times B$ (interaction)	$(a-1)(b-1)$	**11**	$\dfrac{\mathbf{11}}{(a-1)(b-1)}$	$\sigma^2 + \dfrac{n}{(a-1)(b-1)}\sum\limits^{ab}(\alpha\beta)^2$
$Y - \bar{Y}$	Within subgroups	$ab(n-1)$	**12**	$\dfrac{\mathbf{12}}{ab(n-1)}$	σ^2
$Y - \bar{\bar{Y}}$	Total	$abn - 1$	$\overline{\mathbf{7}}$		

Since the present example is a Model I anova for both factors, the expected MS above are correct. Below are the corresponding expressions for other models.

Source of variation	Model II	Mixed model (A fixed, B random)
A	$\sigma^2 + n\sigma_{AB}^2 + nb\sigma_A^2$	$\sigma^2 + n\sigma_{AB}^2 + \dfrac{nb}{a-1}\sum^a \alpha^2$
B	$\sigma^2 + n\sigma_{AB}^2 + na\sigma_B^2$	$\sigma^2 + na\sigma_B^2$
$A \times B$	$\sigma^2 + n\sigma_{AB}^2$	$\sigma^2 + n\sigma_{AB}^2$
Within subgroups	σ^2	σ^2

Anova table

Source of variation	df	SS	MS	F_s
A (columns; species)	1	16.6380	16.638	1.740 ns
B (rows; salinities)	2	181.3210	90.660	9.483**
$A \times B$ (interaction)	2	23.9263	11.963	1.251 ns
Within subgroups (error)	42	401.5213	9.560	
Total	47	623.4066		

$F_{0.05[1,42]} = 4.07$ $F_{0.05[2,42]} = 3.22$ $F_{0.01[2,42]} = 5.15$

Since this is a Model I anova, all mean squares are tested over the error MS. For a discussion of significance tests, see Section 9.2.

Conclusions.—Oxygen consumption does not differ significantly between the two species of limpets but differs with the salinity. At 50% seawater, the O_2 consumption is increased. Salinity appears to affect the two species equally, for there is insufficient evidence of a species × salinity interaction.

TABLE 9.1
Preliminary anova of subgroups in two-way anova. Data from Box 9.1.

Source of variation	df		SS	MS
$\bar{Y} - \bar{\bar{Y}}$ Among subgroups	5	$ab - 1$	221.8853	44.377**
$Y - \bar{Y}$ Within subgroups	42	$ab(n - 1)$	401.5213	9.560
$Y - \bar{\bar{Y}}$ Total	47	$abn -$	623.4066	

quotients we subtract the correction term, computed as quantity **6**. These subtractions are carried out as steps **9** and **10**, respectively. Since the rows and columns are based on equal sample sizes, we do not have to obtain a separate quotient for the square of each row or column sum but carry out a single division after accumulating the squares of the sums.

Let us return for a moment to the preliminary analysis of variance in Table 9.1, which divided the total sum of squares into two parts: the sum of squares among the six subgroups; and that within the subgroups, the error sum of squares. The new sums of squares pertaining to row and column effects clearly are not part of the error, but must contribute to the differences that comprise the sum of squares among the four subgroups. We therefore subtract row and column SS from the subgroup SS. The latter is 221.8853. The row SS is 181.3210, and the column SS is 16.6380. Together they add up to 197.9590, almost but not quite the value of the subgroup sum of squares. The difference represents a third sum of squares, called the *interaction sum of squares*, whose value in this case is 23.9263.

We shall discuss the meaning of this new sum of squares presently. At the moment let us say only that it is almost always present (but not necessarily significant) and generally that it need not be independently computed but may be obtained as illustrated above—by the subtraction of the row SS and the column SS from the subgroup SS. This procedure is shown graphically in Figure 9.1, which illustrates the decomposition of the total sum of squares into the subgroup SS and error SS. The former is subdivided into the row SS, column SS, and interaction SS. The relative magnitudes of these sums of squares will differ from experiment to experiment. In Figure 9.1 they are not shown proportional to their actual values in the limpet experiment; otherwise the area representing the row SS would have to be about 11 times that allotted to the column SS.

Before we can intelligently test for significance in this anova we must understand the meaning of *interaction*. We can best explain interaction in a two-way anova by means of an artificial illustration based on the limpet data we have just studied. If we interchange the readings for 75% and 50% for *A. digitalis* only, we obtain the data table shown in Table 9.2. Only the sums of the subgroups, rows, and columns are shown. We complete the analysis of variance in the manner presented above and note the results at the foot of Table 9.2. The total and error SS are the same as before (Table 9.1). This should not be

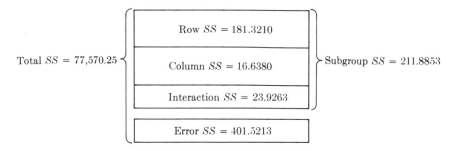

FIGURE 9.1
Diagrammatic representation of the partitioning of the total sums of squares in a two-way orthogonal anova. The areas of the subdivisions are not shown proportional to the magnitudes of the sums of squares.

surprising, since we are using the same data. All that we have done is to interchange the contents of the lower two cells in the right-hand column of the table. When we partition the subgroup SS, we do find some differences. We note that the SS between species (between columns) is unchanged. Since the change we made was within one column, the total for that column was not altered and consequently the column SS did not change. However, the sums

TABLE 9.2

An artificial example to illustrate the meaning of interaction. The readings for 75% and 50% seawater concentrations of *Acmaea digitalis* in Box 9.1 have been interchanged. Only subgroup and marginal totals are given below.

Seawater concentration	Species		
	A. scabra	A. digitalis	$\sum$
100%	84.49	59.43	143.92
75%	63.12	98.61	161.73
50%	97.39	58.70	156.09
$\sum$	245.00	216.74	461.74

Completed anova

Source of variation	df	SS	MS
Species	1	16.6380	16.638 *ns*
Salinities	2	10.3566	5.178 *ns*
Sp × Sal	2	194.8907	97.445**
Error	42	401.5213	9.560
Total	47	623.4066	

of the second and third rows have been altered appreciably as a result of the interchange of the readings for 75% and 50% salinity in *A. digitalis*. The sum for 75% salinity is now very close to that for 50% salinity, and the difference between the salinities, previously quite marked, is now no longer so. By contrast, the interaction *SS*, obtained by subtracting the sums of squares of rows and columns from the subgroup *SS*, is now a large quantity. Remember that the subgroup *SS* is the same in the two examples. In the first example we subtracted sums of squares due to the effects of both species and salinities, leaving only a tiny residual representing the interaction. In the second example these two *main effects* (species and salinities) account only for little of the subgroup sum of squares, leaving the interaction sum of squares as a substantial residual. What is the essential difference between these two examples?

In Table 9.3 we have shown the subgroup and marginal means for the original data from Table 9.1 and for the altered data of Table 9.2. The original results are quite clear: at 75% salinity, oxygen consumption is lower than at the other two salinities, and this is true for both species. We note further that *A. scabra* consumes more oxygen than *A. digitalis* at two of the salinities. Thus our statements about differences due to species or to salinity can be made largely independent of each other. However, if we had to interpret the artificial data (lower half of Table 9.3), we would note that although *A. scabra* still consumes more oxygen than *A. digitalis* (since column sums have not changed), this difference depends greatly on the salinity. At 100% and 50%, *A. scabra* consumes considerably more oxygen than *A. digitalis*, but at 75% this relationship is reversed. Thus, we are no longer able to make an unequivocal statement about the amount of oxygen taken up by the two species. We have to qualify our statement by the seawater concentration at which they are kept. At 100%

TABLE 9.3
Comparison of means of the data in Box 9.1 and Table 9.2.

Seawater concentration	Species		
	A. scabra	A. digitalis	Mean
Original data from Box 9.1			
100%	10.56	7.43	9.00
75%	7.89	7.34	7.61
50%	12.17	12.33	12.25
Mean	10.21	9.03	9.62
Artificial data from Table 9.2			
100%	10.56	7.43	9.00
75%	7.89	12.33	10.11
50%	12.17	7.34	9.76
Mean	10.21	9.03	9.62

and 50%, $\bar{Y}_{\text{scabra}} > \bar{Y}_{\text{digitalis}}$, but at 75%, $\bar{Y}_{\text{scabra}} < \bar{Y}_{\text{digitalis}}$. If we examine the effects of salinity in the artificial example, we notice a mild increase in oxygen consumption at 75%. However, again we have to qualify this statement by the species of the consuming limpet; *scabra* consumes least at 75%, while *digitalis* consumes most at this concentration.

This dependence of the effect of one factor on the level of another factor is called *interaction*. It is a common and fundamental scientific idea. It indicates that the effects of the two factors are not simply additive but that any given combination of levels of factors, such as salinity combined with any one species, contributes a positive or negative increment to the level of expression of the variable. In common biological terminology a large positive increment of this sort is called *synergism*. When drugs act synergistically, the result of the interaction of the two drugs may be above and beyond the sum of the separate effects of each drug. When levels of two factors in combination inhibit each other's effects, we call it *interference*. (Note that "levels" in anova is customarily used in a loose sense to include not only continuous factors, such as the salinity in the present example, but also qualitative factors, such as the two species of limpets.) Synergism and interference will both tend to magnify the interaction *SS*.

Testing for interaction is an important procedure in analysis of variance. If the artificial data of Table 9.2 were real, it would be of little value to state that 75% salinity led to slightly greater consumption of oxygen. This statement would cover up the important differences in the data, which are that *scabra* consumes least at this concentration, while *digitalis* consumes most.

We are now able to write an expression symbolizing the decomposition of a single variate in a two-way analysis of variance in the manner of Expression (7.2) for single-classification anova. The expression below assumes that both factors represent fixed treatment effects, Model I. This would seem reasonable, since species as well as salinity are fixed treatments. Variate Y_{ijk} is the kth item in the subgroup representing the ith group of treatment A and the jth group of treatment B. It is decomposed as follows:

$$Y_{ijk} = \mu + \alpha_i + \beta_j + (\alpha\beta)_{ij} + \epsilon_{ijk} \tag{9.1}$$

where μ equals the parametric mean of the population, α_i is the fixed treatment effect for the ith group of treatment A, β_j is the fixed treatment effect of the jth group of treatment B, $(\alpha\beta)_{ij}$ is the interaction effect in the subgroup representing the ith group of factor A and the jth group of factor B, and ϵ_{ijk} is the error term of the kth item in subgroup ij. We make the usual assumption that ϵ_{ijk} is normally distributed with a mean of 0 and a variance of σ^2. If one or both of the factors represent Model II effects, we replace the α_i and/or β_j in the formula by A_i and/or B_j.

In previous chapters we have seen that each sum of squares represents a sum of squared deviations. What actual deviations does an interaction *SS* represent? We can see this easily by referring back to the anovas of Table 9.1. The variation among subgroups is represented by $(\bar{Y} - \bar{\bar{Y}})$, where $\bar{Y}$ stands for the subgroup mean, and $\bar{\bar{Y}}$ for the grand mean. When we subtract the deviations due to rows $(\bar{R} - \bar{\bar{Y}})$ and those due to columns $(\bar{C} - \bar{\bar{Y}})$ from those due to subgroups, we obtain

$$(\bar{Y} - \bar{\bar{Y}}) - (\bar{R} - \bar{\bar{Y}}) - (\bar{C} - \bar{\bar{Y}}) = \bar{Y} - \bar{\bar{Y}} - \bar{R} + \bar{\bar{Y}} - \bar{C} + \bar{\bar{Y}}$$
$$= \bar{Y} - \bar{R} - \bar{C} + \bar{\bar{Y}}$$

This somewhat involved expression is the deviation due to interaction. When we evaluate one such expression for each subgroup, square it, sum the squares, and multiply the sum by n, we obtain the interaction SS. This partition of the deviations also holds for their squares. This is so because the sums of the products of the separate terms cancel out.

A simple method for revealing the nature of the interaction present in the data is to inspect the means of the original data table. We can do this in Table 9.3. The original data, showing no interaction, yield the following pattern of relative magnitudes:

	Scabra	Digitalis
100%		
	V	V
75%		
	∧	∧
50%		

The relative magnitudes of the means in the lower part of Table 9.3 can be summarized as follows:

	Scabra	Digitalis
100%		
	V	∧
75%		
	∧	V
50%		

When the pattern of signs expressing relative magnitudes is not uniform as in this latter table, interaction is indicated. As long as the pattern of means is consistent, as in the former table, interaction may not be present. However, interaction is often present without change in the *direction* of the differences; sometimes only the relative magnitudes are affected. In any case, the statistical test needs to be performed to test whether the deviations are larger than can be expected from chance alone.

In summary, when the effect of two treatments applied together cannot be predicted from the average responses of the separate factors, statisticians call this phenomenon interaction and test its significance by means of an interaction mean square. This is a very common phenomenon. If we say that the effect of density on the fecundity or weight of a beetle depends on its genotype, we imply that a genotype × density interaction is present. If the success of several alternative surgical procedures depends on the nature of the postoperative treatment, we speak of a procedure × treatment interaction. Or if the effect of temperature on a metabolic process is independent of the effect of oxygen concentration, we say that temperature × oxygen interaction is absent.

Significance testing in a two-way anova will be deferred until the next section. However, we should point out that the computational steps **4** and **9** of Box 9.1 could have been shortened by employing the simplified formula for a sum of squares between *two* groups, illustrated in Section 8.4. In an analysis with only two rows and two columns the interaction *SS* can be computed directly as

$$\frac{(\text{Sum of one diagonal} - \text{sum of other diagonal})^2}{abn}$$

9.2 Two-way anova: Significance testing

Before we can test hypotheses about the sources of variation isolated in Box 9.1, we must become familiar with the expected mean squares for this design. In the anova table of Box 9.1 we first show the expected-mean squares for Model I, both species differences and seawater concentrations being fixed treatment effects. The terms should be familiar in the context of your experience in the previous chapter. The quantities $\Sigma^a \alpha^2$, $\Sigma^b \beta^2$, and $\Sigma^{ab}(\alpha\beta)^2$ represent added components due to treatment for columns, rows, and interaction, respectively. Note that the within-subgroups or error *MS* again estimates the parametric variance of the items, σ^2.

The most important fact to remember about a Model I anova is that the mean square at each level of variation carries only the added effect due to that level of treatment. Except for the parametric variance of the items, it does not contain any term from a lower line. Thus, the expected *MS* of factor *A* contains only the parametric variance of the items plus the added term due to factor *A*, but does not also include interaction effects. In Model I, the significance test is therefore simple and straightforward. Any source of variation is tested by the variance ratio of the appropriate mean square over the error *MS*. Thus, for the appropriate tests we employ variance ratios *A*/**Error**, *B*/**Error** and (*A* × *B*)/**Error**, where each boldface term signifies a mean square. Thus $A = MS_A$, **Error** $= MS_{\text{within}}$.

When we do this in the example of Box 9.1, we find only factor *B*, salinity, significant. Neither factor *A* nor the interaction is significant. We conclude that the differences in oxygen consumption are induced by varying salinities (O_2 consumption responds in a V-shaped manner), and there does not appear to be sufficient evidence for species differences in oxygen consumption. The tabulation of the relative magnitudes of the means in the previous section shows that the pattern of signs in the two lines is identical. However, this may be misleading, since the mean of *A. scabra* is far higher at 100% seawater than at 75%, but that of *A. digitalis* is only very slightly higher. Although the oxygen consumption curves of the two species when graphed appear far from parallel (see Figure 9.2), this suggestion of a species × salinity interaction cannot be shown to be significant when compared with the within-subgroups variance. Finding a significant difference among salinities does not conclude the analysis. The data suggest that at 75% salinity there is a real reduction in oxygen consumption. Whether this is really so could be tested by the methods of Section 8.6.

When we analyze the results of the artificial example in Table 9.2, we find only the interaction *MS* significant. Thus, we would conclude that the response to salinity differs in the two species. This is brought out by inspection of the data, which show that at 75% salinity *A. scabra* consumes least oxygen and *A. digitalis* consumes most.

In the last (artificial) example the mean squares of the two factors (main effects) are not significant, in any case. However, many statisticians would not even test them once they found the interaction mean square to be significant, since in such a case an overall statement for each factor would have little meaning. A simple statement of response to salinity would be unclear. The presence of interaction makes us qualify our statements: "The pattern of response to changes in salinity differed in the two species." We would consequently have to describe separate, nonparallel response curves for the two species. Occasionally, it becomes important to test for overall significance in a Model I anova in spite of the presence of interaction. We may wish to demonstrate the significance of the effect of a drug, regardless of its significant interaction with age of the patient. To support this contention, we might wish to test the mean square among drug concentrations (over the error *MS*), regardless of whether the interaction *MS* is significant.

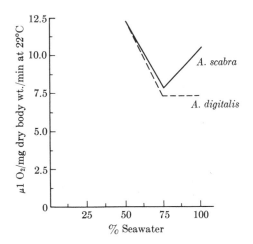

FIGURE 9.2
Oxygen consumption by two species of limpets at three salinities. Data from Box 9.1.

Box 9.1 also lists expected mean squares for a Model II anova and a mixed-model two-way anova. Here, variance components for columns (factor A). for rows (factor B), and for interaction make their appearance, and they are designated σ_A^2, σ_B^2, and σ_{AB}^2, respectively. In the Model II anova note that the two main effects contain the variance component of the interaction as well as their own variance component. In a Model II anova we first test $(A \times B)/\textbf{Error}$. If the interaction is significant, we continue testing $A/(A \times B)$ and $B/(A \times B)$. But when $A \times B$ is not significant, some authors suggest computation of a pooled error $MS = (SS_{A \times B} + SS_{\text{within}})/(df_{A \times B} + df_{\text{within}})$ to test the significance of the main effects. The conservative position is to continue to test the main effects over the interaction MS, and we shall follow this procedure in this book. Only one type of mixed model is shown in Box 9.1, in which factor A is assumed to be fixed and factor B to be random. If the situation is reversed, the expected mean squares change accordingly. In the mixed model, it is the mean square representing the fixed treatment that carries with it the variance component of the interaction, while the mean square representing the random factor contains only the error variance and its own variance component and does not include the interaction component. We therefore test the MS of the random main effect over the error, but test the fixed treatment MS over the interaction.

9.3 Two-way anova without replication

In many experiments there will be no replication for each combination of factors represented by a cell in the data table. In such cases we cannot easily talk of "subgroups," since each cell contains a single reading only. Frequently it may be too difficult or too expensive to obtain more than one reading per cell, or the measurements may be known to be so repeatable that there is little point in estimating their error. As we shall see in the following, a two-way anova without replication can be properly applied only with certain assumptions. For some models and tests in anova we must assume that there is no interaction present.

Our illustration for this design is from a study in metabolic physiology. In Box 9.2 we show levels of a chemical, S-PLP, in the blood serum of eight students before, immediately after, and 12 hours after the administration of an alcohol dose. Each student has been measured only once at each time. What is the appropriate model for this anova?

Clearly, the times are Model I. The eight individuals, however, are not likely to be of specific interest. It is improbable that an investigator would try to ask why student 4 has an S-PLP level so much higher than that of student 3. We would draw more meaningful conclusions from this problem if we considered the eight individuals to be randomly sampled. We could then estimate the variation among individuals with respect to the effect of alcohol over time.

The computations are shown in Box 9.2. They are the same as those in Box 9.1 except that the expressions to be evaluated are considerably simpler. Since $n = 1$, much of the summation can be omitted. The subgroup sum of squares

BOX 9.2

Two-way anova without replication.

Serum-pyridoxal-t-phosphate (S-PLP) content (ng per ml of serum) of blood serum before and after ingestion of alcohol in eight subjects. This is a mixed-model anova.

	Factor A: Time (a = 3)			
Factor B: Individuals (b = 8)	Before alcohol ingestion	Immediately after ingestion	12 hours later	Σ
1	20.00	12.34	17.45	49.79
2	17.62	16.72	18.25	52.59
3	11.77	9.84	11.45	33.06
4	30.78	20.25	28.70	79.73
5	11.25	9.70	12.50	33.45
6	19.17	15.67	20.04	54.88
7	9.33	8.06	10.00	27.39
8	32.96	19.10	30.45	82.51
Σ	152.88	111.68	148.84	413.40

Source: Data from Leinert et al. (1983).

The eight sets of three readings are treated as replications (blocks) in this analysis. Time is a fixed treatment effect, while differences between individuals are considered to be random effects. Hence, this is a mixed-model anova.

Preliminary computations

1. Grand total $= \sum\limits^{a}\sum\limits^{b} Y = 413.40$

2. Sum of the squared observations $= \sum\limits^{a}\sum\limits^{b} Y^2 = (20.00)^2 + \cdots + (30.45)^2 = 8349.4138$

3. Sum of squared column totals divided by sample size of a column $= \dfrac{\sum\limits^{a}\left(\sum\limits^{b} Y\right)^2}{b} = \dfrac{(152.88)^2 + (111.68)^2 + (148.84)^2}{8} = 7249.7578$

4. Sum of squared row totals divided by sample size of a row $= \dfrac{\sum\limits^{b}\left(\sum\limits^{a} Y\right)^2}{a} = \dfrac{(49.79)^2 + \cdots + (82.51)^2}{3} = 8127.8059$

5. Grand total squared and divided by the total sample size $=$ correction term $CT = \dfrac{\left(\sum\limits^{a}\sum\limits^{b} Y\right)^2}{ab}$

$= \dfrac{(\text{quantity } \mathbf{1})^2}{ab} = \dfrac{(413.40)^2}{24} = 7120.8150$

6. $SS_{\text{total}} = \sum\limits^{a}\sum\limits^{b} Y^2 - CT = $ quantity $\mathbf{2}$ $-$ quantity $\mathbf{5} = 8349.4138 - 7120.8150 = 1228.5988$

7. SS_A (SS of columns) $= \dfrac{\sum\limits^{a}\left(\sum\limits^{b} Y\right)^2}{b} - CT = $ quantity $\mathbf{3}$ $-$ quantity $\mathbf{5} = 7249.7578 - 7120.8150 = 128.9428$

8. SS_B (SS of rows) $= \dfrac{\sum\limits^{b}\left(\sum\limits^{a} Y\right)^2}{a} - CT = $ quantity $\mathbf{4}$ $-$ quantity $\mathbf{5} = 8127.8059 - 7120.8150 = 1006.9909$

9. SS_{error} (remainder; discrepance) $= SS_{\text{total}} - SS_A - SS_B = $ quantity $\mathbf{6}$ $-$ quantity $\mathbf{7}$ $-$ quantity $\mathbf{8}$
$= 1228.5988 - 128.9428 - 1006.9909 = 92.6651$

202

BOX 9.2
Continued

Anova table

Source of variation		df	SS	MS	F_s	Expected MS
$\bar{Y}_A - \bar{\bar{Y}}$	A (columns; time)	2	128.9428	64.4714	9.741**	$\sigma^2 + \sigma^2_{AB} + \dfrac{b}{a-1}\sum\alpha^2$
$\bar{Y}_B - \bar{\bar{Y}}$	B (rows; individuals)	7	1006.9909	143.8558	(21.734)**	$\sigma^2 + a\sigma^2_B$
$Y - \bar{Y}_A - \bar{Y}_B + \bar{\bar{Y}}$	Error (remainder; discrepance)	14	92.6651	6.6189		$\sigma^2 + \sigma^2_{AB}$
$Y - \bar{\bar{Y}}$	Total	23	1228.5988			

Conclusions.—Highly significant differences are found with time. There is a sharp decrease in S-PLP level immediately after ingestion of alcohol, but the level returns to near normal after 12 hours. For testing among individuals, we must assume interaction between time and individuals to be zero. From the magnitude of the F_s value it is fairly evident that there are large differences among individuals in S-PLP levels. Inspection of the row totals confirms this conclusion.

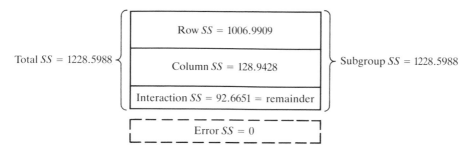

FIGURE 9.3
Diagrammatic representation of the partitioning of the total sums of squares in a two-way ortho-gonal anova without replication. The areas of the subdivisions are not shown proportional to the magnitudes of the sums of squares.

in this example is the same as the total sum of squares. If this is not immediately apparent, consult Figure 9.3, which, when compared with Figure 9.1, illustrates that the error sum of squares based on variation within subgroups is missing in this example. Thus, after we subtract the sum of squares for columns (factor A) and for rows (factor B) from the total SS, we are left with only a single sum of squares, which is the equivalent of the previous interaction SS but which is now the only source for an error term in the anova. This SS is known as the *remainder SS* or the *discrepance*.

If you refer to the expected mean squares for the two-way anova in Box 9.1, you will discover why we made the statement earlier that for some models and tests in a two-way anova without replication we must assume that the inter-action is not significant. If interaction is present, only a Model II anova can be entirely tested, while in a mixed model only the fixed level can be tested over the remainder mean square. But in a pure Model I anova, or for the random factor in a mixed model, it would be improper to test the main effects over the remainder unless we could reliably assume that no added effect due to interaction is present. General inspection of the data in Box 9.2 convinces us that the trends with time for any one individual are faithfully reproduced for the other individuals. Thus, interaction is unlikely to be present. If, for example, some individuals had not responded with a lowering of their S-PLP levels after ingestion of alcohol, interaction would have been apparent, and the test of the mean square among individuals carried out in Box 9.2 would not have been legitimate.

Since we assume no interaction, the row and column mean squares are tested over the error MS. The results are not surprising; casual inspection of the data would have predicted our findings. Differences with time are highly significant, yielding an F_s value of 9.741. The added variance among individuals is also highly significant, assuming there is no interaction.

A common application of two-way anova without replication is the *repeated testing of the same individuals*. By this we mean that the same group of individuals is tested repeatedly over a period of time. The individuals are one factor (usually considered as random and serving as replication), and the time dimension is the second factor, a fixed treatment effect. For example, we might measure growth of a structure in ten individuals at regular intervals. When we test for the presence of an added variance component (due to the random factor), we again must assume that there is no interaction between time and the individuals; that is, the responses of the several individuals are parallel through time. Another use of this design is found in various physiological and psychological experiments in which we test the same group of individuals for the appearance of some response after treatment. Examples include increasing immunity after antigen inoculations, altered responses after conditioning, and measures of learning after a number of trials. Thus, we may study the speed with which ten rats, repeatedly tested on the same maze, reach the end point. The fixed-treatment effect would be the successive trials to which the rats have been subjected. The second factor, the ten rats, is random, presumably representing a random sample of rats from the laboratory population.

One special case, common enough to merit separate discussion, is repeated testing of the same individuals in which only two treatments ($a = 2$) are given. This case is also known as *paired comparisons*, because each observation for one treatment is paired with one for the other treatment. This pair is composed of the same individuals tested twice or of two individuals with common experiences, so that we can legitimately arrange the data as a two-way anova.

Let us elaborate on this point. Suppose we test the muscle tone of a group of individuals, subject them to severe physical exercise, and measure their muscle tone once more. Since the same group of individuals will have been tested twice, we can arrange our muscle tone readings in pairs, each pair representing readings on one individual (before and after exercise). Such data are appropriately treated by a two-way anova without replication, which in this case would be a paired-comparisons test because there are only two treatment classes. This "before and after treatment" comparison is a very frequent design leading to paired comparisons. Another design simply measures two stages in the development of a group of organisms, time being the treatment intervening between the two stages. The example in Box 9.3 is of this nature. It measures lower face width in a group of girls at age five and in the same group of girls when they are six years old. The paired comparison is for each individual girl, between her face width when she is five years old and her face width at six years.

Paired comparisons often result from dividing an organism or other individual unit so that half receives treatment 1 and the other half treatment 2, which may be the control. Thus, if we wish to test the strength of two antigens or allergens we might inject one into each arm of a single individual and measure the diameter of the red area produced. It would not be wise, from the point of view of experimental design, to test antigen 1 on individual 1 and antigen 2 on individual 2. These individuals may be differentially susceptible to these antigens, and we may learn little about the relative potency of the

BOX 9.3
Paired comparisons (randomized blocks with $a = 2$).

Lower face width (skeletal bigonial diameter in cm) for 15 North American white girls measured when 5 and again when 6 years old.

Individuals	(1) 5-year-olds	(2) 6-year-olds	(3) $\sum$	(4) $D = Y_{i2} - Y_{i1}$ (difference)
1	7.33	7.53	14.86	0.20
2	7.49	7.70	15.19	.21
3	7.27	7.46	14.73	.19
4	7.93	8.21	16.14	.28
5	7.56	7.81	15.37	.25
6	7.81	8.01	15.82	.20
7	7.46	7.72	15.18	.26
8	6.94	7.13	14.07	.19
9	7.49	7.68	15.17	.19
10	7.44	7.66	15.10	.22
11	7.95	8.11	16.06	.16
12	7.47	7.66	15.13	.19
13	7.04	7.20	14.24	.16
14	7.10	7.25	14.35	.15
15	7.64	7.79	15.43	.15
$\sum Y$	111.92	114.92	226.84	3.00
$\sum Y^2$	836.3300	881.8304	3435.6992	0.6216

Source: From a larger study by Newman and Meredith (1956).

Two-way anova without replication

Anova table

Source of variation	df	SS	MS	F_s	Expected MS
Ages (columns; factor A)	1	0.3000	0.3000	388.89**	$\sigma^2 + \sigma_{AB}^2 + \dfrac{b}{a-1}\sum \alpha^2$
Individuals (rows; factor B)	14	2.6367	0.188,34	(244.14)**	$\sigma^2 \qquad + a\sigma_B^2$
Remainder	14	0.0108	0.000,771,43		$\sigma^2 + \sigma_{AB}^2$
Total	29	2.9475			

$F_{0.01[1,14]} = 8.86$ $F_{0.01[12,12]} = 4.16$ (Conservative tabled value)

Conclusions.—The variance ratio for ages is highly significant. We conclude that faces of 6-year-old girls are wider than those of 5-year-olds. If we are willing

BOX 9.3
Continued

to assume that the interaction σ_{AB}^2 is zero, we may test for an added variance component among individual girls and would find it significant.

The t *test for paired comparisons*

$$t_s = \frac{\bar{D} - (\mu_1 - \mu_2)}{s_{\bar{D}}}$$

where $\bar{D}$ is the mean difference between the paired observations.

$$\bar{D} = \frac{\Sigma D}{b} = \frac{3.00}{15} = 0.20$$

and $s_{\bar{D}} = s_D/\sqrt{b}$ is the standard error of $\bar{D}$ calculated from the observed differences in column (4):

$$s_D = \sqrt{\frac{\sum D^2 - (\sum D)^2/b}{b-1}} = \sqrt{\frac{0.6216 - (3.00^2/15)}{14}} = \sqrt{\frac{0.0216}{14}}$$

$$= \sqrt{0.001,542,86} = 0.039,279,2$$

and thus

$$s_{\bar{D}} = \frac{s_D}{\sqrt{b}} = \frac{0.039,279,2}{\sqrt{15}} = 0.010,141,9$$

We assume that the true difference between the means of the two groups, $\mu_1 - \mu_2$, equals zero:

$$t_s = \frac{\bar{D} - 0}{s_{\bar{D}}} = \frac{0.20 - 0}{0.010,141,9} = 19.7203 \quad \text{with} \quad b - 1 = 14 \, df.$$

This yields $P \ll 0.01$. Also $t_s^2 = 388.89$, which equals the previous F_s.

antigens, since this would be confounded by the differential responses of the subjects. A much better design would be first to inject antigen 1 into the left arm and antigen 2 into the right arm of a group of n individuals and then to analyze the data as a two-way anova without replication, with n rows (individuals) and 2 columns (treatments). It is probably immaterial whether an antigen is injected into the right or left arm, but if we were designing such an experiment and knew little about the reaction of humans to antigens, we might, as a precaution, randomly allocate antigen 1 to the left or right arm for different subjects, antigen 2 being injected into the opposite arm. A similar example is the testing of certain plant viruses by rubbing a concentration of the virus over the surface of a leaf and counting the resulting lesions. Since different leaves are susceptible in different degrees, a conventional way of measuring the strength of the virus is to wipe it over the half of the leaf on one side of the midrib, rubbing the other half of the leaf with a control or standard solution.

Another design leading to paired comparisons is to apply the treatment to two individuals sharing a common experience, be this genetic or environmental. Thus, a drug or a psychological test might be given to groups of twins or sibs, one of each pair receiving the treatment, the other one not.

Finally, the paired-comparisons technique may be used when the two individuals to be compared share a single experimental unit and are thus subjected to common environmental experiences. If we have a set of rat cages, each of which holds two rats, and we are trying to compare the effect of a hormone injection with a control, we might inject one of each pair of rats with the hormone and use its cage mate as a control. This would yield a $2 \times n$ anova for n cages.

One reason for featuring the paired-comparisons test separately is that it alone among the two-way anovas without replication has an equivalent, alternative method of analysis—the t test for paired comparisons, which is the traditional method of analyzing it.

The paired-comparisons case shown in Box 9.3 analyzes face widths of five- and six-year-old girls, as already mentioned. The question being asked is whether the faces of six-year-old girls are significantly wider than those of five-year-old girls. The data are shown in columns (1) and (2) for 15 individual girls. Column (3) features the row sums that are necessary for the analysis of variance. The computations for the two-way anova without replication are the same as those already shown for Box 9.2 and thus are not shown in detail. The anova table shows that there is a highly significant difference in face width between the two age groups. If interaction is assumed to be zero, there is a large added variance component among the individual girls, undoubtedly representing genetic as well as environmental differences.

The other method of analyzing paired-comparisons designs is the well-known t *test for paired comparisons*. It is quite simple to apply and is illustrated in the second half of Box 9.3. It tests whether the mean of sample differences between pairs of readings in the two columns is significantly different from a hypothetical mean, which the null hypothesis puts at zero. The standard error over which this is tested is the standard error of the mean difference. The difference column has to be calculated and is shown in column (4) of the data table in Box 9.3. The computations are quite straightforward, and the conclusions are the same as for the two-way anova. This is another instance in which we obtain the value of F_s when we square the value of t_s.

Although the paired-comparisons t test is the traditional method of solving this type of problem, we prefer the two-way anova. Its computation is no more time-consuming and has the advantage of providing a measure of the variance component among the rows (blocks). This is useful knowledge, because if there is no significant added variance component among blocks, one might simplify the analysis and design of future, similar studies by employing single classification anova.

Exercises

9.1   Swanson, Latshaw, and Tague (1921) determined soil pH electrometrically for various soil samples from Kansas. An extract of their data (acid soils) is shown below. Do subsoils differ in pH from surface soils (assume that there is no interaction between localities and depth for pH reading)?

County	Soil type	Surface pH	Subsoil pH
Finney	Richfield silt loam	6.57	8.34
Montgomery	Summit silty clay loam	6.77	6.13
Doniphan	Brown silt loam	6.53	6.32
Jewell	Jewell silt loam	6.71	8.30
Jewell	Colby silt loam	6.72	8.44
Shawnee	Crawford silty clay loam	6.01	6.80
Cherokee	Oswego silty clay loam	4.99	4.42
Greenwood	Summit silty clay loam	5.49	7.90
Montgomery	Cherokee silt loam	5.56	5.20
Montgomery	Oswego silt loam	5.32	5.32
Cherokee	Bates silt loam	5.92	5.21
Cherokee	Cherokee silt loam	6.55	5.66
Cherokee	Neosho silt loam	6.53	5.66

ANS. MS between surface and subsoils = 0.6246, $MS_{residual} = 0.6985$, $F_s = 0.849$ which is clearly not significant at the 5% level.

9.2   The following data were extracted from a Canadian record book of purebred dairy cattle. Random samples of 10 mature (five-year-old and older) and 10 two-year-old cows were taken from each of five breeds (honor roll, 305-day class). The average butterfat percentages of these cows were recorded. This gave us a total of 100 butterfat percentages, broken down into five breeds and into two age classes. The 100 butterfat percentages are given below. Analyze and discuss your results. You will note that the tedious part of the calculation has been done for you.

	Ayshire		Canadian		Guerusey		Holstein-Friesian		Jersey	
	Mature	2-yr	Mature	2-yr	Mature	2-yr	Mature	2-yr	Mature	2-yr
	3.74	4.44	3.92	4.29	4.54	5.30	3.40	3.79	4.80	5.75
	4.01	4.37	4.95	5.24	5.18	4.50	3.55	3.66	6.45	5.14
	3.77	4.25	4.47	4.43	5.75	4.59	3.83	3.58	5.18	5.25
	3.78	3.71	4.28	4.00	5.04	5.04	3.95	3.38	4.49	4.76
	4.10	4.08	4.07	4.62	4.64	4.83	4.43	3.71	5.24	5.18
	4.06	3.90	4.10	4.29	4.79	4.55	3.70	3.94	5.70	4.22
	4.27	4.41	4.38	4.85	4.72	4.97	3.30	3.59	5.41	5.98
	3.94	4.11	3.98	4.66	3.88	5.38	3.93	3.55	4.77	4.85
	4.11	4.37	4.46	4.40	5.28	5.39	3.58	3.55	5.18	6.55
	4.25	3.53	5.05	4.33	4.66	5.97	3.54	3.43	5.23	5.72
$\sum Y$	40.03	41.17	43.66	45.11	48.48	50.52	37.21	36.18	52.45	53.40
$\bar{Y}$	4.003	4.117	4.366	4.511	4.848	5.052	3.721	3.618	5.245	5.340

$$\overset{abn}{\sum} Y^2 = 2059.6109$$

9.3   Blakeslee (1921) studied length-width ratios of second seedling leaves of two types of Jimson weed called globe (*G*) and nominal (*N*). Three seeds of each type were planted in 16 pots. Is there sufficient evidence to conclude that globe and nominal differ in length-width ratio?

Pot identification number	Types					
	G			N		
16533	1.67	1.53	1.61	2.18	2.23	2.32
16534	1.68	1.70	1.49	2.00	2.12	2.18
16550	1.38	1.76	1.52	2.41	2.11	2.60
16668	1.66	1.48	1.69	1.93	2.00	2.00
16767	1.38	1.61	1.64	2.32	2.23	1.90
16768	1.70	1.71	1.71	2.48	2.11	2.00
16770	1.58	1.59	1.38	2.00	2.18	2.16
16771	1.49	1.52	1.68	1.94	2.13	2.29
16773	1.48	1.44	1.58	1.93	1.95	2.10
16775	1.28	1.45	1.50	1.77	2.03	2.08
16776	1.55	1.45	1.44	2.06	1.85	1.92
16777	1.29	1.57	1.44	2.00	1.94	1.80
16780	1.36	1.22	1.41	1.87	1.87	2.26
16781	1.47	1.43	1.61	2.24	2.00	2.23
16787	1.52	1.56	1.56	1.79	2.08	1.89
16789	1.37	1.38	1.40	1.85	2.10	2.00

ANS. $MS_{within} = 0.0177$, $MS_{T \times P} = 0.0203$, $MS_{types} = 7.3206$ ($F_s = 360.62**$), $MS_{pots} = 0.0598$ ($F_s = 3.378**$). The effect of pots is considered to be a Model II factor, and types, a Model I factor.

9.4   The following data were extracted from a more entensive study by Sokal and Karten (1964). The data represent mean dry weights (in mg) of three genotypes of beetles, *Tribolium castaneum*, reared at a density of 20 beetles per gram of flour. The four series of experiments represent replications.

Series	Genotypes		
	++	+b	bb
1	0.958	0.986	0.925
2	0.971	1.051	0.952
3	0.927	0.891	0.829
4	0.971	1.010	0.955

Test whether the genotypes differ in mean dry weight.

9.5   The mean length of developmental period (in days) for three strains of house-flies at seven densities is given. (Data by Sullivan and Sokal, 1963.) Do these flies differ in development period with density and among strains? You may assume absence of strain × density interaction.

		Strains	
Density per container	OL	BELL	bwb
60	9.6	9.3	9.3
80	10.6	9.1	9.2
160	9.8	9.3	9.5
320	10.7	9.1	10.0
640	11.1	11.1	10.4
1280	10.9	11.8	10.8
2560	12.8	10.6	10.7

ANS. $MS_{residual} = 0.3426$, $MS_{strains} = 1.3943$ ($F_s = 4.070*$), $MS_{density} = 2.0905$ ($F_s = 6.1019**$).

9.6   The following data are extracted from those of French (1976), who carried out a study of energy utilization in the pocket mouse *Perognathus longimembris* during hibernation at different temperatures. Is there evidence that the amount of food available affects the amount of energy consumed at different temperatures during hibernation?

Restricted food				Ad-libitum food			
8°C		18°C		8°C		18°C	
Animal no.	Energy used (kcal/g)	Animal no.	Energy used (kcal/g)	Animal no.	Energy used (kcal/g)	Animal no.	Energy used (kcal/g)
1	62.69	5	72.60	13	95.73	17	101.19
2	54.07	6	70.97	14	63.95	18	76.88
3	65.73	7	74.32	15	144.30	19	74.08
4	62.98	8	53.02	16	144.30	20	81.40

CHAPTER 10

Assumptions of Analysis of Variance

We shall now examine the underlying assumptions of the analysis of variance, methods for testing whether these assumptions are valid, the consequences for an anova if the assumptions are violated, and steps to be taken if the assumptions cannot be met. We should stress that before you carry out any anova on an actual research problem, you should assure yourself that the assumptions listed in this chapter seem reasonable. If they are not, you should carry out one of several possible alternative steps to remedy the situation.

In Section 10.1 we briefly list the various assumptions of analysis of variance. We describe procedures for testing some of them and briefly state the consequences if the assumptions do not hold, and we give instructions on how to proceed if they do not. The assumptions include random sampling, independence, homogeneity of variances, normality, and additivity.

In many cases, departure from the assumptions of analysis of variance can be rectified by transforming the original data by using a new scale. The rationale behind this is given in Section 10.2, together with some of the common transformations.

When transformations are unable to make the data conform to the assumptions of analysis of variance, we must use other techniques of analysis, analogous to the intended anova. These are the nonparametric or distribution-free techniques, which are sometimes used by preference even when the parametric method (anova in this case) can be legitimately employed. Researchers often like to use the nonparametric methods because the assumptions underlying them are generally simple and because they lend themselves to rapid computation on a small calculator. However, when the assumptions of anova are met, these methods are less efficient than anova. Section 10.3 examines three nonparametric methods in lieu of anova for two-sample cases only.

10.1 The assumptions of anova

Randomness. All anovas require that sampling of individuals be at random. Thus, in a study of the effects of three doses of a drug (plus a control) on five rats each, the five rats allocated to each treatment must be selected at random. If the five rats employed as controls are either the youngest or the smallest or the heaviest rats while those allocated to some other treatment are selected in some other way, it is clear that the results are not apt to yield an unbiased estimate of the true treatment effects. Nonrandomness of sample selection may well be reflected in lack of independence of the items, in heterogeneity of variances, or in nonnormal distribution—all discussed in this section. Adequate safeguards to ensure random sampling during the design of an experiment, or during sampling from natural populations, are essential.

Independence. An assumption stated in each explicit expression for the expected value of a variate (for example, Expression (7.2) was $Y_{ij} = \mu + \alpha_i + \epsilon_{ij}$) is that the error term ϵ_{ij} is a random normal variable. In addition, for completeness we should also add the statement that it is assumed that the ϵ's are independently and identically (as explained below under "Homogeneity of variances") distributed.

Thus, if you arranged the variates within any one group in some logical order independent of their magnitude (such as the order in which the measurements were obtained), you would expect the ϵ_{ij}'s to succeed each other in a random sequence. Consequently, you would assume a long sequence of large positive values followed by an equally long sequence of negative values to be quite unlikely. You would also not expect positive and negative values to alternate with regularity.

How could departures from independence arise? An obvious example would be an experiment in which the experimental units were plots of ground laid out in a field. In such a case it is often found that adjacent plots of ground give rather similar yields. It would thus be important not to group all the plots containing the same treatment into an adjacent series of plots but rather to randomize the allocation of treatments among the experimental plots. The physical process of randomly allocating the treatments to the experimental plots ensures that the ϵ's will be independent.

Lack of independence of the ϵ's can result from correlation in time rather than space. In an experiment we might measure the effect of a treatment by recording weights of ten individuals. Our balance may suffer from a maladjustment that results in giving successive underestimates, compensated for by several overestimates. Conversely, compensation by the operator of the balance may result in regularly alternating over- and underestimates of the true weight. Here again, randomization may overcome the problem of nonindependence of errors. For example, we may determine the sequence in which individuals of the various groups are weighed according to some random procedure.

There is no simple adjustment or transformation to overcome the lack of independence of errors. The basic design of the experiment or the way in which it is performed must be changed. If the ϵ's are not independent, the validity of the usual F test of significance can be seriously impaired.

Homogeneity of variances. In Section 8.4 and Box 8.2, in which we described the t test for the difference between two means, you were told that the statistical test was valid only if we could assume that the variances of the two samples were equal. Although we have not stressed it so far, this assumption that the ϵ_{ij}'s have identical variances also underlies the equivalent anova test for two samples—and in fact any type of anova. *Equality of variances* in a set of samples is an important precondition for several statistical tests. Synonyms for this condition are *homogeneity of variances* and *homoscedasticity*. This latter term is coined from Greek roots meaning equal scatter; the converse condition (inequality of variances among samples) is called *heteroscedasticity*. Because we assume that each sample variance is an estimate of the same parametric error variance, the assumption of homogeneity of variances makes intuitive sense.

We have already seen how to test whether two samples are homoscedastic prior to a t test of the differences between two means (or the mathematically equivalent two-sample analysis of variance): we use an F test for the hypotheses $H_0: \sigma_1^2 = \sigma_2^2$ and $H_1: \sigma_1^2 \neq \sigma_2^2$, as illustrated in Section 7.3 and Box 7.1. For more than two samples there is a "quick and dirty" method, preferred by many because of its simplicity. This is the F_{max} test. This test relies on the tabled cumulative probability distribution of a statistic that is the variance ratio of the largest to the smallest of several sample variances. This distribution is shown in Table **VI**. Let us assume that we have six anthropological samples of 10 bone lengths each, for which we wish to carry out an anova. The variances of the six samples range from 1.2 to 10.8. We compute the maximum variance ratio $s_{max}^2/s_{min}^2 = \frac{10.8}{1.2} = 9.0$ and compare it with $F_{max \, \alpha[\alpha,\nu]}$, critical values of which are found in Table **VI**. For $a = 6$ and $\nu = n - 1 = 9$, F_{max} is 7.80 and 12.1 at the 5% and 1% levels, respectively. We conclude that the variances of the six samples are significantly heterogeneous.

What may cause such heterogeneity? In this case, we suspect that some of the populations are inherently more variable than others. Some races or species are relatively uniform for one character, while others are quite variable for the same character. In an anova representing the results of an experiment, it may well be that one sample has been obtained under less standardized conditions than the others and hence has a greater variance. There are also many cases in which the heterogeneity of variances is a function of an improper choice of measurement scale. With some measurement scales, variances vary as functions of means. Thus, differences among means bring about heterogeneous variances. For example, in variables following the Poisson distribution the variance is in fact equal to the mean, and populations with greater means will therefore have greater variances. Such departures from the assumption of homoscedasticity can often be easily corrected by a suitable transformation, as discussed later in this chapter.

A rapid first inspection for heteroscedasticity is to check for correlation between the means and variances or between the means and the ranges of the samples. If the variances increase with the means (as in a Poisson distribution), the ratios $s^2/\bar{Y}$ or $s/\bar{Y} = V$ will be approximately constant for the samples. If means and variances are independent, these ratios will vary widely.

The consequences of moderate heterogeneity of variances are not too serious for the overall test of significance, but single degree of freedom comparisons may be far from accurate.

If transformation cannot cope with heteroscedasticity, nonparametric methods (Section 10.3) may have to be resorted to.

Normality. We have assumed that the error terms ϵ_{ij} of the variates in each sample will be independent, that the variances of the error terms of the several samples will be equal, and, finally, that the error terms will be normally distributed. If there is serious question about the normality of the data, a graphic test, as illustrated in Section 5.5, might be applied to each sample separately.

The consequences of nonnormality of error are not too serious. Only very skewed distribution would have a marked effect on the significance level of the F test or on the efficiency of the design. The best way to correct for lack of normality is to carry out a transformation that will make the data normally distributed, as explained in the next section. If no simple transformation is satisfactory, a nonparametric test, as carried out in Section 10.3, should be substituted for the analysis of variance.

Additivity. In two-way anova without replication it is necessary to assume that interaction is not present if one is to make tests of the main effects in a Model I anova. This assumption of no interaction in a two-way anova is sometimes also referred to as the assumption of additivity of the main effects. By this we mean that any single observed variate can be decomposed into additive components representing the treatment effects of a particular row and column as well as a random term special to it. If interaction is actually present, then the F test will be very inefficient, and possibly misleading if the effect of the interaction is very large. A check of this assumption requires either more than a single observation per cell (so that an error mean square can be computed) or an independent estimate of the error mean square from previous *comparable* experiments.

Interaction can be due to a variety of causes. Most frequently it means that a given treatment combination, such as level 2 of factor A when combined with level 3 of factor B, makes a variate deviate from the expected value. Such a deviation is regarded as an inherent property of the natural system under study, as in examples of synergism or interference. Similar effects occur when a given replicate is quite aberrant, as may happen if an exceptional plot is included in an agricultural experiment, if a diseased individual is included in a physiological experiment, or if by mistake an individual from a different species is included in a biometric study. Finally, an interaction term will result if the effects of the two factors A and B on the response variable Y are multiplicative rather than additive. An example will make this clear.

In Table 10.1 we show the additive and multiplicative treatment effects in a hypothetical two-way anova. Let us assume that the expected population mean μ is zero. Then the mean of the sample subjected to treatment 1 of factor A and treatment 1 of factor B should be 2, by the conventional additive model. This is so because each factor at level 1 contributes unity to the mean. Similarly, the expected subgroup mean subjected to level 3 for factor A and level 2 for factor B is 8, the respective contributions to the mean being 3 and 5. However, if the process is multiplicative rather than additive, as occurs in a variety of physicochemical and biological phenomena, the expected values will be quite different. For treatment $A_1 B_1$, the expected value equals 1, which is the product of 1 and 1. For treatment $A_3 B_2$, the expected value is 15, the product of 3 and 5. If we were to analyze multiplicative data of this sort by a conventional anova, we would find that the interaction sum of squares would be greatly augmented because of the nonadditivity of the treatment effects. In this case, there is a simple remedy. By transforming the variable into logarithms (Table 10.1), we are able to restore the additivity of the data. The third item in each cell gives the logarithm of the expected value, assuming multiplicative

TABLE 10.1
Illustration of additive and multiplicative effects.

Factor B	Factor A			
	$\alpha_1 = 1$	$\alpha_2 = 2$	$\alpha_3 = 3$	
$\beta_1 = 1$	2	3	4	Additive effects
	1	2	3	Multiplicative effects
	0	0.30	0.48	Log of multiplicative effects
$\beta_2 = 5$	6	7	8	Additive effects
	5	10	15	Multiplicative effects
	0.70	1.00	1.18	Log of multiplicative effects

relations. Notice that the increments are strictly additive again ($SS_{A \times B} = 0$). As a matter of fact, on a logarithmic scale we could simply write $\alpha_1 = 0$, $\alpha_2 = 0.30$, $\alpha_3 = 0.48$, $\beta_1 = 0$, $\beta_2 = 0.70$. Here is a good illustration of how transformation of scale, discussed in detail in Section 10.2, helps us meet the assumptions of analysis of variance.

10.2 Transformations

If the evidence indicates that the assumptions for an analysis of variance or for a t test cannot be maintained, two courses of action are open to us. We may carry out a different test not requiring the rejected assumptions, such as one of the distribution-free tests in lieu of anova, discussed in the next section. A second approach would be to transform the variable to be analyzed in such a manner that the resulting transformed variates meet the assumptions of the analysis.

Let us look at a simple example of what transformation will do. A single variate of the simplest kind of anova (completely randomized, single-classification, Model I) decomposes as follows: $Y_{ij} = \mu + \alpha_i + \epsilon_{ij}$. In this model the components are additive, with the error term ϵ_{ij} normally distributed. However, we might encounter a situation in which the components were multiplicative in effect, so that $Y_{ij} = \mu \alpha_i \epsilon_{ij}$, which is the product of the three terms. In such a case the assumptions of normality and of homoscedasticity would break down. In any one anova, the parametric mean μ is constant but the treatment effect α_i differs from group to group. Clearly, the scatter among the variates Y_{ij} would double in a group in which α_i is twice as great as in another. Assume that $\mu = 1$, the smallest $\epsilon_{ij} = 1$, and the greatest, 3; then if $\alpha_i = 1$, the range of the Y's will be $3 - 1 = 2$. However, when $\alpha_i = 4$, the corresponding range will be four times as wide, from $4 \times 1 = 4$ to $4 \times 3 = 12$, a range of 8. Such data will be heteroscedastic. We can correct this situation simply by transforming our model into logarithms. We would therefore obtain $\log Y_{ij} = \log \mu + \log \alpha_i + \log \epsilon_{ij}$, which is additive and homoscedastic. The entire analysis of variance would then be carried out on the transformed variates.

At this point many of you will feel more or less uncomfortable about what we have done. Transformation seems too much like "data grinding." When you learn that often a statistical test may be made significant after transformation of a set of data, though it would not be so without such a transformation, you may feel even more suspicious. What is the justification for transforming the data? It takes some getting used to the idea, but there is really no scientific necessity to employ the common linear or arithmetic scale to which we are accustomed. You are probably aware that teaching of the "new math" in elementary schools has done much to dispel the naive notion that the decimal system of numbers is the only "natural" one. In a similar way, with some experience in science and in the handling of statistical data, you will appreciate the fact that the linear scale, so familiar to all of us from our earliest experience, occupies a similar position with relation to other scales of measurement as does the decimal system of numbers with respect to the binary and octal numbering systems and others. If a system is multiplicative on a linear scale, it may be much more convenient to think of it as an additive system on a logarithmic scale. Another frequent transformation is the square root of a variable. The square root of the surface area of an organism is often a more appropriate measure of the fundamental biological variable subjected to physiological and evolutionary forces than is the area. This is reflected in the normal distribution of the square root of the variable as compared to the skewed distribution of areas. In many cases experience has taught us to express experimental variables not in linear scale but as logarithms, square roots, reciprocals, or angles. Thus, *pH* values are logarithms and dilution series in microbiological titrations are expressed as reciprocals. As soon as you are ready to accept the idea that the scale of measurement is arbitrary, you simply have to look at the distributions of transformed variates to decide which transformation most closely satisfies the assumptions of the analysis of variance before carrying out an anova.

A fortunate fact about transformations is that very often several departures from the assumptions of anova are simultaneously cured by the same transformation to a new scale. Thus, simply by making the data homoscedastic, we also make them approach normality and ensure additivity of the treatment effects.

When a transformation is applied, tests of significance are performed on the transformed data, but estimates of means are usually given in the familiar untransformed scale. Since the transformations discussed in this chapter are nonlinear, confidence limits computed in the transformed scale and changed back to the original scale would be asymmetrical. Stating the standard error in the original scale would therefore be misleading. In reporting results of research with variables that require transformation, furnish means in the back-transformed scale followed by their (asymmetrical) confidence limits rather than by their standard errors.

An easy way to find out whether a given transformation will yield a distribution satisfying the assumptions of anova is to plot the cumulative distributions of the several samples on probability paper. By changing the scale of the second coordinate axis from linear to logarithmic, square root, or any other one, we can see whether a previously curved line, indicating skewness, straightens out to indicate normality (you may wish to refresh your memory on these graphic techniques studied in Section 5.5). We can look up upper class limits on transformed scales or employ a variety of available probability graph papers whose second axis is in logarithmic, angular, or other scale. Thus, we not only test whether the data become more normal through transformation, but we can also get an estimate of the standard deviation under transformation as measured by the slope of the fitted line. The assumption of homoscedasticity implies that the slopes for the several samples should be the same. If the slopes are very heterogeneous, homoscedasticity has not been achieved. Alternatively, we can examine goodness of fit tests for normality (see Chapter 13) for the samples under various transformations. That transformation yielding the best fit over all samples will be chosen for the anova. It is important that the transformation not be selected on the basis of giving the best anova results, since such a procedure would distort the significance level.

The logarithmic transformation. The most common transformation applied is conversion of all variates into logarithms, usually common logarithms. Whenever the mean is positively correlated with the variance (greater means are accompanied by greater variances), the logarithmic transformation is quite likely to remedy the situation and make the variance independent of the mean. Frequency distributions skewed to the right are often made more symmetrical by transformation to a logarithmic scale. We saw in the previous section and in Table 10.1 that logarithmic transformation is also called for when effects are multiplicative.

The square root transformation. We shall use a square root transformation as a detailed illustration of transformation of scale. When the data are counts, as of insects on a leaf or blood cells in a hemacytometer, we frequently find the square root transformation of value. You will remember that such distributions are likely to be Poisson-distributed rather than normally distributed and that in a Poisson distribution the variance is the same as the mean. Therefore, the mean and variance cannot be independent but will vary identically. Transforming the variates to square roots will generally make the variances independent of the means. When the counts include zero values, it has been found desirable to code all variates by adding 0.5. The transformation then is $\sqrt{Y + \frac{1}{2}}$.

Table 10.2 shows an application of the square root transformation. The sample with the greater mean has a significantly greater variance prior to transformation. After transformation the variances are not significantly different. For reporting means the transformed means are squared again and confidence limits are reported in lieu of standard errors.

The arcsine transformation. This transformation (also known as the *angular transformation*) is especially appropriate to percentages and proportions. You may remember from Section 4.2 that the standard deviation of a binomial distribution is $\sigma = \sqrt{pq/k}$. Since $\mu = p$, $q = 1 - p$, and k is constant for any one problem, it is clear that in a binomial distribution the variance would be a function of the mean. The arcsine transformation preserves the independence of the two.

The transformation finds $\theta = \arcsin \sqrt{p}$, where p is a proportion. The term "arcsin" is synonymous with inverse sine or $\sin^{-1}$, which stands for "the angle whose sine is" the given quantity. Thus, if we compute or look up arcsin $\sqrt{0.431} = 0.6565$, we find $41.03°$, the angle whose sine is 0.6565. The arcsine transformation stretches out both tails of a distribution of percentages or proportions and compresses the middle. When the percentages in the original data fall between 30% and 70%, it is generally not necessary to apply the arcsine transformation.

TABLE 10.2
An application of the square root transformation. The data represent the number of adult *Drosophila* emerging from single-pair cultures for two different medium formulations (medium A contained DDT).

(1) Number of flies emerging Y	(2) Square root of number of flies $\sqrt{Y}$	(3) Medium A f	(4) Medium B f
0	0.00	1	—
1	1.00	5	—
2	1.41	6	—
3	1.73	—	—
4	2.00	3	—
5	2.24	—	—
6	2.45	—	—
7	2.65	—	2
8	2.83	—	1
9	3.00	—	2
10	3.16	—	3
11	3.32	—	1
12	3.46	—	1
13	3.61	—	1
14	3.74	—	1
15	3.87	—	1
16	4.00	—	2
		$\overline{15}$	$\overline{15}$

Untransformed variable

	Medium A	Medium B
$\bar{Y}$	1.933	11.133
s^2	1.495	9.410

Square root transformation

$\sqrt{Y}$	1.299	3.307
$s^2_{\sqrt{Y}}$	0.2634	0.2099

Tests of equality of variances

Untransformed

$$F_s = \frac{s_2^2}{s_1^2} = \frac{9.410}{1.495} = 6.294**$$

$$F_{0.025[14,14]} = 2.98$$

Transformed

$$F_s = \frac{s^2_{\sqrt{Y_1}}}{s^2_{\sqrt{Y_2}}} = \frac{0.2634}{0.2099} = 1.255 \; ns$$

Back-transformed (squared) means

	Medium A	Medium B
$(\sqrt{Y})^2$	1.687	10.937

95% confidence limits

	Medium A	Medium B
$L_1 = \overline{\sqrt{Y}} - t_{0.05}s_{\sqrt{Y}}$	$1.297 - 2.145\sqrt{\frac{0.2634}{15}}$	$3.307 - 2.145\sqrt{\frac{0.2099}{15}}$
	$= 1.015$	$= 3.053$
$L_2 = \overline{\sqrt{Y}} + t_{0.05}s_{\sqrt{Y}}$	1.583	3.561

Back-transformed (squared) confidence limits

	Medium A	Medium B
L_1^2	1.030	9.324
L_2^2	2.507	12.681

Source: Data from unpublished study by R. R. Sokal.

10.3 Nonparametric methods in lieu of anova

If none of the above transformations manage to make our data meet the assumptions of analysis of variance, we may resort to an analogous nonparametric method. These techniques are also called *distribution-free methods*, since they are not dependent on a given distribution (such as the normal in anova), but usually will work for a wide range of different distributions. They are called nonparametric methods because their null hypothesis is not concerned with specific parameters (such as the mean in analysis of variance) but only with the distribution of the variates. In recent years, nonparametric analysis of variance has become quite popular because it is simple to compute and permits freedom from worry about the distributional assumptions of an anova. Yet we should point out that in cases where those assumptions hold entirely or even approximately, the analysis of variance is generally the more efficient statistical test for detecting departures from the null hypothesis.

We shall discuss only nonparametric tests for two samples in this section. For a design that would give rise to a t test or anova with two classes, we employ the nonparametric *Mann-Whitney U test* (Box 10.1). The null hypothesis is that the two samples come from populations having the same distribution. The data in Box 10.1 are measurements of heart (ventricular) function in two groups of patients that have been allocated to their respective groups on the basis of other criteria of ventricular dysfunction. The Mann-Whitney U test as illustrated in Box 10.1 is a semigraphical test and is quite simple to apply. It will be especially convenient when the data are already graphed and there are not too many items in each sample.

Note that this method does not really require that each individual observation represent a precise measurement. So long as you can order the observations, you are able to perform these tests. Thus, for example, suppose you placed some meat out in the open and studied the arrival times of individuals of two species of blowflies. You could record exactly the time of arrival of each individual fly, starting from a point zero in time when the meat was set out. On the other hand, you might simply rank arrival times of the two species, noting that individual 1 of species B came first, 2 individuals from species A next, then 3 individuals of B, followed by the simultaneous arrival of one of each of the two species (a tie), and so forth. While such ranked or ordered data could not be analyzed by the parametric methods studied earlier, the techniques of Box 10.1 are entirely applicable.

The method of calculating the sample statistic U_s for the Mann-Whitney test is straightforward, as shown in Box 10.1. It is desirable to obtain an intuitive understanding of the rationale behind this test. In the Mann-Whitney test we can conceive of two extreme situations: in one case the two samples overlap and coincide entirely; in the other they are quite separate. In the latter case, if we take the sample with the lower-valued variates, there will be no points of the contrasting sample below it; that is, we can go through every observation in the lower-valued sample without having any items of the higher-valued one below

BOX 10.1

Mann-Whitney U test for two samples, ranked observations, not paired.

A measure of heart function (left ventricle ejection fraction) measured in two samples of patients admitted to the hospital under suspicion of heart attack. The patients were classified on the basis of physical examinations during admission into different so-called Killip classes of ventricular dysfunction. We compare the left ventricle ejection fraction for patients classified as Killip classes I and III. The higher Killip class signifies patients with more severe symptons. The findings were already graphed in the source publication, and step **1** illustrates that only a graph of the data is required for the Mann-Whitney U test. Designate the sample size of the larger sample as n_1 and that of the smaller sample as n_2. In this case, $n_1 = 29$, $n_2 = 8$. When the two samples are of equal size it does not matter which is designated as n_1.

1. Graph the two samples as shown below. Indicate the ties by placing dots at the same level.

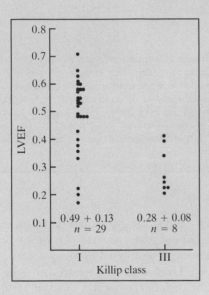

2. For each observation in one sample (it is convenient to use the smaller sample), count the number of observations in the other sample which are lower in value (below it in this graph). Count $\frac{1}{2}$ for each tied observation. For example, there are $1\frac{1}{2}$ observations in class I below the first observation in class III. The half is introduced because of the variate in class I tied with the lowest variate in class III. There are $2\frac{1}{2}$ observations below the tied second and third observations in class III. There are 3 observations below the fourth and fifth variates in class III, 4 observations below the sixth variate, and 6 and 7 observations, respectively, below the seventh and eight variates in class III. The sum of these counts $C = 29\frac{1}{2}$. The Mann-Whitney statistic U_s is the greater of the two quantities C and $(n_1 n_2 - C)$, in this case $29\frac{1}{2}$ and $[(29 \times 8) - 29\frac{1}{2}] = 202\frac{1}{2}$.

Box 10.1

Continued

Testing the significance of U_s

No tied variates in samples (or variates tied within samples only). When $n_1 \leq 20$, compare U_s with critical value for $U_{\alpha[n_1, n_2]}$ in Table **XI**. The null hypothesis is rejected if the observed value is too large.

In cases where $n_1 > 20$, calculate the following quantity

$$t_s = \frac{U_s - n_1 n_2 / 2}{\sqrt{\dfrac{n_1 n_2 (n_1 + n_2 + 1)}{12}}}$$

which is approximately normally distributed. The denominator 12 is a constant. Look up the significance of t_s in Table **III** against critical values of $t_{\alpha[\infty]}$ for a one-tailed or two-tailed test as required by the hypothesis. In our case this would yield

$$t_s = \frac{202.5 - (29)(8)/2}{\sqrt{\dfrac{(29)(8)(29 + 8 + 1)}{12}}} = \frac{86.5}{\sqrt{734.667}} = 3.191$$

A further complication arises from observations tied between the two groups. Our example is a case in point. There is no exact test. For sample sizes $n_1 < 20$, use Table **XI**, which will then be conservative. Larger sample sizes require a more elaborate formula. But it takes a substantial number of ties to affect the outcome of the test appreciably. Corrections for ties increase the t_s value slightly; hence the uncorrected formula is more conservative. We may conclude that the two samples with a t_s value of 3.191 by the uncorrected formula are significantly different at $P < 0.01$.

---

it. Conversely, all the points of the lower-valued sample would be below every point of the higher-valued one if we started out with the latter. Our total count would therefore be the total count of one sample multiplied by every observation in the second sample, which yields $n_1 n_2$. Thus, since we are told to take the greater of the two values, the sum of the counts C or $n_1 n_2 - C$, our result in this case would be $n_1 n_2$. On the other hand, if the two samples coincided completely, then for each point in one sample we would have those points below it plus a half point for the tied value representing that observation in the second sample which is at exactly the same level as the observation under consideration. A little experimentation will show this value to be $[n(n-1)/2] + (n/2) = n^2/2$. Clearly, the range of possible U values must be between this and $n_1 n_2$, and the critical value must be somewhere within this range.

Our conclusion as a result of the tests in Box 10.1 is that the two admission classes characterized by physical examination differ in their ventricular dysfunction as measured by left ventricular ejection fraction. The sample characterized as more severely ill has a lower ejection fraction than the sample characterized as less ill.

The Mann-Whitney U test is based on ranks, and it measures differences in location. A nonparametric test that tests differences between two distributions is the *Kolmogorov-Smirnov two-sample test*. Its null hypothesis is identity in distribution for the two samples, and thus the test is sensitive to differences in location, dispersion, skewness, and so forth. This test is quite simple to carry out. It is based on the unsigned differences between the relative cumulative frequency distributions of the two samples. Expected critical values can be looked up in a table or evaluated approximately. Comparison between observed and expected values leads to decisions whether the maximum difference between the two cumulative frequency distributions is significant.

Box 10.2 shows the application of the method to samples in which both n_1 and $n_2 \leq 25$. The example in this box features morphological measurements

---

BOX 10.2

Kolmogorov-Smirnov two-sample test, testing differences in distributions of two samples of continuous observations. (Both n_1 and $n_2 \leq 25$.)

Two samples of nymphs of the chigger *Trombicula lipovskyi*. Variate measured is length of cheliceral base stated as micrometer units. The sample sizes are $n_1 = 16$, $n_2 = 10$.

Sample A Y	Sample B Y
104	100
109	105
112	107
114	107
116	108
118	111
118	116
119	120
121	121
123	123
125	
126	
126	
128	
128	
128	

Source: Data by D. A. Crossley

Computational steps

1. Form cumulative frequencies F of the items in samples 1 and 2. Thus in column (2) we note that there are 3 measurements in sample A at or below 112.5 micrometer units. By contrast there are 6 such measurements in sample B (column (3)).

2. Compute relative cumulative frequencies by dividing frequencies in columns (2) and (3) by n_1 and n_2, respectively, and enter in columns (4) and (5).

Box 10.2
Continued

3. Compute d, the absolute value of the difference between the relative cumulative frequencies in columns (4) and (5), and enter in column (6).

4. Locate the largest unsigned difference D. It is 0.475.

5. Multiply D by $n_1 n_2$. We obtain $(16)(10)(0.475) = 76$.

6. Compare $n_1 n_2 D$ with its critical value in Table **XIII**, where we obtain a value of 84 for $P = 0.05$. We accept the null hypothesis that the two samples have been taken from populations with the same distribution. The Kolmogorov-Smirnov test is less powerful than the Mann-Whitney U test shown in Box 10.1 with respect to the alternative hypothesis of the latter, i.e., differences in location. However, Kolmogorov-Smirnov tests differences in both shape and location of the distributions and is thus a more comprehensive test.

(1)	(2)	(3)	(4)	(5)	(6)
Y	Sample A F_1	Sample B F_2	$\dfrac{F_1}{n_1}$	$\dfrac{F_2}{n_2}$	$d = \left\| \dfrac{F_1}{n_1} - \dfrac{F_2}{n_2} \right\|$
100		1		0.100	0.100
101	0	1	0	0.100	0.100
102	0	1	0	0.100	0.100
103	0	1	0	0.100	0.100
104	1	1	0.062	0.100	0.038
105	1	2	0.062	0.200	0.138
106	1	2	0.062	0.200	0.138
107	1	4	0.062	0.400	0.338
108	1	5	0.062	0.500	0.438
109	2	5	0.125	0.500	0.375
110	2	5	0.125	0.500	0.375
111	2	6	0.125	0.600	0.475 ← D
112	3	6	0.188	0.600	0.412
113	3	6	0.188	0.600	0.412
114	4	6	0.250	0.600	0.350
115	4	6	0.250	0.600	0.350
116	5	7	0.312	0.700	0.388
117	5	7	0.312	0.700	0.388
118	7	7	0.438	0.700	0.262
119	8	7	0.500	0.700	0.200
120	8	8	0.500	0.800	0.300
121	9	9	0.562	0.900	0.338
122	9	9	0.562	0.900	0.338
123	10	10	0.625	1.000	0.375
124	10	10	0.625	1.000	0.375
125	11	10	0.688	1.000	0.312
126	13	10	0.812	1.000	0.188
127	13	10	0.812	1.000	0.188
128	16	10	1.000	1.000	0

of two samples of chigger nymphs. We use the symbol F for cumulative frequencies, which are summed with respect to the class marks shown in column (1), and we give the cumulative frequencies of the two samples in columns (2) and (3). Relative expected frequencies are obtained in columns (4) and (5) by dividing by the respective sample sizes, while column (6) features the unsigned difference between relative cumulative frequencies. The maximum unsigned difference is $D = 0.475$. It is multiplied by $n_1 n_2$ to yield 76. The critical value for this statistic can be found in Table **XIII**, which furnishes critical values for the two-tailed two-sample Kolmogorov-Smirnov test. We obtain $n_1 n_2 D_{0.10} = 76$ and $n_1 n_2 D_{0.05} = 84$. Thus, there is a 10% probability of obtaining the observed difference by chance alone, and we conclude that the two samples do not differ significantly in their distributions.

When these data are subjected to the Mann-Whitney U test, however, one finds that the two samples are significantly different at $0.05 > P > 0.02$. This contradicts the findings of the Kolmogorov-Smirnov test in Box 10.2. But that is because the two tests differ in their sensitivities to different alternative hypotheses—the Mann-Whitney U test is sensitive to the number of interchanges in rank (shifts in location) necessary to separate the two samples, whereas the Kolmogorov-Smirnov test measures differences in the entire distributions of the two samples and is thus less sensitive to differences in location only.

It is an underlying assumption of all Kolmogorov-Smirnov tests that the variables studied are continuous. Goodness of fit tests by means of this statistic are treated in Chapter 13.

Finally, we shall present a nonparametric method for the paired-comparisons design, discussed in Section 9.3 and illustrated in Box. 9.3. The most widely used method is that of *Wilcoxon's signed-ranks test*, illustrated in Box 10.3. The example to which it is applied has not yet been encountered in this book. It records mean litter size in two strains of guinea pigs kept in large colonies during the years 1916 through 1924. Each of these values is the average of a large number of litters. Note the parallelism in the changes in the variable in the two strains. During 1917 and 1918 (war years for the United States), a shortage of caretakers and of food resulted in a decrease in the number of offspring per litter. As soon as better conditions returned, the mean litter size increased. Notice that a subsequent drop in 1922 is again mirrored in both lines, suggesting that these fluctuations are environmentally caused. It is therefore quite appropriate that the data be treated as paired comparisons, with years as replications and the strain differences as the fixed treatments to be tested.

Column (3) in Box 10.3 lists the differences on which a conventional paired-comparisons t test could be performed. For Wilcoxon's test these differences are ranked *without regard to sign* in column (4), so that the smallest absolute difference is ranked 1 and the largest absolute difference (of the nine differences) is ranked 9. Tied ranks are computed as averages of the ranks; thus if the fourth and fifth difference have the same absolute magnitude they will both be assigned rank 4.5. After the ranks have been computed, the original sign of each difference

BOX 10.3

Wilcoxon's signed-ranks test for two groups, arranged as paired observations.

Mean litter size of two strains of guinea pigs, compared over $n = 9$ years.

Year	(1) Strain B	(2) Strain 13	(3) D	(4) Rank (R)
1916	2.68	2.36	+0.32	+9
1917	2.60	2.41	+0.19	+8
1918	2.43	2.39	+0.04	+2
1919	2.90	2.85	+0.05	+3
1920	2.94	2.82	+0.12	+7
1921	2.70	2.73	−0.03	−1
1922	2.68	2.58	+0.10	+6
1923	2.98	2.89	+0.09	+5
1924	2.85	2.78	+0.07	+4

Absolute sum of negative ranks	1
Sum of positive ranks	44

Source: Data by S. Wright.

Procedure

1. Compute the differences between the n pairs of observations. These are entered in column (3), labeled D.

2. Rank these differences from the smallest to the largest *without regard to sign.*

3. Assign to the ranks the original signs of the differences.

4. Sum the positive and negative ranks separately. The sum that is smaller in absolute value, T_s, is compared with the values in Table **XII** for $n = 9$.

Since $T_s = 1$, which is equal to or less than the entry for one-tailed $\alpha = 0.005$ in the table, our observed difference is significant at the 1% level. Litter size in strain B is significantly different from that of strain 13.

For large samples ($n > 50$) compute

$$t_s = \frac{T_s - \dfrac{n(n+1)}{4}}{\sqrt{\dfrac{n(n+\frac{1}{2})(n+1)}{12}}}$$

where T_s is as defined in step **4** above. Compare the computed value with $t_{\alpha[\infty]}$ in Table **III**.

is assigned to the corresponding rank. The sum of the positive or of the negative ranks, whichever one is smaller in absolute value, is then computed (it is labeled T_s) and is compared with the critical value T in Table **XII** for the corresponding sample size. In view of the significance of the rank sum, it is clear that strain B has a litter size different from that of strain 13.

This is a very simple test to carry out, but it is, of course, not as efficient as the corresponding parametric t test, which should be preferred if the necessary assumptions hold. Note that one needs minimally six differences in order to carry out Wilcoxon's signed-ranks test. With only six paired comparisons, all differences must be of like sign for the test to be significant at the 5% level.

For a large sample an approximation using the normal curve is available, which is given in Box 10.3. Note that the absolute magnitudes of the differences play a role only insofar as they affect the ranks of the differences.

A still simpler test is the *sign test*, in which we count the number of positive and negative signs among the differences (omitting all differences of zero). We then test the hypothesis that the n plus and minus signs are sampled from a population in which the two kinds of signs are present in equal proportions, as might be expected if there were no true difference between the two paired samples. Such sampling should follow the binomial distribution, and the test of the hypothesis that the parametric frequency of the plus signs is $\hat{p} = 0.5$ can be made in a number of ways. Let us learn these by applying the sign test to the guinea pig data of Box 10.3. There are nine differences, of which eight are positive and one is negative. We could follow the methods of Section 4.2 (illustrated in Table 4.3) in which we calculate the expected probability of sampling one minus sign in a sample of nine on the assumption of $\hat{p} = \hat{q} = 0.5$. The probability of such an occurrence and all "worse" outcomes equals 0.0195. Since we have no a priori notions that one strain should have a greater litter size than the other, this is a two-tailed test, and we double the probability to 0.0390. Clearly, this is an improbable outcome, and we reject the null hypothesis that $\hat{p} = \hat{q} = 0.5$.

Since the computation of the exact probabilities may be quite tedious if no table of cumulative binomial probabilities is at hand, we may take a second approach, using Table **IX**, which furnishes confidence limits for $\hat{p}$ for various sample sizes and sampling outcomes. Looking up sample size 9 and $Y = 1$ (number showing the property), we find the 95% confidence limits to be 0.0028 and 0.4751 by interpolation, thus excluding the value $\hat{p} = \hat{q} = 0.5$ postulated by the null hypothesis. At least at the 5% significance level we can conclude that it is unlikely that the number of plus and minus signs is equal. The confidence limits imply a two-tailed distribution; if we intend a one-tailed test, we can infer a 0.025 significance level from the 95% confidence limits and a 0.005 level from the 99% limits. Obviously, such a one-tailed test would be carried out only if the results were in the direction of the alternative hypothesis. Thus, if the alternative hypothesis were that strain 13 in Box 10.3 had greater litter size than strain B, we would not bother testing this example at all, since the observed proportion of years showing this relation is less than half. For larger samples, we can use the normal approximation to the binomial distribution as follows: $t_s = (Y - \mu)/\sigma_Y = (Y - kp)/\sqrt{kpq}$, where we substitute the mean and standard deviation of the binomial distribution learned in Section 4.2. In our case, we let n stand for k and assume that $\hat{p} = \hat{q} = 0.5$. Therefore, $t_s = (Y - \frac{1}{2}n)/\sqrt{\frac{1}{4}n} = (Y - \frac{1}{2}n)/\frac{1}{2}\sqrt{n}$. The value of t_s is then compared with $t_{\alpha[\infty]}$ in Table **III**, using one tail or two tails of the distribution as warranted. When the sample size $n \geq 12$, this is a satisfactory approximation.

A third approach we can use is to test the departure from the expectation that $\hat{p} = \hat{q} = 0.5$ by one of the methods of Chapter 13.

Exercises

10.1 Allee and Bowen (1932) studied survival time of goldfish (in minutes) when placed in colloidal silver suspensions. Experiment no. 9 involved 5 replications, and experiment no. 10 involved 10 replicates. Do the results of the two experiments differ? Addition of urea, NaCl, and Na_2S to a third series of suspensions apparently prolonged the life of the fish.

	Colloidal silver		*Urea and salts added*
Experiment no. 9	*Experiment no. 10*	
210	150	330
180	180	300
240	210	300
210	240	420
210	240	360
	120	270
	180	360
	240	360
	120	300
	150	120

Analyze and interpret. Test equality of variances. Compare anova results with those obtained using the Mann-Whitney U test for the two comparisons under study. To test the effect of urea it might be best to pool Experiments 9 and 10, if they prove not to differ significantly. ANS. Test for homogeneity of Experiments 9 and 10, $U_s = 33$, ns. For the comparison of Experiments 9 and 10 versus urea and salts, $U_s = 136$, $P < 0.001$.

10.2 In a study of flower color in Butterflyweed (*Asclepias tuberosa*), Woodson (1964) obtained the following results:

Geographic region	$\bar{Y}$	n	s
C1	29.3	226	4.59
SW2	15.8	94	10.15
SW3	6.3	23	1.22

The variable recorded was a color score (ranging from 1 for pure yellow to 40 for deep orange-red) obtained by matching flower petals to sample colors in Maerz and Paul's *Dictionary of Color*. Test whether the samples are homoscedastic.

10.3 Test for a difference in surface and subsoil pH in the data of Exercise 9.1, using Wilcoxon's signed-ranks test. ANS. $T_s = 38$; $P > 0.10$.

10.4 Number of bacteria in 1 cc of milk from three cows counted at three periods (data from Park, Williams, and Krumwiede, 1924):

Cow no.	At time of milking	After 24 hours	After 48 hours
1	12,000	14,000	57,000
2	13,000	20,000	65,000
3	21,500	31,000	106,000

(a) Calculate means and variances for the three periods and examine the relation between these two statistics. Transform the variates to logarithms and compare means and variances based on the transformed data. Discuss.

(b) Carry out an anova on transformed and untransformed data. Discuss your results.

10.5 Analyze the measurements of the two samples of chigger nymphs in Box 10.2 by the Mann-Whitney U test. Compare the results with those shown in Box 10.2 for the Kolmogorov-Smirnov test. ANS. $U_s = 123.5$, $P < 0.05$.

10.6 Allee et al. (1934) studied the rate of growth of *Ameiurus melas* in conditioned and unconditioned well water and obtained the following results for the gain in average length of a sample fish. Although the original variates are not available, we may still test for differences between the two treatment classes. Use the sign test to test for differences in the paired replicates.

	Average gain in length (in millimeters)	
Replicate	Conditioned water	Unconditioned water
1	2.20	1.06
2	1.05	0.06
3	3.25	3.55
4	2.60	1.00
5	1.90	1.10
6	1.50	0.60
7	2.25	1.30
8	1.00	0.90
9	−0.09	−0.59
10	0.83	0.58

CHAPTER 11

Regression

We now turn to the simultaneous analysis of two variables. Even though we may have considered more than one variable at a time in our studies so far (for example, seawater concentration and oxygen consumption in Box 9.1, or age of girls and their face widths in Box 9.3), our actual analyses were of only one variable. However, we frequently measure two or more variables on each individual, and we consequently would like to be able to express more precisely the nature of the relationships between these variables. This brings us to the subjects of *regression* and *correlation*. In regression we estimate the relationship of one variable with another by expressing the one in terms of a linear (or a more complex) function of the other. We also use regression to predict values of one variable in terms of the other. In correlation analysis, which is sometimes confused with regression, we estimate the degree to which two variables vary together. Chapter 12 deals with correlation, and we shall postpone our effort to clarify the relation and distinction between regression and correlation until then. The variables involved in regression and correlation are either continuous or meristic; if meristic, they are treated as though they were continuous. When variables are qualitative (that is, when they are attributes), the methods of regression and correlation cannot be used.

In Section 11.1 we review the notion of mathematical functions and introduce the new terminology required for regression analysis. This is followed in Section 11.2 by a discussion of the appropriate statistical models for regression analysis. The basic computations in simple linear regression are shown in Section 11.3 for the case of one dependent variate for each independent variate. The case with several dependent variates for each independent variate is treated in Section 11.4. Tests of significance and computation of confidence intervals for regression problems are discussed in Section 11.5.

Section 11.6 serves as a summary of regression and discusses the various uses of regression analysis in biology. How transformation of scale can straighten out curvilinear relationships for ease of analysis is shown in Section 11.7. When transformation cannot linearize the relation between variables, an alternative approach is by a nonparametric test for regression. Such a test is illustrated in Section 11.8.

11.1 Introduction to regression

Much scientific thought concerns the relations between pairs of variables hypothesized to be in a cause-and-effect relationship. We shall be content with establishing the form and significance of *functional relationships* between two variables, leaving the demonstration of cause-and-effect relationships to the established procedures of the scientific method. A *function* is a mathematical relationship enabling us to predict what values of a variable Y correspond to given values of a variable X. Such a relationship, generally written as $Y = f(X)$, is familiar to all of us.

A typical linear regression is of the form shown in Figure 11.1, which illustrates the effect of two drugs on the blood pressure of two species of

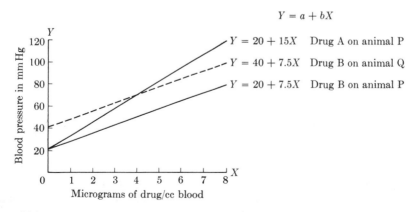

FIGURE 11.1
Blood pressure of an animal in mmHg as a function of drug concentration in μg per cc of blood.

animals. The relationships depicted in this graph can be expressed by the formula $Y = a + bX$. Clearly, Y is a function of X. We call the variable Y the *dependent variable*, while X is called the *independent variable*. The magnitude of blood pressure Y depends on the amount of the drug X and can therefore be predicted from the independent variable, which presumably is free to vary. Although a cause would always be considered an independent variable and an effect a dependent variable, a functional relationship observed in nature may actually be something other than a cause-and-effect relationship. The highest line is of the relationship $Y = 20 + 15X$, which represents the effect of drug A on animal P. The quantity of drug is measured in micrograms, the blood pressure in millimeters of mercury. Thus, after 4 μg of the drug have been given, the blood pressure would be $Y = 20 + (15)(4) = 80$ mmHg. The independent variable X is multiplied by a coefficient b, the slope factor. In the example chosen, $b = 15$; that is, for an increase of one microgram of the drug, the blood pressure is raised by 15 mm.

In biology, such a relationship can clearly be appropriate over only a limited range of values of X. Negative values of X are meaningless in this case; it is also unlikely that the blood pressure will continue to increase at a uniform rate. Quite probably the slope of the functional relationship will flatten out as the drug level rises. But, for a limited portion of the range of variable X (micrograms of the drug), the linear relationship $Y = a + bX$ may be an adequate description of the functional dependence of Y on X.

By this formula, when the independent variable equals zero, the dependent variable equals a. This point is the interesection of the function line with the Y axis. It is called the Y *intercept*. In Figure 11.1, when $X = 0$, the function just studied will yield a blood pressure of 20 mmHg, which is the normal blood pressure of animal P in the absence of the drug.

The two other functions in Figure 11.1 show the effects of varying both a, the Y intercept, and b, the slope. In the lowest line, $Y = 20 + 7.5X$, the Y intercept remains the same but the slope has been halved. We visualize this as the effect of a different drug, B, on the same organism P. Obviously, when no drug is administered, the blood pressure should be at the same Y intercept, since the identical organism is being studied. However, a different drug is likely to exert a different hypertensive effect, as reflected by the different slope. The third relationship also describes the effect of drug B, which is assumed to remain the same, but the experiment is carried out on a different species, Q, whose normal blood pressure is assumed to be 40 mmHg. Thus, the equation for the effect of drug B on species Q is written as $Y = 40 + 7.5X$. This line is parallel to that corresponding to the second equation.

From your knowledge of analytical geometry you will have recognized the slope factor b as the *slope* of the function $Y = a + bX$, generally symbolized by m. In calculus, b is the *derivative* of that same function ($dY/dX = b$). In biostatistics, b is called the *regression coefficient*, and the function is called a *regression equation*. When we wish to stress that the regression coefficient is of variable Y on variable X, we write $b_{Y \cdot X}$.

11.2 Models in regression

In any real example, observations would not lie perfectly along a regression line but would scatter along both sides of the line. This scatter is usually due to inherent, natural variation of the items (genetically and environmentally caused) and also due to measurement error. Thus, in regression a functional relationship does not mean that given an X the value of Y must be $a + bX$, but rather that the mean (or expected value) of Y is $a + bX$.

The appropriate computations and significance tests in regression relate to the following two models. The more common of these, *Model I regression*, is especially suitable in experimental situations. It is based on four assumptions.

1. The independent variable X is measured without error. We therefore say that the X's are "fixed." We mean by this that whereas Y, the dependent variable, is a random variable, X does not vary at random but is under the control of the investigator. Thus, in the example of Figure 11.1 we have varied dose of drug at will and studied the response of the random variable blood presssure. We can manipulate X in the same way that we were able to manipulate the treatment effect in a Model I anova. As a matter of fact, as you shall see later, there is a very close relationship between Model I anova and Model I regression.
2. The expected value for the variable Y for any given value. X is described by the linear function $\mu_Y = \alpha + \beta X$. This is the same relation we have just encountered, but we use Greek letters instead of a and b, since we are describing a parametric relationship. Another way of stating this assumption is that the parametric means μ_Y of the values of Y are a function of X and lie on a straight line described by this equation.
3. For any given value X_i of X, the Y's are independently and normally distributed. This can be represented by the equation $Y_i = \alpha + \beta X_i + \epsilon_i$, where the ϵ_i's are assumed to be normally distributed error terms with a mean of zero. Figure 11.2 illustrates this concept with a regression line similar to the ones in Figure 11.1. A given experiment can be repeated several times. Thus, for instance, we could administer 2, 4, 6, 8, and 10 μg of the drug to each of 20 individuals of an animal species and obtain a

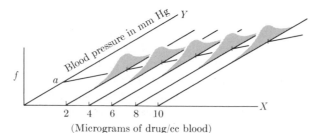

(Micrograms of drug/cc blood)

FIGURE 11.2
Blood pressure of an animal in mmHg as a function of drug concentration in μg per cc of blood. Repeated sampling for a given drug concentration.

frequency distribution of blood pressure responses Y to the independent variates $X = 2, 4, 6, 8$, and 10 μg. In view of the inherent variability of biological material, the responses to each dosage would not be the same in every individual; you would obtain a frequency distribution of values of Y (blood pressure) around the expected value. Assumption 3 states that these sample values would be independently and normally distributed. This is indicated by the normal curves which are superimposed about several points in the regression line in Figure 11.2. A few are shown to give you an idea of the scatter about the regression line. In actuality there is, of course, a continuous scatter, as though these separate normal distributions were stacked right next to each other, there being, after all, an infinity of possible intermediate values of X between any two dosages. In those rare cases in which the independent variable is discontinuous, the distributions of Y would be physically separate from each other and would occur only along those points of the abscissa corresponding to independent variates. An example of such a case would be weight of offspring (Y) as a function of number of offspring (X) in litters of mice. There may be three or four offspring per litter but there would be no intermediate value of X representing 3.25 mice per litter.

Not every experiment will have more than one reading of Y for each value of X. In fact, the basic computations we shall learn in the next section are for only one value of Y per value of X, this being the more common case. However, you should realize that even in such instances the basic assumption of Model I regression is that the single variate of Y corresponding to the given value of X is a sample from a population of independently and normally distributed variates.

4. The final assumption is a familiar one. We assume that these samples along the regression line are homoscedastic; that is, that they have a common variance σ^2, which is the variance of the ϵ's in the expression in item 3. Thus, we assume that the variance around the regression line is constant and independent of the magnitude of X or Y.

Many regression analyses in biology do not meet the assumptions of Model I regression. Frequently both X and Y are subject to natural variation and/or measurement error. Also, the variable X is sometimes not fixed, that is, under control of the investigator. Suppose we sample a population of female flies and measure wing length and total weight of each individual. We might be interested in studying wing length as a function of weight or we might wish to predict wing length for a given weight. In this case the weight, which we treat as an independent variable, is not fixed and certainly not the "cause" of differences in wing length. The weights of the flies will vary for genetic and environmental reasons and will also be subject to measurement error. The general case where both variables show random variation is called *Model II regression*. Although, as will be discussed in the next chapter, cases of this sort are much better analyzed by the methods of correlation analysis, we sometimes wish to describe the functional relationship between such variables. To do so, we need to resort to the special techniques of Model II regression. In this book we shall limit ourselves to a treatment of Model I regression.

11.3 The linear regression equation

To learn the basic computations necessary to carry out a Model I linear regression, we shall choose an example with only one Y value per independent variate X, since this is computationally simpler. The extension to a sample of values of Y for each X is shown in Section 11.4. Just as in the case of the previous analyses, there are also simple computational formulas, which will be presented at the end of this section.

The data on which we shall learn regression come from a study of water loss in *Tribolium confusum*, the confused flour beetle. Nine batches of 25 beetles were weighed (individual beetles could not be weighed with available equipment), kept at different relative humidities, and weighed again after six days of starvation. Weight loss in milligrams was computed for each batch. This is clearly a Model I regression, in which the weight loss is the dependent variable Y and the relative humidity is the independent variable X, a fixed treatment effect under the control of the experimenter. The purpose of the analysis is to establish whether the relationship between relative humidity and weight loss can be adequately described by a linear regression of the general form $Y = a + bX$.

The original data are shown in columns (1) and (2) of Table 11.1. They are plotted in Figure 11.3, from which it appears that a negative relationship exists between weight loss and humidity; as the humidity increases, the weight loss decreases. The means of weight loss and relative humidity, $\bar{Y}$ and $\bar{X}$, respectively, are marked along the coordinate axes. The average humidity is 50.39%, and the average weight loss is 6.022 mg. How can we fit a regression line to these data, permitting us to estimate a value of Y for a given value of X? Unless the actual observations lie exactly on a straight line, we will need a criterion for determining the best possible placing of the regression line. Statisticians have generally followed the principle of least squares, which we first encountered in Chapter 3 when learning about the arithmetic mean and the variance. If we were to draw a horizontal line through $\bar{X}$, $\bar{Y}$ (that is, a line parallel to the X axis at the level of $\bar{Y}$), then deviations to that line drawn parallel to the Y axis would represent the deviations from the mean for these observations with respect to variable Y (see Figure 11.4). We learned in Chapter 3 that the sum of these observations $\Sigma (Y - \bar{Y}) = \Sigma y = 0$. The sum of squares of these deviations, $\Sigma (Y - \bar{Y})^2 = \Sigma y^2$, is less than that from any other horizontal line. Another way of saying this is that the arithmetic mean of Y represents the least squares horizontal line. Any horizontal line drawn through the data at a point other than $\bar{Y}$ would yield a sum of deviations other than zero and a sum of deviations squared greater than Σy^2. Therefore, a mathematically correct but impractical

TABLE 11.1
Basic computations in regression. Weight loss (in mg) of nine batches of 25 *Tribolium* beetles after six days of starvation at nine different humidities.

(1) Percent relative humidity X	(2) Weight loss (in mg) Y	(3) $x = (X - \bar{X})$	(4) $y = (Y - \bar{Y})$	(5) x^2	(6) xy	(7) y^2	(8) $\hat{Y}$	(9) $d_{Y \cdot x} = Y - \hat{Y}$	(10) $d_{Y \cdot x}^2$	(11) $\hat{y} = \hat{Y} - \bar{Y}$	(12) $\hat{y}^2$
0	8.98	−50.39	2.958	2539.1521	−149.0536	8.7498	8.7038	0.2762	0.0763	2.6818	7.1921
12	8.14	−38.39	2.118	1473.7921	−81.3100	4.4859	8.0652	0.0748	0.0056	2.0432	4.1747
29.5	6.67	−20.89	0.648	436.3921	−13.5367	0.4199	7.1338	−0.4638	0.2151	1.1118	1.2361
43	6.08	−7.39	0.058	54.6121	−0.4286	0.0034	6.4153	−0.3353	0.1124	0.3933	0.1547
53	5.90	2.61	−0.122	6.8121	−0.3184	0.0149	5.8831	0.0169	0.0003	−0.1389	0.0193
62.5	5.83	12.11	−0.192	146.6521	−2.3251	0.0369	5.3776	0.4524	0.2047	−0.6444	0.4153
75.5	4.68	25.11	−1.342	630.5121	−33.6976	1.8010	4.6857	−0.0057	0.0000	−1.3363	1.7857
85	4.20	34.61	−1.822	1197.8521	−63.0594	3.3197	4.1801	0.0199	0.0004	−1.8419	3.3926
93	3.72	42.61	−2.302	1815.6121	−98.0882	5.2992	3.7543	−0.0343	0.0012	−2.2677	5.1425
Sum 453.5	54.20	−0.01	0.002	8301.3889	−441.8176	24.1307	54.1989	0.0011	0.6160	0.0009	23.5130
Mean 50.39	6.022						6.022				
Sum/(n − 1)				1037.6736	−55.2272	3.0163			0.0880[a]		

Source: Nelson (1964).

[a] Sum divided by $n - 2$.

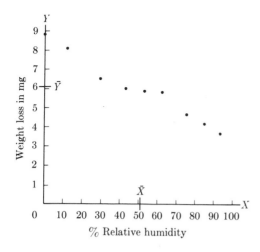

FIGURE 11.3
Weight loss (in mg) of nine batches of 25 *Tribolium* beetles after six days of starvation at nine different relative humidities. Data from Table 11.1, after Nelson (1964).

method for finding the mean of *Y* would be to draw a series of horizontal lines across a graph, calculate the sum of squares of deviations from it, and choose that line yielding the smallest sum of squares.

In linear regression, we still draw a straight line through our observations, but it is no longer necessarily horizontal. A sloped regression line will indicate for each value of the independent variable X_i an estimated value of the dependent variable. We should distinguish the estimated value of Y_i, which we shall hereafter designate as $\hat{Y}_i$, (read: Y-hat or Y-caret), and the observed values, conventionally designated as Y_i. The regression equation therefore should read

$$\hat{Y} = a + bX \tag{11.1}$$

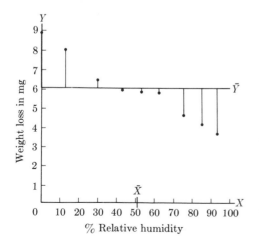

FIGURE 11.4
Deviations from the mean (of *Y*) for the data of Figure 11.3.

which indicates that for given values of X, this equation calculates estimated values $\hat{Y}$ (as distinct from the observed values Y in any actual case). The deviation of an observation Y_i from the regression line is $(Y_i - \hat{Y}_i)$ and is generally symbolized as $d_{Y \cdot X}$. These deviations can still be drawn parallel to the Y axis, but they meet the sloped regression line at an angle (see Figure 11.5). The sum of these deviations is again zero ($\Sigma d_{Y \cdot X} = 0$), and the sum of their squares yields a quantity $\Sigma (Y - \hat{Y})^2 = \Sigma d_{Y \cdot X}^2$ analogous to the sum of squares Σy^2. For reasons that will become clear later, $\Sigma d_{Y \cdot X}^2$ is called the unexplained sum of squares. The least squares *linear regression line* through a set of points is defined as that straight line which results in the smallest value of $\Sigma d_{Y \cdot X}^2$. Geometrically, the basic idea is that one would prefer using a line that is in some sense close to as many points as possible. For purposes of ordinary Model I regression analysis, it is most useful to define closeness in terms of the vertical distances from the points to a line, and to use the line that makes the sum of the squares of these deviations as small as possible. A convenient consequence of this criterion is that the line must pass through the point $\bar{X}, \bar{Y}$. Again, it would be possible but impractical to calculate the correct regression slope by pivoting a ruler around the point $\bar{X}, \bar{Y}$ and calculating the unexplained sum of squares $\Sigma d_{Y \cdot X}^2$ for each of the innumerable possible positions. Whichever position gave the smallest value of $\Sigma d_{Y \cdot X}^2$ would be the least squares regression line.

The formula for the slope of a line based on the minimum value of $\Sigma d_{Y \cdot X}^2$ is obtained by means of the calculus. It is

$$b_{Y \cdot X} = \frac{\Sigma xy}{\Sigma x^2} \tag{11.2}$$

Let us calculate $b = \Sigma xy / \Sigma x^2$ for our weight loss data.

We first compute the deviations from the respective means of X and Y, as shown in columns (3) and (4) of Table 11.1. The sums of these deviations,

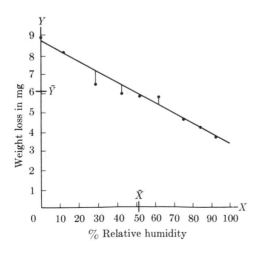

FIGURE 11.5
Deviations from the regression line for the data of Figure 11.3.

Σx and Σy, are slightly different from their expected value of zero because of rounding errors. The squares of these deviations yield sums of squares and variances in columns (5) and (7). In column (6) we have computed the products xy, which in this example are all negative because the deviations are of unlike sign. An increase in humidity results in a decrease in weight loss. The sum of these products $\Sigma^n xy$ is a new quantity, called the *sum of products*. This is a poor but well-established term, referring to Σxy, the sum of the products of the deviations rather than ΣXY, the sum of the products of the variates. You will recall that Σy^2 is called the sum of squares, while ΣY^2 is the sum of the squared variates. The sum of products is analogous to the sum of squares. When divided by the degrees of freedom, it yields the *covariance*, by analogy with the variance resulting from a similar division of the sum of squares. You may recall first having encountered covariances in Section 7.4. Note that the sum of products can be negative as well as positive. If it is negative, this indicates a negative slope of the regression line: as X increases, Y decreases. In this respect it differs from a sum of squares, which can only be positive. From Table 11.1 we find that $\Sigma xy = -441.8176$, $\Sigma x^2 = 8301.3889$, and $b = \Sigma xy/\Sigma x^2 = -0.053,22$. Thus, for a one-unit increase in X, there is a decrease of 0.053,22 units of Y. Relating it to our actual example, we can say that for a 1% increase in relative humidity, there is a reduction of 0.053,22 mg in weight loss.

You may wish to convince yourself that the formula for the regression coefficient is intuitively reasonable. It is the ratio of the sum of *products* of deviations for X and Y to the sum of *squares* of deviations for X. If we look at the product for X_i, a single value of X, we obtain x_iy_i. Similarly, the squared deviation for X_i would be x_i^2, or x_ix_i. Thus the ratio x_iy_i/x_ix_i reduces to y_i/x_i. Although $\Sigma xy/\Sigma x^2$ only approximates the average of y_i/x_i for the n values of X_i, the latter ratio indicates the direction and magnitude of the change in Y for a unit change in X. Thus, if y_i on the average equals x_i, b will equal 1. When $y_i = -x_i$, $b = -1$. Also, when $|y_i| > |x_i|$, $b > |1|$; and conversely, when $|y_i| < |x_i|$, $b < |1|$.

How can we complete the equation $Y = a + bX$? We have stated that the regression line will go through the point $\bar{X}$, $\bar{Y}$. At $\bar{X} = 50.39$, $\hat{\bar{Y}} = 6.022$; that is, we use $\bar{Y}$, the observed mean of Y, as an estimate $\hat{\bar{Y}}$ of the mean. We can substitute these means into Expression (11.1):

$$\hat{Y} = a + bX$$

$$\bar{Y} = a + b\bar{X}$$

$$a = \bar{Y} - b\bar{X}$$

$$a = 6.022 - (-0.053,22)50.39$$

$$= 8.7038$$

Therefore,

$$\hat{Y} = 8.7038 - 0.053,22X$$

This is the equation that relates weight loss to relative humidity. Note that when X is zero (humidity zero), the estimated weight loss is greatest. It is then equal to $a = 8.7038$ mg. But as X increases to a maximum of 100, the weight loss decreases to 3.3818 mg.

We can use the regression formula to draw the regression line: simply estimate $\hat{Y}$ at two convenient points of X, such as $X = 0$ and $X = 100$, and draw a straight line between them. This line has been added to the observed data and is shown in Figure 11.6. Note that it goes through the point $\bar{X}, \bar{Y}$. In fact, for drawing the regression line, we frequently use the intersection of the two means and one other point.

Since

$$a = \bar{Y} - b\bar{X}$$

we can write Expression (11.1), $\hat{Y} = a + bX$, as

$$\hat{Y} = (\bar{Y} - b\bar{X}) + bX$$
$$= \bar{Y} + b(X - \bar{X})$$

Therefore,

$$\hat{Y} = \bar{Y} + bx$$

Also,

$$\hat{Y} - \bar{Y} = bx$$

$$\hat{y} = bx \tag{11.3}$$

where $\hat{y}$ is defined as the deviation $\hat{Y} - \bar{Y}$. Next, using Expression (11.1), we estimate $\hat{Y}$ for every one of our given values of X. The estimated values $\hat{Y}$ are shown in column (8) of Table 11.1. Compare them with the observed values

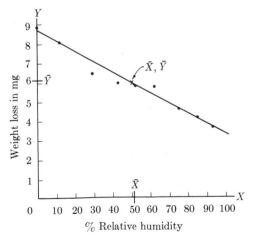

FIGURE 11.6
Linear regression fitted to data of Figure 11.3.

of Y in column (2). Overall agreement between the two columns of values is good. Note that except for rounding errors, $\Sigma \hat{Y} = \Sigma Y$ and hence $\bar{\hat{Y}} = \bar{Y}$. However, our actual Y values usually are different from the estimated values $\hat{Y}$. This is due to individual variation around the regression line. Yet, the regression line is a better base from which to compute deviations than the arithmetic average $\bar{Y}$, since the value of X has been taken into account in constructing it.

When we compute deviations of each observed Y value from its estimated value $(Y - \hat{Y}) = d_{Y \cdot X}$ and list these in column (9), we notice that these deviations exhibit one of the properties of deviations from a mean: they sum to zero except for rounding errors. Thus $\Sigma d_{Y \cdot X} = 0$, just as $\Sigma y = 0$. Next, we compute in column (10) the squares of these deviations and sum them to give a new sum of squares, $\Sigma d_{Y \cdot X}^2 = 0.6160$. When we compare $\Sigma (Y - \bar{Y})^2 = \Sigma y^2 = 24.1307$ with $\Sigma (Y - \hat{Y})^2 = \Sigma d_{Y \cdot X}^2 = 0.6160$, we note that the new sum of squares is much less than the previous old one. What has caused this reduction? Allowing for different magnitudes of X has eliminated most of the variance of Y from the sample. Remaining is the *unexplained sum of squares* $\Sigma d_{Y \cdot X}^2$, which expresses that portion of the total SS of Y that is not accounted for by differences in X. It is unexplained with respect to X. The difference between the total SS, Σy^2, and the unexplained SS, $\Sigma d_{Y \cdot X}^2$, is not surprisingly called the *explained sum of squares*, $\Sigma \hat{y}^2$, and is based on the deviations $\hat{y} = \hat{Y} - \bar{Y}$. The computation of these deviations and their squares is shown in columns (11) and (12). Note that $\Sigma \hat{y}$ approximates zero and that $\Sigma \hat{y}^2 = 23.5130$. Add the unexplained SS (0.6160) to this and you obtain $\Sigma y^2 = \Sigma \hat{y}^2 + \Sigma d_{Y \cdot X}^2 = 24.1290$, which is equal (except for rounding errors) to the independently calculated value of 24.1307 in column (7). We shall return to the meaning of the unexplained and explained sums of squares in later sections.

We conclude this section with a discussion of calculator formulas for computing the regression equation in cases where there is a single value of Y for each value of X. The regression coefficient $\Sigma xy / \Sigma x^2$ can be rewritten as

$$b_{Y \cdot X} = \frac{\sum_{}^{n}(X - \bar{X})(Y - \bar{Y})}{\sum_{}^{n}(X - \bar{X})^2} \qquad (11.4)$$

The denominator of this expression is the sum of squares of X. Its computational formula, as first encountered in Section 3.9, is $\Sigma x^2 = \Sigma X^2 - (\Sigma X)^2/n$. We shall now learn an analogous formula for the numerator of Expression (11.4), the sum of products. The customary formula is

$$\sum_{}^{n} xy = \sum_{}^{n} XY - \frac{\left(\sum_{}^{n} X\right)\left(\sum_{}^{n} Y\right)}{n} \qquad (11.5)$$

The quantity ΣXY is simply the accumulated product of the two variables. Expression (11.5) is derived in Appendix A1.5. The actual computations for a regression equation (single value of Y per value of X) are illustrated in Box 11.1, employing the weight loss data of Table 11.1.

To compute regression statistics, we need six quantities initially. These are n, ΣX, ΣX^2, ΣY, ΣY^2, and ΣXY. From these the regression equation is calculated as shown in Box 11.1, which also illustrates how to compute the explained

---

BOX 11.1
Computation of regression statistics. Single value of Y for each value of X.

Data from Table 11.1.

Weight loss in mg (Y)	8.98	8.14	6.67	6.08	5.90	5.83	4.68	4.20	3.72
Percent relative humidity (X)	0	12.0	29.5	43.0	53.0	62.5	75.5	85.0	93.0

Basic computations

1. Compute sample size, sums, sums of the squared observations, and the sum of the XY's.

$$n = 9 \qquad \sum X = 453.5 \qquad \sum Y = 54.20$$

$$\sum X^2 = 31,152.75 \qquad \sum Y^2 = 350.5350 \qquad \sum XY = 2289.260$$

2. The means, sums of squares, and sum of products are

$$\bar{X} = 50.389 \qquad \bar{Y} = 6.022$$

$$\sum x^2 = 8301.3889 \qquad \sum y^2 = 24.1306$$

$$\sum xy = \sum XY - \frac{(\sum X)(\sum Y)}{n}$$

$$= 2289.260 - \frac{(453.5)(54.20)}{9} = -441.8178$$

3. The regression coefficient is

$$b_{Y \cdot X} = \frac{\sum xy}{\sum x^2} = \frac{-441.8178}{8301.3889} = -0.053,22$$

4. The Y intercept is

$$a = \bar{Y} - b_{Y \cdot X}\bar{X} = 6.022 - (-0.053,22)(50.389) = 8.7037 \checkmark$$

5. The explained sum of squares is

$$\sum \hat{y}^2 = \frac{(\sum xy)^2}{\sum x^2} = \frac{(-441.8178)^2}{8301.3889} = 23.5145$$

6. The unexplained sum of squares is

$$\sum d_{Y \cdot X}^2 = \sum y^2 - \sum \hat{y}^2 = 24.1306 - 23.5145 = 0.6161$$

sum of squares $\Sigma \hat{y}^2 = \Sigma (\hat{Y} - \bar{Y})^2$ and the unexplained sum of squares $\Sigma d^2_{Y \cdot X} = \Sigma (Y - \hat{Y})^2$. That

$$\Sigma d^2_{Y \cdot X} = \Sigma y^2 - \frac{(\Sigma xy)^2}{\Sigma x^2} \tag{11.6}$$

is demonstrated in Appendix A1.6. The term subtracted from Σy^2 is obviously the explained sum of squares, as shown in Expression (11.7) below:

$$\Sigma \hat{y}^2 = \Sigma b^2 x^2 = b^2 \Sigma x^2 = \frac{(\Sigma xy)^2}{(\Sigma x^2)^2} \Sigma x^2 \tag{11.7}$$

$$\Sigma \hat{y}^2 = \frac{(\Sigma xy)^2}{\Sigma x^2}$$

11.4 More than one value of Y for each value of X

We now take up Model I regression as originally defined in Section 11.2 and illustrated by Figure 11.2. For each value of the treatment X we sample Y repeatedly, obtaining a sample distribution of Y values at each of the chosen points of X. We have selected an experiment from the laboratory of one of us (Sokal) in which *Tribolium* beetles were reared from eggs to adulthood at four different densities. The percentage survival to adulthood was calculated for varying numbers of replicates at these densities. Following Section 10.2, these percentages were given arcsine transformations, which are listed in Box 11.2. These transformed values are more likely to be normal and homoscedastic than are percentages. The arrangement of these data is very much like that of a single-classification model I anova. There are four different densities and several survival values at each density. We now would like to determine whether there are differences in survival among the four groups, and also whether we can establish a regression of survival on density.

A first approach, therefore, is to carry out an analysis of variance, using the methods of Section 8.3 and Table 8.1. Our aim in doing this is illustrated in Figure 11.7 (see page 247). If the analysis of variance were not significant, this would indicate, as shown in Figure 11.7A, that the means are not significantly different from each other, and it would be unlikely that a regression line fitted to these data would have a slope significantly different from zero. However, although both the analysis of variance and linear regression test the same null hypothesis—equality of means—the regression test is more powerful (less type II error; see Section 6.8) against the alternative hypothesis that there is a linear relationship between the group means and the independent variable X. Thus, when the means increase or decrease slightly as X increases it may be that they are not different enough for the mean square among groups to be significant by anova but that a significant regression can still be found. When we find a marked regression of the means on X, as shown in Figure 11.7B, we usually will find a significant difference among the means by an anova. However, we cannot turn

BOX 11.2
Computation of regression with more than one value of Y per value of X.

The variates Y are arcsine transformations of the percentage survival of the bettle *Tribolium castaneum* at 4 densities (X = number of eggs per gram of flour medium).

	Density = X (a = 4)			
	5/g	20/g	50/g	100/g
Survival; in degrees	61.68	68.21	58.69	53.13
	58.37	66.72	58.37	49.89
	69.30	63.44	58.37	49.82
	61.68	60.84		
	69.30			
$\sum^{n_i} Y$	320.33	259.21	175.43	152.84
n_i	5	4	3	3
$\bar{Y}_i$	64.07	64.80	58.48	50.95

$$\sum^{a} n_i = 15$$

$$\sum^{a}\sum^{n_i} Y = 907.81$$

Source: Data by Sokal (1967).

The anova computations are carried out as in Table 8.1.

Anova table

Source of variation	df	SS	MS	F_s
$\bar{Y} - \bar{\bar{Y}}$ Among groups	3	423.7016	141.2339	11.20**
$Y - \bar{Y}$ Within groups	11	138.6867	12.6079	
$Y - \bar{\bar{Y}}$ Total	14	562.3883		

The groups differ significantly with respect to survival.

We proceed to test whether the differences among the survival values can be accounted for by linear regression on density. If $F_s < [1/(a-1)]\, F_{\alpha[1, \Sigma^a n_i - a]}$, it is impossible for regression to be significant.

Computation for regression analysis

1. Sum of X weighted by sample size $= \sum^{a} n_i X$

$$= 5(5) + 4(20) + 3(50) + 3(100)$$

$$= 555$$

BOX 11.2
Continued

2. Sum of X^2 weighted by sample size $= \sum\limits^{a} n_i X^2$

$$= 5(5)^2 + 4(20)^2 + 3(50)^2 + 3(100)^2$$

$$= 39,225$$

3. Sum of products of X and $\bar{Y}$ weighted by sample size

$$= \sum\limits^{a} n_i X \bar{Y} = \sum\limits^{a} X \left(\sum\limits^{n_i} Y \right) = 5(320.33) + \cdots + 100(152.84)$$

$$= 30,841.35$$

4. Correction term for $X = CT_X = \dfrac{\left(\sum\limits^{a} n_i X \right)^2}{\sum\limits^{a} n_i}$

$$= \dfrac{(\text{quantity } 1)^2}{\sum\limits^{a} n_i} = \dfrac{(555)^2}{15} = 20,535.00$$

5. Sum of squares of $X = \sum x^2 = \sum\limits^{a} n_i X^2 - CT_X$

$$= \text{quantity } 2 - \text{quantity } 4 = 39,225 - 20,535$$

$$= 18,690$$

6. Sum of products $= \sum xy$

$$= \sum\limits^{a} X \left(\sum\limits^{n_i} Y \right) - \dfrac{\left(\sum\limits^{a} n_i X \right) \left(\sum\limits^{a} \sum\limits^{n_i} Y \right)}{\sum\limits^{a} n_i}$$

$$= \text{quantity } 3 - \dfrac{\text{quantity } 1 \times \sum\limits^{a} \sum\limits^{n_i} Y}{\sum\limits^{a} n_i}$$

$$= 30,841.35 - \dfrac{(555)(907.81)}{15} = -2747.62$$

7. Explained sum of squares $= \sum \hat{y}^2 = \dfrac{(\sum xy)^2}{\sum x^2}$

$$= \dfrac{(\text{quantity } 6)^2}{\text{quantity } 5} = \dfrac{(-2747.62)^2}{18,690} = 403.9281$$

8. Unexplained sum of squares $= \sum d^2_{Y \cdot X} = SS_{\text{groups}} - \sum \hat{y}^2$

$$= SS_{\text{groups}} - \text{quantity } 7$$

$$= 423.7016 - 403.9281 = 19.7735$$

BOX 11.2
Continued

Completed anova table with regression

	Source of variation	df	SS	MS	F_s
$\bar{Y} - \bar{\bar{Y}}$	Among densities (groups)	3	423.7016	141.2339	11.20**
$\hat{Y} - \bar{\bar{Y}}$	Linear regression	1	403.9281	403.9281	40.86*
$\bar{Y} - \hat{Y}$	Deviations from regression	2	19.7735	9.8868	<1 ns
$Y - \bar{Y}$	Within groups	11	138.6867	12.6079	
$Y - \bar{\bar{Y}}$	Total	14	562.3883		

In addition to the familiar mean squares, MS_{groups} and MS_{within}, we now have the mean square due to linear regression, $MS_{\hat{Y}}$, and the mean square for deviations from regression, $MS_{Y \cdot X} (= s^2_{Y \cdot X})$. To test if the deviations from linear regression are significant, compare the ratio $F_s = MS_{Y \cdot X}/MS_{within}$ with $F_{\alpha[a-2, \Sigma^a n_i - a]}$. Since we find $F_s < 1$, we accept the null hypothesis that the deviations from linear regression are zero.

To test for the presence of linear regression, we therefore tested $MS_{\hat{Y}}$ over the mean square of deviations from regression $s^2_{Y \cdot X}$ and, since $F_s = 403.9281/9.8868 = 40.86$ is greater than $F_{0.05[1,2]} = 18.5$, we clearly reject the null hypothesis that there is no regression, or that $\beta = 0$.

9. Regression coefficient (slope of regression line)

$$= b_{Y \cdot X} = \frac{\sum xy}{\sum x^2} = \frac{\text{quantity 6}}{\text{quantity 5}} = \frac{-2747.62}{18,690} = -0.147,01$$

10. Y intercept $= a = \bar{\bar{Y}} - b_{Y \cdot X}\bar{X}$

$$= \frac{\sum\limits^{a}\sum\limits^{n_i} Y}{\sum\limits^{a} n_i} - \frac{\text{quantity 9} \times \text{quantity 1}}{\sum\limits^{a} n_i}$$

$$= \frac{907.81}{15} - \frac{(-0.147,01)555}{15} = 60.5207 + 5.4394 = 65.9601$$

Hence, the regression equation is $\hat{Y} = 65.9601 - 0.147,01X$.

this argument around and say that a significant difference among means as shown by an anova necessarily indicates that a significant linear regression can be fitted to these data. In Figure 11.7C, the means follow a **U**-shaped function (a parabola). Though the means would likely be significantly different from each other, clearly a straight line fitted to these data would be a horizontal line halfway between the upper and the lower points. In such a set of data, *linear* regression can explain only little of the variation of the dependent variable. However, a curvilinear parabolic regression would fit these data and remove

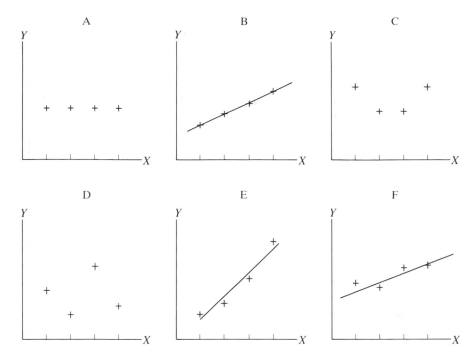

FIGURE 11.7
Differences among means and linear regression. General trends only are indicated by these figures. Significance of any of these would depend on the outcomes of appropriate tests.

most of the variance of Y. A similar case is shown in Figure 11.7D, in which the means describe a periodically changing phenomenon, rising and falling alternatingly. Again the regression line for these data has slope zero. A curvilinear (cyclical) regression could also be fitted to such data, but our main purpose in showing this example is to indicate that there could be heterogeneity among the means of Y apparently unrelated to the magnitude of X. Remember that in real examples you will rarely ever get a regression as clear-cut as the linear case in 11.7B, or the curvilinear one in 11.7C, not will you necessarily get heterogeneity of the type shown in 11.7D, in which any straight line fitted to the data would be horizontal. You are more likely to get data in which linear regression can be demonstrated, but which will not fit a straight line well. Sometimes the residual deviations of the means around linear regression can be removed by changing from linear to curvilinear regression (as is suggested by the pattern of points in Figure 11.7E), and sometimes they may remain as inexplicable residual heterogeneity around the regression line, as indicated in Figure 11.7F.

We carry out the computations following the by now familiar outline for analysis of variance and obtain the anova table shown in Box 11.2. The three degrees of freedom among the four groups yield a mean square that would be highly significant if tested over the within-groups mean square. The additional steps for the regression analysis follow in Box 11.2. We compute the sum of squares of X, the sum of products of X and Y, the explained sum of squares of Y, and the unexplained sum of squares of Y. The formulas will look unfamiliar because of the complication of the several Y's per value of X. The computations for the sum of squares of X involve the multiplication of X by the number of items in the study. Thus, though there may appear to be only four densities, there are, in fact, as many densities (although of only four magnitudes) as there are values of Y in the study. Having completed the computations, we again present the results in the form of an anova table, as shown in Box 11.2. Note that the major quantities in this table are the same as in a single-classification anova, but in addition we now have a sum of squares representing linear regression, which is always based on one degree of freedom. This sum of squares is subtracted from the SS among groups, leaving a residual sum of squares (of two degrees of freedom in this case) representing the deviations from linear regression.

We should understand what these sources of variation represent. The linear model for regression with replicated Y per X is derived directly from Expression (7.2), which is

$$Y_{ij} = \mu + \alpha_i + \epsilon_{ij}$$

The treatment effect $\alpha_i = \beta x_i + D_i$, where βx is the component due to linear regression and D_i is the deviation of the mean $\bar{Y}_i$ from regression, which is assumed to have a mean of zero and a variance of σ_D^2. Thus we can write

$$Y_{ij} = \mu + \beta x_i + D_i + \epsilon_{ij}$$

The SS due to linear regression represents that portion of the SS among groups that is explained by linear regression on X. The SS due to deviations from regression represents the residual variation or scatter around the regression line as illustrated by the various examples in Figure 11.7. The SS within groups is a measure of the variation of the items around each group mean.

We first test whether the mean square for deviations from regression $(MS_{Y \cdot X} = s_{Y \cdot X}^2)$ is significant by computing the variance ratio of $MS_{Y \cdot X}$ over the within-groups MS. In our case, the deviations from regression are clearly not significant, since the mean square for deviations is less than that within groups. We now test the mean square for regression, MS_Y, over the mean square for deviations from regression and find it to be significant. Thus linear regression on density has clearly removed a significant portion of the variation of survival values. Significance of the mean square for deviations from regression could mean either that Y is a curvilinear function of X or that there is a large amount of random heterogeneity around the regression line (as already discussed in connection with Figure 11.7; actually a mixture of both conditions may prevail).

Some workers, when analyzing regression examples with several Y variates at each value of X, proceed as follows when the deviations from regression are not significant. They add the sum of squares for deviations and that within groups as well as their degrees of freedom. Then they calculate a pooled error mean square by dividing the pooled sums of squares by the pooled degrees of freedom. The mean square for regression is then tested over the pooled error mean square, which, since it is based on more degrees of freedom, will be a better estimator of the error variance and should permit more sensitive tests. Other workers prefer never to pool, arguing that pooling the two sums of squares confounds the pseudoreplication of having several Y variates at each value of X with the true replication of having more X points to determine the slope of the regression line. Thus if we had only three X points but one hundred Y variates at each, we would be able to estimate the mean value of Y for each of the three X values very well, but we would be estimating the slope of the line on the basis of only three points, a risky procedure. The second attitude, forgoing pooling, is more conservative and will decrease the likelihood that a nonexistent regression will be declared significant.

We complete the computation of the regression coefficient and regression equation as shown at the end of Box 11.2. Our conclusions are that as density increases, survival decreases, and that this relationship can be expressed by a significant linear regression of the form $\hat{Y} = 65.9601 - 0.147,01X$, where X is density per gram and $\hat{Y}$ is the arcsine transformation of percentage survival. This relation is graphed in Figure 11.8.

The sums of products and regression slopes of both examples discussed so far have been negative, and you may begin to believe that this is always so. However, it is only an accident of choice of these two examples. In the exercises at the end of this chapter a positive regression coefficient will be encountered.

When we have equal sample sizes of Y values for each value of X, the computations become simpler. First we carry out the anova in the manner of Box 8.1. Steps 1 through 8 in Box 11.2 become simplified because the unequal sample sizes n_i are replaced by a constant sample size n, which can generally be factored out of the various expressions. Also, $\Sigma^a n_i = an$. Significance tests applied to such cases are also simplified.

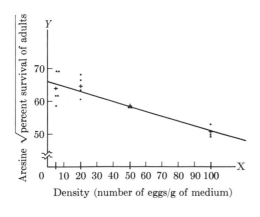

FIGURE 11.8
Linear regression fitted to data of Box 11.2. Sample means are identified by plus signs.

11.5 Tests of significance in regression

We have so far interpreted regression as a method for providing an estimate, $\hat{Y}_1$, given a value of X_1. Another interpretation is as a method for explaining some of the variation of the dependent variable Y in terms of the variation of the independent variable X. The SS of a sample of Y values, $\sum y^2$, is computed by summing and squaring deviations $y = Y - \bar{Y}$. In Figure 11.9 we can see that the deviation y can be decomposed into two parts, $\hat{y}$ and $d_{Y \cdot X}$. It is also clear from Figure 11.9 that the deviation $\hat{y} = \hat{Y} - \bar{Y}$ represents the deviation of the estimated value $\hat{Y}$ from the mean of Y. The height of $\hat{y}$ is clearly a function of x. We have already seen that $\hat{y} = bx$ (Expression (11.3)). In analytical geometry this is called the point-slope form of the equation. If b, the slope of the regression line, were steeper, $\hat{y}$ would be relatively larger for a given value of x. The remaining portion of the deviation y is $d_{Y \cdot X}$. It represents the residual variation of the variable Y after the explained variation has been subtracted. We can see that $y = \hat{y} + d_{Y \cdot X}$ by writing out these deviations explicitly as $Y - \bar{Y} = (\hat{Y} - \bar{Y}) + (Y - \hat{Y})$.

For each of these deviations we can compute a corresponding sum of squares. Appendix A1.6 gives the calculator formula for the unexplained sum of squares,

$$\sum d_{Y \cdot X}^2 = \sum y^2 - \frac{(\sum xy)^2}{\sum x^2}$$

Transposed, this yields

$$\sum y^2 = \frac{(\sum xy)^2}{\sum x^2} + \sum d_{Y \cdot X}^2$$

Of course, $\sum y^2$ corresponds to y, $\sum d_{Y \cdot X}^2$ to $d_{Y \cdot X}$, and

$$\sum \hat{y}^2 = \frac{(\sum xy)^2}{\sum x^2}$$

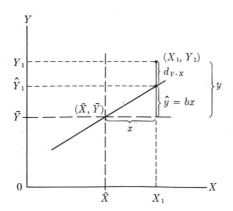

FIGURE 11.9
Schematic diagram to show relations involved in partitioning the sum of squares of the dependent variable.

corresponds to $\hat{y}$ (as shown in the previous section). Thus we are able to partition the sum of squares of the dependent variable in regression in a way analogous to the partition of the total SS in analysis of variance. You may wonder how the additive relation of the deviations can be matched by an additive relation of their squares without the presence of any cross products. Some simple algebra in Appendix A1.7 will show that the cross products cancel out. The magnitude of the unexplained deviation $d_{Y \cdot X}$ is independent of the magnitude of the explained deviation $\hat{y}$, just as in anova the magnitude of the deviation of an item from the sample mean is independent of the magnitude of the deviation of the sample mean from the grand mean. This relationship between regression and analysis of variance can be carried further. We can undertake an analysis of variance of the partitioned sums of squares as follows:

Source of variation		df	SS	MS	Expected MS
$\hat{Y} - \bar{Y}$	Explained (estimated Y from mean of Y)	1	$\sum \hat{y}^2 = \dfrac{(\sum xy)^2}{\sum x^2}$	$s_{\hat{Y}}^2$	$\sigma_{Y \cdot X}^2 + \beta^2 \sum x^2$
$Y - \hat{Y}$	Unexplained, error (observed Y from estimated Y)	$n - 2$	$\sum d_{Y \cdot X}^2 = \sum y^2 - \sum \hat{y}^2$	$s_{Y \cdot X}^2$	$\sigma_{Y \cdot X}^2$
$Y - \bar{Y}$	Total (observed Y from mean of Y)	$n - 1$	$\sum y^2 = \sum Y^2 - \dfrac{(\sum Y)^2}{n}$	s_Y^2	

The *explained mean square*, or *mean square due to linear regression*, measures the amount of variation in Y accounted for by variation of X. It is tested over the *unexplained mean square*, which measures the residual variation and is used as an error MS. The mean square due to linear regression, $s_{\hat{Y}}^2$, is based on one degree of freedom, and consequently $(n - 2)$ df remain for the error MS since the total sum of squares possesses $n - 1$ degrees of freedom. The test is of the null hypothesis $H_0: \beta = 0$. When we carry out such an anova on the weight loss data of Box 11.1, we obtain the following results:

Source of variation	df	SS	MS	F_s
Explained—due to linear regression	1	23.5145	23.5145	267.18**
Unexplained—error around regression line	7	0.6161	0.08801	
Total	8	24.1306		

The significance test is $F_s = s_{\hat{Y}}^2 / s_{Y \cdot X}^2$. It is clear from the observed value of F_s that a large and significant portion of the variance of Y has been explained by regression on X.

BOX 11.3
Standard errors of regression statistics and their degrees of freedom.

For explanation of this box, see Section 11.5: ν identifies degrees of freedom; a = number of values of X when the number of Y values for each X is n_i; n = sample size when there is a single Y value for each value of X.

Statistic	s	More than one Y value for each value of X		Single Y value for each value of X	
$b_{Y \cdot X}$ (Regression coefficient)	s_b	$\sqrt{\dfrac{s^2_{Y \cdot x}}{\sum x^2}}$	$\nu = a - 2$	$\sqrt{\dfrac{s^2_{Y \cdot x}}{\sum x^2}}$	$\nu = n - 2$
$\bar{Y}_i$ (Sample mean)	$s_{\bar{Y}}$	At any value X_i, $\sqrt{\dfrac{MS_{within}}{n_i}}$	$\nu = \sum n_i - a$	At $\bar{X}$, $\sqrt{\dfrac{s^2_{Y \cdot x}}{n}}$	$\nu = n - 2$
$\hat{Y}_i$ (Estimated Y for a given value X_i)	$s_{\hat{Y}}$	$\sqrt{s^2_{Y \cdot x}\left[\dfrac{1}{\sum n_i} + \dfrac{(X_i - \bar{\bar{X}})^2}{\sum x^2}\right]}$ $(\nu = \sum n_i - 2$ If pooled $s^2_{Y \cdot x}$ is employed)	$\nu = a - 2$	$\sqrt{s^2_{Y \cdot x}\left[\dfrac{1}{n} + \dfrac{(X_i - \bar{X})^2}{\sum x^2}\right]}$	$\nu = n - 2$

We now proceed to the standard errors for various regression statistics, their employment in tests of hypotheses, and the computation of confidence limits. Box 11.3 lists these standard errors in two columns. The right-hand column is for the case with a single Y value for each value of X. The first row of the table gives the *standard error of the regression coefficient*, which is simply the square root of the ratio of the unexplained variance to the sum of squares of X. Note that the unexplained variance $s_{Y \cdot X}^2$ is a fundamental quantity that is a part of all standard errors in regression. The standard error of the regression coefficient permits us to test various hypotheses and to set confidence limits to our sample estimate of b. The computation of s_b is illustrated in step **1** of Box 11.4, using the weight loss example of Box 11.1.

BOX 11.4

Significance tests and computation of confidence limits of regression statistics. Single value of Y for each value of X.

Based on standard errors and degrees of freedom of Box 11.3; using example of Box 11.1.

$$n = 9 \qquad \bar{X} = 50.389 \qquad \bar{Y} = 6.022$$

$$b_{Y \cdot X} = -0.053,22 \qquad \sum x^2 = 8301.3889$$

$$s_{Y \cdot X}^2 = \frac{\sum d_{Y \cdot X}^2}{(n-2)} = \frac{0.6161}{7} = 0.088,01$$

1. Standard error of the regression coefficient:

$$s_b = \sqrt{\frac{s_{Y \cdot X}^2}{\sum x^2}} = \sqrt{\frac{0.088,01}{8301.3889}} = \sqrt{0.000,010,602} = 0.003,256,1$$

2. Testing significance of the regression coefficent:

$$t_s = \frac{b - 0}{s_b} = \frac{-0.053,22}{0.003,256,1} = -16.345$$

$$t_{0.001[7]} = 5.408 \qquad P < 0.001$$

3. 95% confidence limits for regression coefficient:

$$t_{0.05[7]}s_b = 2.365(0.003,256,1) = 0.007,70$$

$$L_1 = b - t_{0.05[7]}s_b = -0.053,22 - 0.007,70 = -0.060,92$$

$$L_2 = b + t_{0.05[7]}s_b = -0.053,22 + 0.007,70 = -0.045,52$$

4. Standard error of the sampled mean $\bar{Y}$ (at $\bar{X}$):

$$s_{\bar{Y}} = \sqrt{\frac{s_{Y \cdot X}^2}{n}} = \sqrt{\frac{0.088,01}{9}} = 0.098,888,3$$

BOX 11.4

Continued

5. 95% confidence limits for the mean μ_Y corresponding to $\bar{X}(\bar{Y} = 6.022)$:

$$t_{0.05[7]}s_{\bar{Y}} = 2.365(0.098,888,3) = 0.233,871$$

$$L_1 = \bar{Y} - t_{0.05[7]}s_{\bar{Y}} = 6.022 - 0.2339 = 5.7881$$

$$L_2 = \bar{Y} + t_{0.05[7]}s_{\bar{Y}} = 6.022 + 0.2339 = 6.2559$$

6. Standard error of $\hat{Y}_i$, an estimated Y for a given value of X_i:

$$s_{\hat{Y}_i} = \sqrt{s_{Y \cdot X}^2 \left[\frac{1}{n} + \frac{(X_i - \bar{X})^2}{\sum x^2} \right]}$$

For example, for $X_i = 100\%$ relative humidity

$$s_{\hat{Y}} = \sqrt{0.088,01 \left[\frac{1}{9} + \frac{(100 - 50.389)^2}{8301.3889} \right]}$$

$$= \sqrt{0.088,01(0.407,60)} = \sqrt{0.035,873} = 0.189,40$$

7. 95% confidence limits for μ_{Y_i} corresponding to the estimate $\hat{Y}_i = 3.3817$ at $X_i = 100\%$ relative humidity:

$$t_{0.05[7]}s_{\hat{Y}} = 2.365(0.189,40) = 0.447,93$$

$$L_1 = \hat{Y}_i - t_{0.05[7]}s_{\hat{Y}} = 3.3817 - 0.4479 = 2.9338$$

$$L_2 = \hat{Y}_i + t_{0.05[7]}s_{\hat{Y}} = 3.3817 + 0.4479 = 3.8296$$

The significance test illustrated in step **2** tests the *"significance" of the regression coefficient*; that is, it tests the null hypothesis that the sample value of b comes from a population with a parametric value $\beta = 0$ for the regression coefficient. This is a t test, the appropriate degrees of freedom being $n - 2 = 7$. If we cannot reject the null hypothesis, there is no evidence that the regression is significantly deviant from zero in either the positive or negative direction. Our conclusions for the weight loss data are that a highly significant negative regression is present. We saw earlier (Section 8.4) that $t^2 = F$. When we square $t_s = -16.345$ from Box 11.4, we obtain 267.16, which (within rounding error) equals the value of F_s found in the anova earlier in this section. The significance test in step **2** of Box 11.4 could, of course, also be used to test whether b is significantly different from a parametric value β other than zero.

Setting confidence limits to the regression coefficient presents no new features, since b is normally distributed. The computation is shown in step **3** of Box 11.4. In view of the small magnitude of s_b, the confidence interval is quite narrow. The confidence limits are shown in Figure 11.10 as dashed lines representing the 95% bounds of the slope. Note that the regression line as well as its

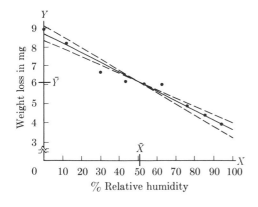

FIGURE 11.10
95% confidence limits to regression line of Figure 11.6.

confidence limits passes through the means of X and Y. Variation in b therefore rotates the regression line about the point $\bar{X}$, $\bar{Y}$.

Next, we calculate a *standard error for the observed sample mean* $\bar{Y}$. You will recall from Section 6.1 that $s_{\bar{Y}}^2 = s_Y^2/n$. However, now that we have regressed Y on X, we are able to account for (that is, hold constant) some of the variation of Y in terms of the variation of X. The variance of Y around the point $\bar{X}$, $\bar{Y}$ on the regression line is less than s_Y^2; it is $s_{Y \cdot X}^2$. At $\bar{X}$ we may therefore compute confidence limits of $\bar{Y}$, using as a standard error of the mean $s_{\bar{Y}} = \sqrt{s_{Y \cdot X}^2/n}$ with $n - 2$ degrees of freedom. This standard error is computed in step **4** of Box 11.4, and 95% confidence limits for the sampled mean $\bar{Y}$ at $\bar{X}$ are calculated in step **5**. These limits (5.7881–6.2559) are considerably narrower than the confidence limits for the mean based on the conventional standard error $s_{\bar{Y}}$, which would be from 4.687 to 7.357. Thus, knowing the relative humidity greatly reduces much of the uncertainty in weight loss.

The standard error for $\bar{Y}$ is only a special case of the *standard error for any estimated value $\hat{Y}$ along the regression line*. A new factor, whose magnitude is in part a function of the distance of a given value X_i from its mean $\bar{X}$, now enters the error variance. Thus, the farther away X_i is from its mean, the greater will be the error of estimate. This factor is seen in the third row of Box 11.3 as the deviation $X_i - \bar{X}$, squared and divided by the sum of squares of X. The standard error for an estimate $\hat{Y}_i$ for a relative humidity $X_i = 100\%$ is given in step **6** of Box 11.4. The 95% confidence limits for $\mu_{\hat{Y}_i}$, the parametric value corresponding to the estimate $\hat{Y}_i$, are shown in step **7** of that box. Note that the width of the confidence interval is $3.8296 - 2.9338 = 0.8958$, considerably wider than the confidence interval at $\bar{X}$ calculated in step **5**, which was $6.2559 - 5.7881 = 0.4678$. If we calculate a series of confidence limits for different values of X_i, we obtain a biconcave confidence belt as shown in Figure 11.11. The farther we get away from the mean, the less reliable are our estimates of Y, because of the uncertainty about the true slope, β, of the regression line.

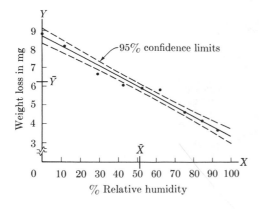

FIGURE 11.11
95% confidence limits to regression estimates
for data of Figure 11.6.

Furthermore, the linear regressions that we fit are often only rough approximations to the more complicated functional relationships between biological variables. Very often there is an approximately linear relation along a certain range of the independent variable, beyond which range the slope changes rapidly. For example, heartbeat of a poikilothermic animal will be directly proportional to temperature over a range of tolerable temperatures, but beneath and above this range the heartbeat will eventually decrease as the animal freezes or suffers heat prostration. Hence common sense indicates that one should be very cautious about extrapolating from a regression equation if one has any doubts about the linearity of the relationship.

The confidence limits for α, the parametric value of a, are a special case of those for $\mu_{\hat{Y}_i}$ at $X_i = 0$, and the standard error of a is therefore

$$ s_a = \sqrt{s_{Y \cdot X}^2 \left[\frac{1}{n} + \frac{\bar{X}^2}{\sum x^2} \right]} $$

Tests of significance in regression analyses where there is more than one variate Y per value of X are carried out in a manner similar to that of Box 11.4, except that the standard errors in the left-hand column of Box 11.3 are employed.

Another significance test in regression is a test of the differences between two regression lines. Why would we be interested in testing differences between regression slopes? We might find that different toxicants yield different dosage-mortality curves or that different drugs yield different relationships between dosage and response (see, for example, Figure 11.1). Or genetically differing cultures might yield different responses to increasing density, which would be important for understanding the effect of natural selection in these cultures. The regression slope of one variable on another is as fundamental a statistic of a sample as is the mean or the standard deviation, and in comparing samples it may be as important to compare regression coefficients as it is to compare these other statistics.

The test for the difference between two regression coefficients can be carried out as an F test. We compute

$$F_s = \frac{(b_1 - b_2)^2}{\dfrac{\sum x_1^2 + \sum x_2^2}{(\sum x_1^2)(\sum x_2^2)} \bar{s}_{Y \cdot X}^2}$$

where $\bar{s}_{Y \cdot X}^2$ is the weighted average $s_{Y \cdot X}^2$ of the two groups. Its formula is

$$\bar{s}_{Y \cdot X}^2 = \frac{(\sum d_{Y \cdot X}^2)_1 + (\sum d_{Y \cdot X}^2)_2}{v_2}$$

For one Y per value of X, $v_2 = n_1 + n_2 - 4$, but when there is more than one variate Y per value of X, $v_2 = a_1 + a_2 - 4$. Compare F_s with $F_{\alpha[1, v_2]}$.

11.6 The uses of regression

We have been so busy learning the mechanics of regression analysis that we have not had time to give much thought to the various applications of regression. We shall take up four more or less distinct applications in this section. All are discussed in terms of Model I regression.

First, we might mention the *study of causation*. If we wish to know whether variation in a variable Y is caused by changes in another variable X, we manipulate X in an experiment and see whether we can obtain a significant regression of Y on X. The idea of causation is a complex, philosophical one that we shall not go into here. You have undoubtedly been cautioned from your earliest scientific experience not to confuse concomitant variation with causation. Variables may vary together, yet this covariation may be accidental or both may be functions of a common cause affecting them. The latter cases are usually examples of Model II regression with both variables varying freely. When we manipulate one variable and find that such manipulations affect a second variable, we generally are satisfied that the variation of the independent variable X is the cause of the variation of the dependent variable Y (not the cause of the variable!). However, even here it is best to be cautious. When we find that heartbeat rate in a cold-blooded animal is a function of ambient temperature, we may conclude that temperature is one of the causes of differences in hearbeat rate. There may well be other factors affecting rate of heartbeat. A possible mistake is to invert the cause-and-effect relationship. It is unlikely that anyone would suppose that hearbeat rate affects the temperature of the general environment, but we might be mistaken about the cause-and-effect relationships between two chemical substances in the blood, for instance. Despite these cautions, regression analysis is a commonly used device for screening out causal relationships. While a significant regression of Y on X does not prove that changes in X are the cause of variations in Y, the converse statement is true. When we find no significant regression of Y on X, we can in all but the most complex cases infer quite safely (allowing for the possibility of type II error) that deviations of X do not affect Y.

The *description of scientific laws* and *prediction* are a second general area of application of regression analysis. Science aims at mathematical description of relations between variables in nature, and Model I regression analysis permits us to estimate functional relationships between variables, one of which is subject to error. These functional relationships do not always have clearly interpretable biological meaning. Thus, in many cases it may be difficult to assign a biological interpretation to the statistics a and b, or their corresponding parameters α and β. When we can do so, we speak of a *structural mathematical model*, one whose component parts have clear scientific meaning. However, mathematical curves that are not structural models are also of value in science. Most regression lines are *empirically fitted curves*, in which the functions simply represent the best mathematical fit (by a criterion such as least squares) to an observed set of data.

Comparison of dependent variates is another application of regression. As soon as it is established that a given variable is a function of another one, as in Box 11.2, where we found survival of beetles to be a function of density, one is bound to ask to what degree any observed difference in survival between two samples of beetles is a function of the density at which they have been raised. It would be unfair to compare beetles raised at very high density (and expected to have low survival) with those raised under optimal conditions of low density. This is the same point of view that makes us disinclined to compare the mathematical knowledge of a fifth-grader with that of a college student. Since we could undoubtedly obtain a regression of mathematical knowledge on years of schooling in mathematics, we should be comparing how far a given individual deviates from his or her expected value based on such a regression. Thus, relative to his or her classmates and age group, the fifth-grader may be far better than is the college student relative to his or her peer group. This suggests that we calculate *adjusted Y values* that allow for the magnitude of the independent variable X. A conventional way of calculating such adjusted Y values is to estimate the Y value one would expect if the independent variable were equal to its mean $\bar{X}$ and the observation retained its observed deviation $(d_{Y \cdot X})$ from the regression line. Since $\hat{Y} = \bar{Y}$ when $X = \bar{X}$, the adjusted Y value can be computed as

$$Y_{\text{adj}} = \bar{Y} + d_{Y \cdot X} = Y - bx \tag{11.8}$$

Statistical control is an application of regression that is not widely known among biologists and represents a scientific philosophy that is not well established in biology outside agricultural circles. Biologists frequently categorize work as either descriptive or experimental, with the implication that only the latter can be analytical. However, statistical approaches applied to descriptive work can, in a number of instances, take the place of experimental techniques quite adequately—occasionally they are even to be preferred. These approaches are attempts to substitute statistical manipulation of a concomitant variable for control of the variable by experimental means. An example will clarify this technique.

Let us assume that we are studying the effects of various diets on blood pressure in rats. We find that the variability of blood pressure in our rat population is considerable, even before we introduce differences in diet. Further study reveals that the variability is largely due to differences in age among the rats of the experimental population. This can be demonstrated by a significant linear regression of blood pressure on age. To reduce the variability of blood pressure in the population, we should keep the age of the rats constant. The reaction of most biologists at this point will be to repeat the experiment using rats of only one age group; this is a valid, commonsense approach, which is part of the experimental method. An alternative approach is superior in some cases, when it is impractical or too costly to hold the variable constant. We might continue to use rats of variable ages and simply record the age of each rat as well as its blood pressure. Then we regress blood pressure on age and use an adjusted mean as the basic blood pressure reading for each individual. We can now evaluate the effect of differences in diet on these adjusted means. Or we can analyze the effects of diet on unexplained deviations, $d_{Y \cdot X}$, after the experimental blood pressures have been regressed on age (which amounts to the same thing).

What are the advantages of such an approach? Often it will be impossible to secure adequate numbers of individuals all of the same age. By using regression we are able to utilize all the individuals in the population. The use of statistical control assumes that it is relatively easy to record the independent variable X and, of course, that this variable can be measured without error, which would be generally true of such a variable as age of a laboratory animal. Statistical control may also be preferable because we obtain information over a wider range of both Y and X and also because we obtain added knowledge about the relations between these two variables, which would not be so if we restricted ourselves to a single age group.

11.7 Residuals and transformations in regression

An examination of regression residuals, $d_{Y \cdot X}$, may detect outliers in a sample. Such outliers may reveal systematic departures from regression that can be adjusted by transformation of scale, or by the fitting of a curvilinear regression line. When it is believed that an outlier is due to an observational or recording error, or to contamination of the sample studied, removal of such an outlier may improve the regression fit considerably. In examining the magnitude of residuals, we should also allow for the corresponding deviation from $\bar{X}$. Outlying values of Y_i that correspond to deviant variates X_i will have a greater influence in determining the slope of the regression line than will variates close to $\bar{X}$. We can examine the residuals in column (9) of Table 11.1 for the weight loss data. Although several residuals are quite large, they tend to be relatively close to $\bar{Y}$. Only the residual for 0% relative humidity is suspiciously large and, at the same time, is the single most deviant observation from $\bar{X}$. Perhaps the reading at this extreme relative humidity does not fit into the generally linear relations described by the rest of the data.

In transforming either or both variables in regression, we aim at simplifying a curvilinear relationship to a linear one. Such a procedure generally increases the proportion of the variance of the dependent variable explained by the independent variable, and the distribution of the deviations of points around the regression line tends to become normal and homoscedastic. Rather than fit a complicated curvilinear regression to points plotted on an arithmetic scale, it is far more expedient to compute a simple linear regression for variates plotted on a transformed scale. A general test of whether transformation will improve linear regression is to graph the points to be fitted on ordinary graph paper as well as on other graph paper in a scale suspected to improve the relationship. If the function straightens out and the systematic deviation of points around a visually fitted line is reduced, the transformation is worthwhile.

We shall briefly discuss a few of the transformations commonly applied in regression analysis. Square root and arcsine transformations (Section 10.2) are not mentioned below, but they are also effective in regression cases involving data suited to such transformations.

The *logarithmic transformation* is the most frequently used. Anyone doing statistical work is therefore well advised to keep a supply of semilog paper handy. Most frequently we transform the dependent variable Y. This transformation is indicated when *percentage* changes in the dependent variable vary directly with changes in the independent variable. Such a relationship is indicated by the equation $Y = ae^{bX}$, where a and b are constants and e is the base of the natural logarithm. After the transformation, we obtain $\log Y = \log a + b(\log e)X$. In this expression $\log e$ is a constant which when multiplied by b yields a new constant factor b' which is equivalent to a regression coefficient. Similarly, $\log a$ is a new Y intercept, a'. We can then simply regress $\log Y$ on X to obtain the function $\widehat{\log Y} = a' + b'X$ and obtain all our prediction equations and confidence intervals in this form. Figure 11.12 shows an example of transforming the dependent variate to logarithmic form, which results in considerable straightening of the response curve.

A logarithmic transformation of the independent variable in regression is effective when proportional changes in the independent variable produce linear responses in the dependent variable. An example might be the decline in weight of an organism as density increases, where the successive increases in density need to be in a constant ratio in order to effect equal decreases in weight. This belongs to a well-known class of biological phenomena, another example of which is the Weber-Fechner law in physiology and psychology, which states that a stimulus has to be increased by a constant proportion in order to produce a constant increment in response. Figure 11.13 illustrates how logarithmic

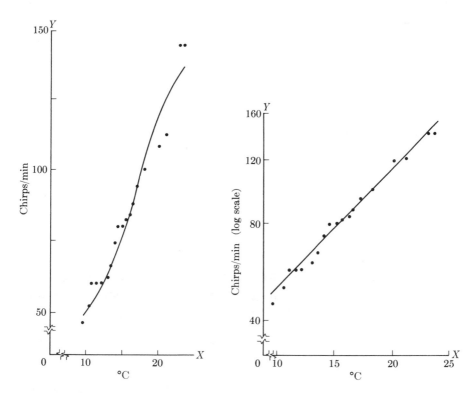

FIGURE 11.12

Logarithmic transformation of a dependent variable in regression. Chirp-rate as a function of temperature in males of the tree cricket *Oecanthus fultoni*. Each point represents the mean chirp rate/min for all observations at a given temperature in °C. Original data in left panel, Y plotted on logarithmic scale in right panel. (Data from Block, 1966.)

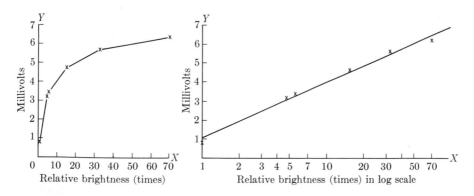

FIGURE 11.13

Logarithmic transformation of the independent variable in regression. This illustrates size of electrical response to illumination in the cephalopod eye. Ordinate, millivolts; abscissa, relative brightness of illumination. A proportional increase in X (relative brightness) produces a linear electrical response Y. (Data in Fröhlich, 1921.)

transformation of the independent variable results in the straightening of the regression line. For computations one would transform X into logarithms.

Logarithmic transformation for both variables is applicable in situations in which the true relationship can be described by the formula $\hat{Y} = aX^b$. The regression equation is rewritten as $\log \hat{Y} = \log a + b \log X$ and the computation is done in the conventional manner. Examples are the greatly disproportionate growth of various organs in some organisms, such as the sizes of antlers of deer or horns of stage beetles, with respect to their general body sizes. A double logarithmic transformation is indicated when a plot on log-log graph paper results in a straight-line graph.

Reciprocal transformation. Many rate phenomena (a given performance per unit of time or per unit of population), such as wing beats per second or number of eggs laid per female, will yield hyperbolic curves when plotted in original measurement scale. Thus, they form curves described by the general mathematical equations $bXY = 1$ or $(a + bX)Y = 1$. From these we can derive $1/Y = bX$ or $1/Y = a + bX$. By transforming the dependent variable into its reciprocal, we can frequently obtain straight-line regressions.

Finally, some cumulative curves can be straightened by the *probit transformation.* Refresh your memory on the cumulative normal curve shown in Figure 5.5. Remember that by changing the ordinate of the cumulative normal into probability scale we were able to straighten out this curve. We do the same thing here except that we graduate the probability scale in standard deviation units. Thus, the 50% point becomes 0 standard deviations, the 84.13% point becomes $+1$ standard deviation, and the 2.27% point becomes -2 standard deviations. Such standard deviations, corresponding to a cumulative percentage, are called *normal equivalent deviates* (*NED*). If we use ordinary graph paper and mark the ordinate in *NED* units, we obtain a straight line when plotting the cumulative normal curve against it. *Probits* are simply normal equivalent deviates coded by the addition of 5.0, which will avoid negative values for most deviates. Thus, the probit value 5.0 corresponds to a cumulative frequency of 50%, the probit value 6.0 corresponds to a cumulative frequency of 84.13%, and the probit value 3.0 corresponds to a cumulative frequency of 2.27%.

Figure 11.14 shows an example of mortality percentages for increasing doses of an insecticide. These represent differing points of a cumulative frequency distribution. With increasing dosages an ever greater proportion of the sample dies until at a high enough dose the entire sample is killed. It is often found that if the doses of toxicants are transformed into logarithms, the tolerances of many organisms to these poisons are approximately normally distributed. These transformed doses are often called *dosages.* Increasing dosages lead to cumulative normal distributions of mortalities, often called *dosage-mortality curves.* These curves are the subject matter of an entire field of biometric analysis, *bioassay,* to which we can refer only in passing here. The most common technique in this field is *probit analysis.* Graphic approximations can be carried out on so-called *probit paper,* which is probability graph paper in which the

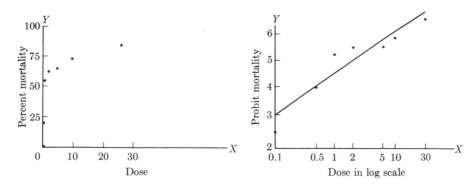

FIGURE 11.14
Dosage mortality data illustrating an application of the probit transformation. Data are mean mortalities for two replicates. Twenty *Drosophila melanogaster* per replicate were subjected to seven doses of an "unknown" insecticide in a class experiment. The point at dose 0.1 which yielded 0% mortality has been assigned a probit value of 2.5 in lieu of $-\infty$, which cannot be plotted.

abscissa has been transformed into logarithmic scale. A regression line is fitted to dosage-mortality data graphed on probit paper (see Figure 11.14). From the regression line the 50% lethal does is estimated by a process of inverse prediction, that is, we estimate the value of X (dosage) corresponding to a kill of probit 5.0, which is equivalent to 50%.

11.8 A nonparametric test for regression

When transformations are unable to linearize the relationship between the dependent and independent variables, the research worker may wish to carry out a simpler, nonparametric test in lieu of regression analysis. Such a test furnishes neither a prediction equation nor a functional relationship, but it does test whether the dependent variable Y is a monotonically increasing (or decreasing) function of the independent variable X. The simplest such test is the *ordering test*, which is equivalent to computing Kendall's rank correlation coefficient (see Box 12.3) and can be carried out most easily as such. In fact, in such a case the distinction between regression and correlation, which will be discussed in detail in Section 12.1, breaks down. The test is carried out as follows.

Rank variates X and Y. Arrange the independent variable X in increasing order of ranks and calculate the Kendall rank correlation of Y with X. The computational steps for the procedure are shown in Box 12.3. If we carry out this computation for the weight loss data of Box 11.1 (reversing the order of percent relative humidity, X, which is negatively related to weight loss, Y), we obtain a quantity $N = 72$, which is significant at $P < 0.01$ when looked up in Table **XIV**. There is thus a significant trend of weight loss as a function of relative humidity. The ranks of the weight losses are a perfect monotonic function

on the ranks of the relative humidities. The minimum number of points required
for significance by the rank correlation method is 5.

Exercises

11.1 The following temperatures (Y) were recorded in a rabbit at various times (X)
after it was inoculated with rinderpest virus (data from Carter and Mitchell, 1958).

Time after injection (h)	Temperature (°F)
24	102.8
32	104.5
48	106.5
56	107.0
72	103.9
80	103.2
96	103.1

Graph the data. Clearly, the last three data points represent a different phenomenon from the first four pairs. *For the first four points:* (a) Calculate b. (b)
Calculate the regression equation and draw in the regression line. (c) Test the
hypothesis that $\beta = 0$ and set 95% confidence limits. (d) Set 95% confidence
limits to your estimate of the rabbit's temperature 50 hours after the injection.
ANS. $a = 100$, $b = 0.1300$, $F_s = 59.4288$, $P < 0.05$, $\hat{Y}_{50} = 106.5$.

11.2 The following table is extracted from data by Sokoloff (1955). Adult weights
of female *Drosophila persimilis* reared at 24°C are affected by their density as
larvae. Carry out an anova among densities. Then calculate the regression of
weight on density and partition the sums of squares among groups into that
explained and unexplained by linear regression. Graph the data with the regression line fitted to the means. Interpret your results.

Larval density	Mean weight of adults (in mg)	s of weights (not $s_{\bar{Y}}$)	n
1	1.356	0.180	9
3	1.356	0.133	34
5	1.284	0.130	50
6	1.252	0.105	63
10	0.989	0.130	83
20	0.664	0.141	144
40	0.475	0.083	24

11.3 Davis (1955) reported the following results in a study of the amount of energy
metabolized by the English sparrow, *Passer domesticus*, under various constant
temperature conditions and a ten-hour photoperiod. Analyze and interpret
ANS. $MS_{\hat{Y}} = 657.5043$, $MS_{Y \cdot X} = 8.2186$, $MS_{\text{within}} = 3.9330$, deviations are not
significant.

Temperature (°C)	Calories $\bar{Y}$	n	s
0	24.9	6	1.77
4	23.4	4	1.99
10	24.2	4	2.07
18	18.7	5	1.43
26	15.2	7	1.52
34	13.7	7	2.70

11.4 Using the *complete* data given in Exercise 11.1, calculate the regression equation and compare it with the one you obtained for the first four points. Discuss the effect of the inclusion of the last three points in the analysis. Compute the residuals from regression.

11.5 The following results were obtained in a study of oxygen consumption (microliters/mg dry weight per hour) in *Heliothis zea* by Phillips and Newsom (1966) under controlled temperatures and photoperiods.

Temperature (°C)	Photoperiod (h)	
	10	14
18	0.51	1.61
21	0.53	1.64
24	0.89	1.73

Compute regression for each photoperiod separately and test for homogeneity of slopes. ANS. For 10 hours: $b = 0.0633$, $s^2_{Y \cdot X} = 0.019,267$. For 14 hours: $b = 0.020,00$, $s^2_{Y \cdot X} = 0.000,60$.

11.6 Length of developmental period (in days) of the potato leafhopper, *Empoasca fabae*, from egg to adult at various constant temperatures (Kouskolekas and Decker, 1966). The original data were weighted means, but for purposes of this analysis we shall consider them as though they were single observed values.

Temperature (°F)	Mean length of developmental period in days $\bar{Y}$
59.8	58.1
67.6	27.3
70.0	26.8
70.4	26.3
74.0	19.1
75.3	19.0
78.0	16.5
80.4	15.9
81.4	14.8
83.2	14.2
88.4	14.4
91.4	14.6
92.5	15.3

Analyze and interpret. Compute deviations from the regression line $(\bar{Y}_i - \hat{\bar{Y}}_i)$ and plot against temperature.

11.7   The experiment cited in Exercise 11.3 was repeated using a 15-hour photoperiod, and the following results were obtained:

Temperature (°C)	Calories $\bar{Y}$	n	s
0	24.3	6	1.93
10	25.1	7	1.98
18	22.2	8	3.67
26	13.8	10	4.01
34	16.4	6	2.92

Test for the equality of slopes of the regression lines for the 10-hour and 15-hour photoperiod. ANS. $F_s = 0.003$.

11.8   Carry out a nonparametric test for regression in Exercises 11.1 and 11.6.

11.9   Water temperature was recorded at various depths in Rot Lake on August 1, 1952, by Vollenweider and Frei (1953).

Depth (m)	0	1	2	3	4	5	6	9	12	15.5
Temperature (°C)	24.8	23.2	22.2	21.2	18.8	13.8	9.6	6.3	5.8	5.6

Plot the data and then compute the regression line. Compute the deviations from regression. Does temperature vary as a linear function of depth? What do the residuals suggest? ANS. $a = 23.384$, $b = -1.435$, $F_s = 45.2398$, $P < 0.01$.

CHAPTER 12

Correlation

In this chapter we continue our discussion of bivariate statistics. In Chapter 11, on regression, we dealt with the functional relation of one variable upon the other; in the present chapter, we treat the measurement of the amount of association between two variables. This general topic is called *correlation analysis.*

It is not always obvious which type of analysis—regression or correlation—one should employ in a given problem. There has been considerable confusion in the minds of investigators and also in the literature on this topic. We shall try to make the distinction between these two approaches clear at the outset in Section 12.1. In Section 12.2 you will be introduced to the product-moment correlation coefficient, the common correlation coefficient of the literature. We shall derive a formula for this coefficient and give you something of its theoretical background. The close mathematical relationship between regression and correlation analysis will be examined in this section. We shall also compute a product-moment correlation coefficient in this section. In Section 12.3 we will talk about various tests of significance involving correlation coefficients. Then, in Section 12.4, we will introduce some of the applications of correlation coefficients.

Section 12.5 contains a nonparametric method that tests for association. It is to be used in those cases in which the necessary assumptions for tests involving correlation coefficients do not hold, or where quick but less than fully efficient tests are preferred for reasons of speed in computation or for convenience.

12.1 Correlation and regression

There has been much confusion on the subject matter of correlation and regression. Quite frequently correlation problems are treated as regression problems in the scientific literature, and the converse is equally true. There are several reasons for this confusion. First of all, the mathematical relations between the two methods of analysis are quite close, and mathematically one can easily move from one to the other. Hence, the temptation to do so is great. Second, earlier texts did not make the distinction between the two approaches sufficiently clear, and this problem has still not been entirely overcome. At least one textbook synonymizes the two, a step that we feel can only compound the confusion. Finally, while an investigator may with good reason intend to use one of the two approaches, the nature of the data may be such as to make only the other approach appropriate.

Let us examine these points at some length. The many and close mathematical relations between regression and correlation will be detailed in Section 12.2. It suffices for now to state that for any given problem, the majority of the computational steps are the same whether one carries out a regression or a correlation analysis. You will recall that the fundamental quantity required for regression analysis is the sum of products. This is the very same quantity that serves as the base for the computation of the correlation coefficient. There are some simple mathematical relations between regression coefficients and correlation coefficients for the same data. Thus the temptation exists to compute a correlation coefficient corresponding to a given regression coefficient. Yet, as we shall see shortly, this would be wrong unless our intention at the outset were to study association and the data were appropriate for such a computation.

Let us then look at the intentions or purposes behind the two types of analyses. In regression we intend to describe the dependence of a variable Y on an independent variable X. As we have seen, we employ regression equations for purposes of lending support to hypotheses regarding the possible causation of changes in Y by changes in X; for purposes of prediction, of variable Y given a value of variable X; and for purposes of explaining some of the variation of Y by X, by using the latter variable as a statistical control. Studies of the effects of temperature on heartbeat rate, nitrogen content of soil on growth rate in a plant, age of an animal on blood pressure, or dose of an insecticide on mortality of the insect population are all typical examples of regression for the purposes named above.

In correlation, by contrast, we are concerned largely whether two variables are interdependent, or *covary*—that is, vary together. We do not express one as a function of the other. There is no distinction between independent and dependent variables. It may well be that of the pair of variables whose correlation is studied, one is the cause of the other, but we neither know nor assume this. A more typical (but not essential) assumption is that the two variables are both effects of a common cause. What we wish to estimate is the degree to which these variables vary together. Thus we might be interested in the correlation between amount of fat in diet and incidence of heart attacks in human populations, between foreleg length and hind leg length in a population of mammals, between body weight and egg production in female blowflies, or between age and number of seeds in a weed. Reasons why we would wish to demonstrate and measure association between pairs of variables need not concern us yet. We shall take this up in Section 12.4. It suffices for now to state that when we wish to establish the degree of association between pairs of variables in a population sample, correlation analysis is the proper approach.

Thus a correlation coefficient computed from data that have been properly analyzed by Model I regression is meaningless as an estimate of any population correlation coefficient. Conversely, suppose we were to evaluate a regression coefficient of one variable on another in data that had been properly computed as correlations. Not only would construction of such a functional dependence for these variables not meet our intentions, but we should point out that a conventional regression coefficient computed from data in which both variables are measured with error—as is the case in correlation analysis— furnishes biased estimates of the functional relation.

Even if we attempt the correct method in line with our purposes we may run afoul of the nature of the data. Thus we may wish to establish cholesterol content of blood as a function of weight, and to do so we may take a random sample of men of the same age group, obtain each individual's cholesterol content and weight, and regress the former on the latter. However, both these variables will have been measured with error. Individual variates of the supposedly independent variable X will not have been deliberately chosen or controlled by the experimenter. The underlying assumptions of Model I regression do not hold, and fitting a Model I regression to the data is not legitimate, although you will have no difficulty finding instances of such improper practices in the published research literature. If it is really an equation describing the dependence of Y on X that we are after, we should carry out a Model II regression. However, if it is the degree of association between the variables (interdependence) that is of interest, then we should carry out a correlation analysis, for which these data are suitable. The converse difficulty is trying to obtain a correlation coefficient from data that are properly computed as a regression—that is, are computed when X is fixed. An example would be heartbeats of a poikilotherm as a function of temperature, where several temperatures have been applied in an experiment. Such a correlation coefficient is easily obtained mathematically but would simply be a numerical value, not an estimate

TABLE **12.1**

The relations between correlation and regression. This table indicates the correct computation for any combination of purposes and variables, as shown.

Purpose of investigator	Nature of the two variables	
	Y random, X fixed	Y_1, Y_2 both random
Establish and estimate dependence of one variable upon another. (Describe functional relationship and/or predict one in terms of the other.)	Model I regression.	Model II regression. (Not treated in this book.)
Establish and estimate association (interdependence) between two variables.	Meaningless for this case. If desired, an estimate of the proportion of the variation of Y explained by X can be obtained as the square of the correlation coefficient between X and Y.	Correlation coefficient. (Significance tests entirely appropriate only if Y_1, Y_2 are distributed as bivariate normal variables.)

of a parametric measure of correlation. There is an interpretation that can be given to the square of the correlation coefficient that has some relevance to a regression problem. However, it is not in any way an estimate of a parametric correlation.

This discussion is summarized in Table 12.1, which shows the relations between correlation and regression. The two columns of the table indicate the two conditions of the pair of variables: in one case one random and measured with error, the other variable fixed; in the other case, both variables random. In this text we depart from the usual convention of labeling the pair of variables Y and X or X_1, X_2 for both correlation and regression analysis. In regression we continue the use of Y for the dependent variable and X for the independent variable, but in correlation both of the variables are in fact random variables, which we have throughout the text designated as Y. We therefore refer to the two variables as Y_1 and Y_2. The rows of the table indicate the intention of the investigator in carrying out the analysis, and the four quadrants of the table indicate the appropriate procedures for a given combination of intention of investigator and nature of the pair of variables.

12.2 The product-moment correlation coefficient

There are numerous correlation coefficients in statistics. The most common of these is called the *product-moment correlation coefficient*, which in its current formulation is due to Karl Pearson. We shall derive its formula through an intuitive approach.

You have seen that the sum of products is a measure of covariation, and it is therefore likely that this will be the basic quantity from which to obtain a formula for the correlation coefficient. We shall label the variables whose correlation is to be estimated as Y_1 and Y_2. Their sum of products will therefore be $\sum y_1 y_2$ and their covariance $[1/(n-1)] \sum y_1 y_2 = s_{12}$. The latter quantity is analogous to a variance, that is, a sum of squares divided by its degrees of freedom.

A standard deviation is expressed in original measurement units such as inches, grams, or cubic centimeters. Similarly, a regression coefficient is expressed as so many units of Y per unit of X, such as 5.2 grams/day. However, a measure of association should be independent of the original scale of measurement, so that we can compare the degree of association in one pair of variables with that in another. One way to accomplish this is to divide the covariance by the standard deviations of variables Y_1 and Y_2. This results in dividing each deviation y_1 and y_2 by its proper standard deviation and making it into a standardized deviate. The expression now becomes the sum of the products of standardized deviates divided by $n - 1$:

$$r_{Y_1 Y_2} = \frac{\sum y_1 y_2}{(n-1)s_{Y_1} s_{Y_2}} \tag{12.1}$$

This is the formula for the product-moment correlation coefficient $r_{Y_1 Y_2}$ between variables Y_1 and Y_2. We shall simplify the symbolism to

$$r_{12} = \frac{\sum y_1 y_2}{(n-1)s_1 s_2} = \frac{s_{12}}{s_1 s_2} \tag{12.2}$$

Expression (12.2) can be rewritten in another common form. Since

$$s\sqrt{n-1} = \sqrt{s^2(n-1)} = \sqrt{\frac{\sum y^2}{n-1}(n-1)} = \sqrt{\sum y^2}$$

Expression (12.2) can be rewritten as

$$r_{12} = \frac{\sum y_1 y_2}{\sqrt{\sum y_1^2 \sum y_2^2}} \tag{12.3}$$

To state Expression (12.2) more generally for variables Y_j and Y_k, we can write it as

$$r_{jk} = \frac{\sum y_j y_k}{(n-1)s_j s_k} \tag{12.4}$$

The correlation coefficient r_{jk} can range from $+1$ for perfect association to -1 for perfect negative association. This is intuitively obvious when we consider the correlation of a variable Y_j with itself. Expression (12.4) would then yield $r_{jj} = \sum y_j y_j / \sqrt{\sum y_j^2 \sum y_j^2} = \sum y_j^2 / \sum y_j^2 = 1$, which yields a perfect correlation of $+1$. If deviations in one variable were paired with opposite but equal deviations representing another variable, this would yield a correlation of -1, because the sum of products in the numerator would be negative. Proof that the correlation coefficient is bounded by $+1$ and -1 will be given shortly.

If the variates follow a specified distribution, the *bivariate normal distribution*, the correlation coefficient r_{jk} will estimate a parameter of that distribution symbolized by ρ_{jk}.

Let us approach the distribution empirically. Suppose you have sampled a hundred items and measured two variables on each item, obtaining two samples of 100 variates in this manner. If you plot these 100 items on a graph in which the variables Y_1 and Y_2 are the coordinates, you will obtain a scattergram of points as in Figure 12.3A. Let us assume that both variables, Y_1 and Y_2, are normally distributed and that they are quite independent of each other, so that the fact that one individual happens to be greater than the mean in character Y_1 has no effect whatsoever on its value for variable Y_2. Thus this same individual may be greater or less than the mean for variable Y_2. If there is absolutely no relation between Y_1 and Y_2 and if the two variables are standardized to make their scales comparable, you will find that the outline of the scattergram is roughly circular. Of course, for a sample of 100 items, the circle will be only imperfectly outlined; but the larger the sample, the more clearly you will be able to discern a circle with the central area around the intersection $\bar{Y}_1$, $\bar{Y}_2$ heavily darkened because of the aggregation there of many points. If you keep sampling, you will have to superimpose new points upon previous points, and if you visualize these points in a physical sense, such as grains of sand, a mound peaked in a bell-shaped fashion will gradually accumulate. This is a three-dimensional realization of a normal distribution, shown in perspective in Figure 12.1. Regarded from either coordinate axis, the mound will present a two-dimensional appearance, and its outline will be that of a normal distribution curve, the two perspectives giving the distributions of Y_1 and Y_2, respectively.

If we assume that the two variables Y_1 and Y_2 are not independent but are positively correlated to some degree, then if a given individual has a large value of Y_1, it is more likely than not to have a large value of Y_2 as well. Similarly, a small value of Y_1 will likely be associated with a small value of Y_2. Were you to sample items from such a population, the resulting scattergram (shown in

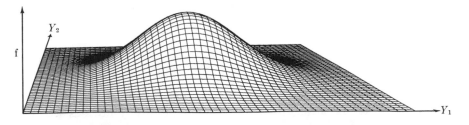

FIGURE 12.1
Bivariate normal frequency distribution. The parametric correlation ρ between variables Y_1 and Y_2 equals zero. The frequency distribution may be visualized as a bell-shaped mound.

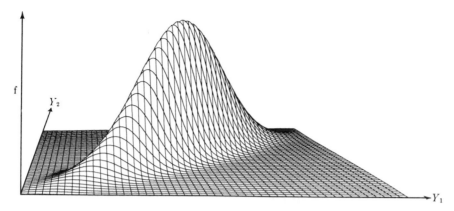

FIGURE 12.2

Bivariate normal frequency distribution. The parametric correlation ρ between variables Y_1 and Y_2 equals 0.9. The bell-shaped mound of Figure 12.1 has become elongated.

Figure 12.3D) would become elongated in the form of an ellipse. This is so because those parts of the circle that formerly included individuals high for one variable and low for the other (and vice versa), are now scarcely represented. Continued sampling (with the sand grain model) yields a three-dimensional elliptic mound, shown in Figure 12.2. If correlation is perfect, all the data will fall along a single regression line (the identical line would describe the regression of Y_1 on Y_2 and of Y_2 on Y_1), and if we let them pile up in a physical model, they will result in a flat, essentially two-dimensional normal curve lying on this regression line.

The circular or elliptical shape of the outline of the scattergram and of the resulting mound is clearly a function of the degree of correlation between the two variables, and this is the parameter ρ_{jk} of the bivariate normal distribution. By analogy with Expression (12.2), the parameter ρ_{jk} can be defined as

$$\rho_{jk} = \frac{\sigma_{jk}}{\sigma_j \sigma_k} \tag{12.5}$$

where σ_{jk} is the parametric covariance of variables Y_j and Y_k and σ_j and σ_k are the parametric standard deviations of variables Y_j and Y_k, as before. When two variables are distributed according to the bivariate normal, a sample correlation coefficient r_{jk} estimates the parametric correlation coefficient ρ_{jk}. We can make some statements about the sampling distribution of ρ_{jk} and set confidence limits to it.

Regrettably, the elliptical shape of scattergrams of correlated variables is not usually very clear unless either very large samples have been taken or the parametric correlation ρ_{jk} is very high. To illustrate this point, we show in Figure 12.3 several graphs illustrating scattergrams resulting from samples of 100 items from bivariate normal populations with differing values of ρ_{jk}. Note

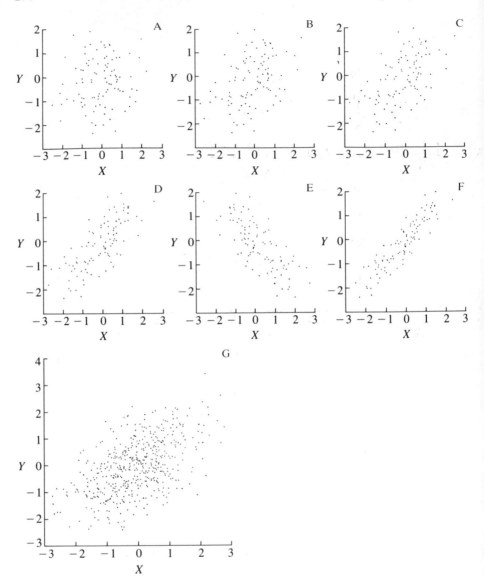

FIGURE 12.3
Random samples from bivariate normal distributions with varying values of the parametric correlation coefficient ρ. Sample sizes $n = 100$ in all graphs except G, which has $n = 500$. (A) $\rho = 0.4$. (B) $\rho = 0.3$. (C) $\rho = 0.5$. (D) $\rho = 0.7$. (E) $\rho = -0.7$. (F) $\rho = 0.9$. (G) $\rho = 0.5$.

that in the first graph (Figure 12.3A), with $\rho_{jk} = 0$, the circular distribution is only very vaguely outlined. A far greater sample is required to demonstrate the circular shape of the distribution more clearly. No substantial difference is noted in Figure 12.3B, based on $\rho_{jk} = 0.3$. Knowing that this depicts a positive correlation, one can visualize a positive slope in the scattergram; but without prior knowledge this would be difficult to detect visually. The next graph (Figure 12.3C, based on $\rho_{jk} = 0.5$) is somewhat clearer, but still does not exhibit an unequivocal trend. In general, correlation cannot be inferred from inspection of scattergrams based on samples from populations with ρ_{jk} between -0.5 and $+0.5$ unless there are numerous sample points. This point is illustrated in the last graph (Figure 12.3G), also sampled from a population with $\rho_{jk} = 0.5$ but based on a sample of 500. Here, the positive slope and elliptical outline of the scattergram are quite evident. Figure 12.3D, based on $\rho_{jk} = 0.7$ and $n = 100$, shows the trend more clearly than the first three graphs. Note that the next graph (Figure 12.3E), based on the same magnitude of ρ_{jk} but representing negative correlation, also shows the trend but is more strung out than Figure 12.3D. The difference in shape of the ellipse has no relation to the negative nature of the correlation; it is simply a function of sampling error, and the comparison of these two figures should give you some idea of the variability to be expected on random sampling from a bivariate normal distribution. Finally, Figure 12.3F, representing a correlation of $\rho_{jk} = 0.9$, shows a tight association between the variables and a reasonable approximation to an ellipse of points.

Now let us return to the expression for the sample correlation coefficient shown in Expression (12.3). Squaring this expression results in

$$r_{12}^2 = \frac{(\sum y_1 y_2)^2}{\sum y_1^2 \sum y_2^2}$$

$$= \frac{(\sum y_1 y_2)^2}{\sum y_1^2} \cdot \frac{1}{\sum y_2^2}$$

Look at the left term of the last expression. It is the square of the sum of products of variables Y_1 and Y_2, divided by the sum of squares of Y_1. If this were a regression problem, this would be the formula for the explained sum of squares of variable Y_2 on variable Y_1, $\sum \hat{y}_2^2$. In the symbolism of Chapter 11, on regression, it would be $\sum \hat{y}^2 = (\sum xy)^2/\sum x^2$. Thus, we can write

$$r_{12}^2 = \frac{\sum \hat{y}_2^2}{\sum y_2^2} \tag{12.6}$$

The square of the correlation coefficient, therefore, is the ratio formed by the explained sum of squares of variable Y_2 divided by the total sum of squares of variable Y_2. Equivalently,

$$r_{12}^2 = \frac{\sum \hat{y}_1^2}{\sum y_1^2} \tag{12.6a}$$

which can be derived just as easily. (Remember that since we are not really regressing one variable on the other, it is just as legitimate to have Y_1 explained by Y_2 as the other way around.) The ratio symbolized by Expressions (12.6) and (12.6a) is a proportion ranging from 0 to 1. This becomes obvious after a little contemplation of the meaning of this formula. The explained sum of squares of any variable must be smaller than its total sum of squares or, maximally, if all the variation of a variable has been explained, it can be as great as the total sum of squares, but certainly no greater. Minimally, it will be zero if none of the variable can be explained by the other variable with which the covariance has been computed. Thus, we obtain an important measure of the proportion of the variation of one variable determined by the variation of the other. This quantity, the square of the correlation coefficient, r_{12}^2, is called the *coefficient of determination*. It ranges from zero to 1 and must be positive regardless of whether the correlation coefficient is negative or positive. Incidentally, here is proof that the correlation coefficient cannot vary beyond -1 and $+1$. Since its square is the coefficient of determination and we have just shown that the bounds of the latter are zero to 1, it is obvious that the bounds of its square root will be ± 1.

The coefficient of determination is useful also when one is considering the relative importance of correlations of different magnitudes. As can be seen by a reexamination of Figure 12.3, the rate at which the scatter diagrams go from a distribution with a circular outline to one resembling an ellipse seems to be more directly proportional to r^2 than to r itself. Thus, in Figure 12.3B, with $\rho^2 = 0.09$, it is difficult to detect the correlation visually. However, by the time we reach Figure 12.3D, with $\rho^2 = 0.49$, the presence of correlation is very apparent.

The coefficient of determination is a quantity that may be useful in regression analysis also. You will recall that in a regression we used anova to partition the total sum of squares into explained and unexplained sums of squares. Once such an analysis of variance has been carried out, one can obtain the ratio of the explained sums of squares over the total SS as a measure of the proportion of the total variation that has been explained by the regression. However, as already discussed in Section 12.1, it would not be meaningful to take the square root of such a coefficient of determination and consider it as an estimate of the parametric correlation of these variables.

We shall now take up a mathematical relation between the coefficients of correlation and regression. At the risk of being repetitious, we should stress again that though we can easily convert one coefficient into the other, this does not mean that the two types of coefficients can be used interchangeably on the same sort of data. One important relationship between the correlation coefficient and the regression coefficient can be derived as follows from Expression (12.3):

$$r_{12} = \frac{\sum y_1 y_2}{\sqrt{\sum y_1^2 \sum y_2^2}} = \frac{\sum y_1 y_2}{\sqrt{\sum y_1^2}} \cdot \frac{1}{\sqrt{\sum y_2^2}}$$

Multiplying numerator and denominator of this expression by $\sqrt{\sum y_1^2}$, we obtain

$$r_{12} = \frac{\sum y_1 y_2}{\sqrt{\sum y_1^2}\sqrt{\sum y_1^2}} \cdot \frac{\sqrt{\sum y_1^2}}{\sqrt{\sum y_2^2}} = \frac{\sum y_1 y_2}{\sum y_1^2} \cdot \frac{\sqrt{\sum y_1^2}}{\sqrt{\sum y_2^2}}$$

Dividing numerator and denominator of the right term of this expression by $\sqrt{n-1}$, we obtain

$$r_{12} = \frac{\sum y_1 y_2}{\sum y_1^2} \cdot \frac{\sqrt{\dfrac{\sum y_1^2}{n-1}}}{\sqrt{\dfrac{\sum y_2^2}{n-1}}} = b_{2 \cdot 1}\frac{s_1}{s_2} \tag{12.7}$$

Similarly, we could demonstrate that

$$r_{12} = b_{1 \cdot 2}\frac{s_2}{s_1} \tag{12.7a}$$

and hence

$$b_{2 \cdot 1} = r_{12}\frac{s_2}{s_1} \qquad b_{1 \cdot 2} = r_{12}\frac{s_1}{s_2} \tag{12.7b}$$

In these expressions $b_{2 \cdot 1}$ is the regression coefficient for variable Y_2 on Y_1. We see, therefore, that the correlation coefficient is the regression slope multiplied by the ratio of the standard deviations of the variables. The correlation coefficient may thus be regarded as a standardized regression coefficient. If the two standard deviations are identical, both regression coefficients and the correlation coefficient will be identical in value.

Now that we know about the coefficient of correlation, some of the earlier work on paired comparisons (see Section 9.3) can be put into proper perspective. In Appendix A1.8 we show for the corresponding parametric expressions that the variance of a sum of two variables is

$$s_{(Y_1 + Y_2)}^2 = s_1^2 + s_2^2 + 2r_{12}s_1 s_2 \tag{12.8}$$

where s_1 and s_2 are standard deviations of Y_1 and Y_2, respectively, and r_{12} is the correlation coefficient between these variables. Similarly, for a difference between two variables, we obtain

$$s_{(Y_1 - Y_2)}^2 = s_1^2 + s_2^2 - 2r_{12}s_1 s_2 \tag{12.9}$$

What Expression (12.8) indicates is that if we make a new composite variable that is the sum of two other variables, the variance of this new variable will be the sum of the variances of the variables of which it is composed plus an added term, which is a function of the standard deviations of these two variables and of the correlation between them. It is shown in Appendix A1.8 that this added term is twice the covariance of Y_1 and Y_2. When the two variables

BOX 12.1
Computation of the product-moment correlation coefficient.

Relationships between gill weight and body weight in the crab *Pachygrapsus crassipes*. $n = 12$.

(1) Y_1 Gill weight in milligrams	(2) Y_2 Body weight in grams
159	14.40
179	15.20
100	11.30
45	2.50
384	22.70
230	14.90
100	1.41
320	15.81
80	4.19
220	15.39
320	17.25
210	9.52

Source: Unpublished data by L. Miller.

Computation

1. $\sum Y_1 = 159 + \cdots + 210 = 2347$

2. $\sum Y_1^2 = 159^2 + \cdots + 210^2 = 583,403$

3. $\sum Y_2 = 14.40 + \cdots + 9.52 = 144.57$

4. $\sum Y_2^2 = (14.40)^2 + \cdots + (9.52)^2 = 2204.1853$

5. $\sum Y_1 Y_2 = 14.40(159) + \cdots + 9.52(210) = 34,837.10$

6. Sum of squares of $Y_1 = \sum y_1^2 = \sum Y_1^2 - \dfrac{(\sum Y_1)^2}{n}$

$$= \text{quantity } 2 - \frac{(\text{quantity } 1)^2}{n} = 583,403 - \frac{(2347)^2}{12}$$

$$= 124,368.9167$$

7. Sum of squares of $Y_2 = \sum y_2^2 = \sum Y_2^2 - \dfrac{(\sum Y_2)^2}{n}$

$$= \text{quantity } 4 - \frac{(\text{quantity } 3)^2}{n} = 2204.1853 - \frac{(144.57)^2}{12}$$

$$= 462.4782$$

BOX 12.1
Continued

8. Sum of products $= \sum y_1 y_2 = \sum Y_1 Y_2 - \dfrac{(\sum Y_1)(\sum Y_2)}{n}$

$$= \text{quantity } \mathbf{5} - \frac{\text{quantity } \mathbf{1} \times \text{quantity } \mathbf{3}}{n}$$

$$= 34{,}837.10 - \frac{(2347)(144.57)}{12} = 6561.6175$$

9. Product-moment correlation coefficient (by Expression (12.3)):

$$r_{12} = \frac{\sum y_1 y_2}{\sqrt{\sum y_1^2 \sum y_2^2}} = \frac{\text{quantity } \mathbf{8}}{\sqrt{\text{quantity } \mathbf{6} \times \text{quantity } \mathbf{7}}}$$

$$= \frac{6561.6175}{\sqrt{(124{,}368.9167)(462.4782)}} = \frac{6561.6175}{\sqrt{57{,}517{,}912.7314}}$$

$$= \frac{6561.6175}{7584.0565} = 0.8652 \approx 0.87$$

being summed are uncorrelated, this added covariance term will be zero, and the variance of the sum will simply be the sum of variances of the two variables. This is the reason why, in an anova or in a t test of the difference between the two means, we had to assume the independence of the two variables to permit us to add their variances. Otherwise we would have had to allow for a covariance term. By contrast, in the paired-comparisons technique we expect correlation between the variables, since the members in each pair share a common experience. The paired-comparisons test automatically subtracts a covariance term, resulting in a smaller standard error and consequently in a larger value of t_s, since the numerator of the ratio remains the same. Thus, whenever correlation between two variables is positive, the variance of their differences will be considerably smaller than the sum of their variances; this is the reason why the paired-comparisons test has to be used in place of the t test for difference of means. These considerations are equally true for the corresponding analyses of variance, single-classification and two-way anova.

The computation of a product-moment correlation coefficient is quite simple. The basic quantities needed are the same six required for computation of the regression coefficient (Section 11.3). Box 12.1 illustrates how the coefficient should be computed. The example is based on a sample of 12 crabs in which gill weight Y_1 and body weight Y_2 have been recorded. We wish to know whether there is a correlation between the weight of the gill and that of the body, the latter representing a measure of overall size. The existence of a positive correlation might lead you to conclude that a bigger-bodied crab with its resulting greater amount of metabolism would require larger gills in order to

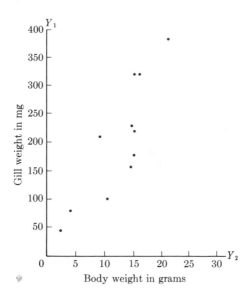

FIGURE 12.4
Scatter diagram for crab data of Box 12.1.

provide the necessary oxygen. The computations are illustrated in Box 12.1. The correlation coefficient of 0.87 agrees with the clear slope and narrow elliptical outline of the scattergram for these data in Figure 12.4.

12.3 Significance tests in correlation

The most common significance test is whether it is possible for a sample correlation coefficient to have come from a population with a parametric correlation coefficient of zero. The null hypothesis is therefore $H_0: \rho = 0$. This implies that the two variables are uncorrelated. If the sample comes from a bivariate normal distribution and $\rho = 0$, the standard error of the correlation coefficient is $s_r = \sqrt{(1 - r^2)/(n - 2)}$. The hypothesis is tested as a t test with $n - 2$ degrees of freedom, $t_s = (r - 0)/\sqrt{(1 - r^2)/(n - 2)} = r\sqrt{(n - 2)/(1 - r^2)}$. We should emphasize that this standard error applies only when $\rho = 0$, so that it cannot be applied to testing a hypothesis that ρ is a specific value other than zero. The t test for the significance of r is mathematically equivalent to the t test for the significance of b, in either case measuring the strength of the association between the two variables being tested. This is somewhat analogous to the situation in Model I and Model II single-classification anova, where the same F test establishes the significance regardless of the model.

Significance tests following this formula have been carried out systematically and are tabulated in Table **VIII**, which permits the direct inspection of a sample correlation coefficient for significance without further computation. Box 12.2 illustrates tests of the hypothesis $H_0: \rho = 0$, using Table **VIII** as well as the t test discussed at first.

BOX 12.2
Tests of significance and confidence limits for correlation coefficients.

Test of the null hypothesis H_0: $\rho = 0$ versus H_1: $\rho \neq 0$

The simplest procedure is to consult Table **VIII**, where the critical values of r are tabulated for $df = n - 2$ from 1 to 1000. If the absolute value of the observed r is greater than the tabulated value in the column for two variables, we reject the null hypothesis.

Examples. In Box 12.1 we found the correlation between body weight and gill weight to be 0.8652, based on a sample of $n = 12$. For 10 degrees of freedom the critical values are 0.576 at the 5% level and 0.708 at the 1% level of significance. Since the observed correlation is greater than both of these, we can reject the null hypothesis, H_0: $\rho = 0$, at $P < 0.01$.

Table **VIII** is based upon the following test, which may be carried out when the table is not available or when an exact test is needed at significance levels or at degrees of freedom other than those furnished in the table. The null hypothesis is tested by means of the t distribution (with $n - 2$ df) by using the standard error of r. When $\rho = 0$,

$$s_r = \sqrt{\frac{(1 - r^2)}{(n - 2)}}$$

Therefore,

$$t_s = \frac{(r - 0)}{\sqrt{(1 - r^2)/(n - 2)}} = r\sqrt{\frac{(n - 2)}{(1 - r^2)}}$$

For the data of Box 12.1, this would be

$$t_s = 0.8652\sqrt{(12 - 2)/(1 - 0.8652^2)} = 0.8652\sqrt{10/0.25143}$$
$$= 0.8652\sqrt{39.7725} = 0.8652(6.3065) = 5.4564 > t_{0.001[10]}$$

For a one-tailed test the 0.10 and 0.02 values of t should be used for 5% and 1% significance tests, respectively. Such tests would apply if the alternative hypothesis were H_1: $\rho > 0$ or H_1: $\rho < 0$, rather than H_1: $\rho \neq 0$.

When n is greater than 50, we can also make use of the z transformation described in the text. Since $\sigma_z = 1/\sqrt{n - 3}$, we test

$$t_s = \frac{z - 0}{1/\sqrt{n - 3}} = z\sqrt{n - 3}$$

Since z is normally distributed and we are using a parametric standard deviation, we compare t_s with $t_{\alpha[\infty]}$ or employ Table **II**, "Areas of the normal curve." If we had a sample correlation of $r = 0.837$ between length of right- and left-wing veins of bees based on $n = 500$, we would find $z = 1.2111$ in Table **X**. Then

$$t_s = 1.2111\sqrt{497} = 26.997$$

This value, when looked up in Table **II**, yields a very small probability ($< 10^{-6}$).

Test of the null hypothesis H_0: $\rho = \rho_1$, where $\rho_1 \neq 0$

To test this hypothesis we cannot use Table **VIII** or the t test given above, but must make use of the z transformation.

BOX 12.2
Continued

Suppose we wish to test the null hypothesis H_0: $\rho = +0.5$ versus H_1: $\rho \neq$ $+0.5$ for the case just considered. We would use the following expression:

$$t_s = \frac{z - \zeta}{1/\sqrt{n-3}} = (z - \zeta)\sqrt{n-3}$$

where z and ζ are the z transformations of r and ρ, respectively. Again we compare t_s with $t_{\alpha[\infty]}$ or look it up in Table **II**. From Table **VIII** we find

For $r = 0.837$ $z = 1.2111$

For $\rho = 0.500$ $\zeta = 0.5493$

Therefore

$$t_s = (1.2111 - 0.5493)(\sqrt{497}) = 14.7538$$

The probability of obtaining such a value of t_s by random sampling is $P < 10^{-6}$ (see Table **II**). It is most unlikely that the parametric correlation between right- and left-wing veins is 0.5.

Confidence limits

If $n > 50$, we can set confidence limits to r using the z transformation. We first convert the sample r to z, set confidence limits to this z, and then transform these limits back to the r scale. We shall find 95% confidence limits for the above wing vein length data.

For $r = 0.837$, $z = 1.2111$, $\alpha = 0.05$,

$$L_1 = z - t_{\alpha[\infty]}\sigma_z = z - \frac{t_{0.05[\infty]}}{\sqrt{n-3}} = 1.2111 - \frac{1.960}{22.2953}$$

$$= 1.2111 - 0.0879 = 1.1232$$

$$L_2 = z + \frac{t_{0.05[\infty]}}{\sqrt{n-3}} = 1.2111 + 0.0879 = 1.2990$$

We retransform these z values to the r scale by finding the corresponding arguments for the z function in Table **X**.

$$L_1 \approx 0.808 \quad \text{and} \quad L_2 \approx 0.862$$

are the 95% confidence limits around $r = 0.837$.

Test of the difference between two correlation coefficients

For two correlation coefficients we may test H_0: $\rho_1 = \rho_2$ versus H_1: $\rho_1 \neq \rho_2$ as follows:

$$t_s = \frac{z_1 - z_2}{\sqrt{\dfrac{1}{n_1 - 3} + \dfrac{1}{n_2 - 3}}}$$

BOX 12.2
Continued

Since $z_1 - z_2$ is normally distributed and we are using a parametric standard deviation, we compare t_s with $t_{\alpha[\infty]}$ or employ Table **II**, "Areas of the normal curve."

For example, the correlation between body weight and wing length in *Drosophila pseudoobscura* was found by Sokoloff (1966) to be 0.552 in a sample of $n_1 = 39$ at the Grand Canyon and 0.665 in a sample of $n_2 = 20$ at Flagstaff, Arizona.

Grand Canyon: $z_1 = 0.6213$ Flagstaff: $z_2 = 0.8017$

$$t_s = \frac{0.6213 - 0.8017}{\sqrt{\frac{1}{36} + \frac{1}{17}}} = \frac{-0.1804}{\sqrt{0.086,601}} = \frac{-0.1804}{0.294,28} = -0.6130$$

By linear interpolation in Table **II**, we find the probability that a value of t_s will be between ± 0.6130 to be about $2(0.229,41) = 0.458,82$, so we clearly have no evidence on which to reject the null hypothesis.

When ρ is close to ± 1.0, the distribution of sample values of r is markedly asymmetrical, and, although a standard error is available for r in such cases, it should not be applied unless the sample is very large ($n > 500$), a most infrequent case of little interest. To overcome this difficulty, we transform r to a function z, developed by Fisher. The formula for z is

$$z = \frac{1}{2} \ln \left(\frac{1+r}{1-r} \right) \tag{12.10}$$

You may recognize this as $z = \tanh^{-1} r$, the formula for the inverse hyperbolic tangent of r. This function has been tabulated in Table **X**, where values of z corresponding to absolute values of r are given. Inspection of Expression (12.10) will show that when $r = 0$, z will also equal zero, since $\frac{1}{2} \ln 1$ equals zero. However, as r approaches ± 1, $(1 + r)/(1 - r)$ approaches $\pm \infty$ and 0; consequently, z approaches $\pm$ infinity. Therefore, substantial differences between r and z occur at the higher values for r. Thus, when r is 0.115, $z = 0.1155$. For $r = -0.531$, we obtain $z = -0.5915$; $r = 0.972$ yields $z = 2.1273$. Note by how much z exceeds r in this last pair of values. By finding a given value of z in Table **X**, we can also obtain the corresponding value of r. Inverse interpolation may be necessary. Thus, $z = 0.70$ corresponds to $r = 0.604$, and a value of $z = -2.76$ corresponds to $r = -0.992$. Some pocket calculators have built-in hyperbolic and inverse hyperbolic functions. Keys for such functions would obviate the need for Table **X**.

The advantage of the z transformation is that while correlation coefficients are distributed in skewed fashion for values of $\rho \neq 0$, the values of z are approximately normally distributed for any value of its parameter, which we call

ζ (zeta), following the usual convention. The expected variance of z is

$$\sigma_z^2 = \frac{1}{n-3} \tag{12.11}$$

This is an approximation adequate for sample sizes $n \geq 50$ and a tolerable approximation even when $n \geq 25$. An interesting aspect of the variance of z evident from Expression (12.11) is that it is independent of the magnitude of r, but is simply a function of sample size n.

As shown in Box 12.2, for sample sizes greater than 50 we can also use the z transformation to test the significance of a sample r employing the hypothesis $H_0: \rho = 0$. In the second section of Box 12.2 we show the test of a null hypothesis that $\rho \neq 0$. We may have a hypothesis that the true correlation between two variables is a given value ρ different from zero. Such hypotheses about the expected correlation between two variables are frequent in genetic work, and we may wish to test observed data against such a hypothesis. Although there is no a priori reason to assume that the true correlation between right and left sides of the bee wing vein lengths in Box 12.2 is 0.5, we show the test of such a hypothesis to illustrate the method. Corresponding to $\rho = 0.5$, there is ζ, the parametric value of z. It is the z transformation of ρ. We note that the probability that the sample r of 0.837 could have been sampled from a population with $\rho = 0.5$ is vanishingly small.

Next, in Box 12.2 we see how to set confidence limits to a sample correlation coefficient r. This is done by means of the z transformation; it will result in asymmetrical confidence limits when these are retransformed to the r scale, as when setting confidence limits with variables subjected to square root or logarithmic transformations.

A test for the significance of the difference between two sample correlation coefficients is the final example illustrated in Box 12.2. A standard error for the difference is computed and tested against a table of areas of the normal curve. In the example the correlation between body weight and wing length in two *Drosophila* populations was tested, and the difference in correlation coefficients between the two populations was found not significant. The formula given is an acceptable approximation when the smaller of the two samples is greater than 25. It is frequently used with even smaller sample sizes, as shown in our example in Box 12.2.

12.4 Applications of correlation

The purpose of correlation analysis is to measure the intensity of association observed between any pair of variables and to test whether it is greater than could be expected by chance alone. Once established, such an association is likely to lead to reasoning about causal relationships between the variables. Students of statistics are told at an early stage not to confuse significant correlation with causation. We are also warned about so-called nonsense correlations, a well-known case being the positive correlation between the number of Baptist ministers and the per capita liquor consumption in cities with populations of over 10,000 in the United States. Individual cases of correlation must be carefully analyzed before inferences are drawn from them. It is useful to distinguish correlations in which one variable is the entire or, more likely, the partial cause of another from others in which the two correlated variables have a common cause and from more complicated situations involving both direct influence and common causes. The establishment of a significant correlation does not tell us which of many possible structural models is appropriate. Further analysis is needed to discriminate between the various models.

The traditional distinction of real versus nonsense or illusory correlation is of little use. In supposedly legitimate correlations, causal connections are known or at least believed to be clearly understood. In so-called illusory correlations, no reasonable connection between the variables can be found; or if one is demonstrated, it is of no real interest or may be shown to be an artifact of the sampling procedure. Thus, the correlation between Baptist ministers and liquor consumption is simply a consequence of city size. The larger the city, the more Baptist ministers it will contain on the average and the greater will be the liquor consumption. The correlation is of little interest to anyone studying either the distribution of Baptist ministers or the consumption of alcohol. Some correlations have time as the common factor, and processes that change with time are frequently likely to be correlated, not because of any functional biological reasons but simply because the change with time in the two variables under consideration happens to be in the same direction. Thus, size of an insect population building up through the summer may be correlated with the height of some weeds, but this may simply be a function of the passage of time. There may be no ecological relation between the plant and the insects. Another situation in which the correlation might be considered an artifact is when one of the variables is in part a mathematical function of the other. Thus, for example, if $Y = Z/X$ and we compute the correlation of X with Y, the existing relation will tend to produce a negative correlation.

Perhaps the only correlations properly called nonsense or illusory are those assumed by popular belief or scientific intuition which, when tested by proper statistical methodology using adequate sample sizes, are found to be not significant. Thus, if we can show that there is no significant correlation between amount of saturated fats eaten and the degree of atherosclerosis, we can consider this to be an illusory correlation. Remember also that when testing significance of correlations at conventional levels of significance, you must allow for type I error, which will lead to your judging a certain percentage of correlations significant when in fact the parametric value of $\rho = 0$.

Correlation coefficients have a history of extensive use and application dating back to the English biometric school at the beginning of the twentieth century. Recent years have seen somewhat less application of this technique as increasing segments of biological research have become experimental. In experiments in which one factor is varied and the response of another variable to the deliberate variation of the first is examined, the method of regression is more appropriate, as has already been discussed. However, large areas of biology and of other sciences remain where the experimental method is not suitable because variables cannot be brought under control of the investigator. There are many areas of medicine, ecology, systematics, evolution, and other fields in which experimental methods are difficult to apply. As yet, the weather cannot be controlled, nor can historical evolutionary factors be altered. Epidemiological variables are generally not subject to experimental manipulation. Nevertheless, we need an understanding of the scientific mechanisms underlying these phenomena as much as of those in biochemistry or experimental embryology. In such cases, correlation analysis serves as a first descriptive technique estimating the degrees of association among the variables involved.

12.5 Kendall's coefficient of rank correlation

Occasionally data are known not to follow the bivariate normal distribution, yet we wish to test for the significance of association between the two variables. One method of analyzing such data is by ranking the variates and calculating a coefficient of rank correlation. This approach belongs to the general family of nonparametric methods we encountered in Chapter 10, where we learned methods for analyses of ranked variates paralleling anova. In other cases especially suited to ranking methods, we cannot measure the variable on an absolute scale, but only on an ordinal scale. This is typical of data in which we estimate relative performance, as in assigning positions in a class. We can say that A is the best student, B is the second-best student, C and D are equal to each other and next-best, and so on. If two instructors independently rank a group of students, we can then test whether the two sets of rankings are independent (which we would not expect if the judgments of the instructors are based on objective evidence). Of greater biological and medical interest are the following examples. We might wish to correlate order of emergence in a sample of insects with a ranking in size, or order of germination in a sample of plants with rank order of flowering. An epidemiologist may wish to associate rank order of occurrence (by time) of an infectious disease within a community, on the one hand, with its severity as measured by an objective criterion, on the other.

We present in Box 12.3 *Kendall's coefficient of rank correlation*, generally symbolized by τ (tau), although it is a sample statistic, not a parameter. The formula for Kendall's coefficient of rank correlation is $\tau = N/n(n-1)$, where n is the conventional sample size and N is a count of ranks, which can be obtained in a variety of ways. A second variable Y_2, if it is perfectly correlated with the first variable Y_1, should be in the same order as the Y_1 variates. However, if the correlation is less than perfect, the order of the variates Y_2 will not entirely correspond to that of Y_1. The quantity N measures how well the second variable corresponds to the order of the first. It has a maximal value of $n(n-1)$ and a minimal value of $-n(n-1)$. The following small example will make this clear.

BOX 12.3
Kendall's coefficient of rank correlation, τ.

Computation of a rank correlation coefficient between the blood neutrophil counts (Y_1; $\times 10^{-3}$ per μl) and total marrow neutrophil mass (Y_2; $\times 10^9$ per kg) in 15 patients with nonhematological tumors; $n = 15$ pairs of observations.

(1) Patient	(2) Y_1	(3) R_1	(4) Y_2	(5) R_2	(1) Patient	(2) Y_1	(3) R_1	(4) Y_2	(5) R_2
1	4.9	6	4.34	1	8	7.1	9	7.12	5
2	4.6	5	9.64	9	9	2.3	1	9.75	10
3	5.5	7	7.39	6	10	3.6	2	8.65	8
4	9.1	11	13.97	12	11	18.0	15	15.34	14
5	16.3	14	20.12	15	12	3.7	3	12.33	11
6	12.7	13	15.01	13	13	7.3	10	5.99	2
7	6.4	8	6.93	4	14	4.4	4	7.66	7
					15	9.8	12	6.07	3

Source: Data extracted from Liu, Kesfeld, and Koo (1983).

Computational steps

1. Rank variables Y_1 and Y_2 separately and then replace the original variates with the ranks (assign tied ranks if necessary so that for both variables you will always have n ranks for n variates). These ranks are listed in columns (3) and (5) above.

2. Write down the n ranks of one of the two variables in order, paired with the rank values assigned for the other variable (as shown below). If only one variable has ties, order the pairs by the variable without ties. If both variables have ties, it does not matter which of the variables is ordered.

3. Obtain a sum of the counts C_i, as follows. Examine the first value in the column of ranks paired with the ordered column. In our case, this is rank 10. Count all ranks subsequent to it which are higher than the rank being considered. Thus, in this case, count all ranks greater than 10. There are fourteen ranks following the 10 and five of them are greater than 10. Therefore, we count a score of $C_1 = 5$. Now we look at the next rank (rank 8) and find that six of the thirteen subsequent ranks are greater than it; therefore, C_2 is equal to 6. The third rank is 11, and four following ranks are higher than it. Hence, $C_3 = 4$. Continue in this manner, taking each rank of the variable in turn and counting the number of higher ranks subsequent to it. This can usually be done in one's head, but we show it explicitly below so that the method will be entirely clear. Whenever a subsequent rank is tied in value with the pivotal rank R_2, count $\frac{1}{2}$ instead of 1.

BOX 12.3
Continued

R_1	R_2	Subsequent ranks greater than pivotal rank R_2	Counts C_i
1	10	11, 12, 13, 15, 14	5
2	8	11, 9, 12, 13, 15, 14	6
3	11	12, 13, 15, 14	4
4	7	9, 12, 13, 15, 14	5
5	9	12, 13, 15, 14	4
6	1	6, 4, 5, 2, 12, 3, 13, 15, 14	9
7	6	12, 13, 15, 14	4
8	4	5, 12, 13, 15, 14	5
9	5	12, 13, 15, 14	4
10	2	12, 3, 13, 15, 14	5
11	12	13, 15, 14	3
12	3	13, 15, 14	3
13	13	15, 14	2
14	15		0
15	14		0

$$\sum_{}^{n} C_i = 59$$

We then need the following quantity:

$$N = 4 \sum_{}^{n} C_i - n(n-1) = 4(59) - 15(14) = 236 - 210 = 26$$

4. The Kendall coefficient of rank correlation, τ, can be found as follows:

$$\tau = \frac{N}{n(n-1)} = \frac{26}{15(14)} = 0.124$$

When there are ties, the coefficient is computed as follows:

$$\tau = \frac{N}{\sqrt{\left[n(n-1) - \sum^{m} T_1 \right]\left[n(n-1) - \sum^{m} T_2 \right]}}$$

where $\Sigma^m T_1$ and $\Sigma^m T_2$ are the sums of correction terms for ties in the ranks of variable Y_1 and Y_2, respectively, defined as follows. A T value equal to $t(t-1)$ is computed for each group of t tied variates and summed over m such groups. Thus if variable Y_2 had had two sets of ties, one involving $t = 2$ variates and a second involving $t = 3$ variates, one would have computed $\Sigma^m T_2 = 2(2-1) + 3(3-1) = 8$. It has been suggested that if the ties are due to lack of precision rather than being real, the coefficient should be computed by the simpler formula.

BOX 12.3
Continued

5. To test significance for sample sizes >40, we can make use of a normal approximation to test the null hypothesis that the true value of $\tau = 0$:

$$t_s = \frac{\tau}{\sqrt{2(2n + 5)/9n(n - 1)}} \qquad \text{compared with} \qquad t_{\alpha[\infty]}$$

When $n \leq 40$, this approximation is not accurate, and Table **XIV** must be consulted. The table gives various (two-tailed) critical values of τ for $n = 4$ to 40. The minimal significant value of the coefficient at $P = 0.05$ is 0.390. Hence the observed value of τ is not significantly different from zero.

Suppose we have a sample of five individuals that have been arrayed by rank of variable Y_1 and whose rankings for a second variable Y_2 are entered paired with the ranks for Y_1:

$$
\begin{array}{cccccc}
Y_1 & 1 & 2 & 3 & 4 & 5 \\
Y_2 & 1 & 3 & 2 & 5 & 4
\end{array}
$$

Note that the ranking by variable Y_2 is not totally concordant with that by Y_1. The technique employed in Box 12.3 is to count the number of higher ranks following any given rank, sum this quantity for all ranks, multiply the sum $\Sigma^n C_i$ by 4, and subtract from the result a correction factor $n(n - 1)$ to obtain a statistic N. If, for purposes of illustration, we undertake to calculate the correlation of variable Y_1 with itself, we will find $\Sigma^n C_i = 4 + 3 + 2 + 1 + 0 = 10$. Then we compute $N = 4 \Sigma^n C_i - n(n - 1) = 40 - 5(4) = 20$, to obtain the maximum possible score $N = n(n - 1) = 20$. Obviously, Y_1, being ordered, is always perfectly concordant with itself. However, for Y_2 we obtain only $\Sigma^n C_i = 4 + 2 + 2 + 0 + 0 = 8$, and so $N = 4(8) - 5(4) = 12$. Since the maximum score of N for Y_2 (the score we would have if the correlation were perfect) is $n(n - 1) = 20$ and the observed score 12, an obvious coefficient suggests itself as $N/n(n - 1) = [4 \Sigma^n C_i - n(n - 1)]/n(n - 1) = 12/20 = 0.6$. Ties between individuals in the ranking process present minor complications that are dealt with in Box 12.3. The correlation in that box is between blood neutrophil counts and total marrow neutrophil mass in 15 cancer patients. The authors note that there is a product-moment correlation of 0.69 between these two variables, but when the data are analyzed by Kendall's rank correlation coefficient, the association between the two variables is low and nonsignificant. Examination of the data reveals that there is marked skewness in both variables. The data cannot, therefore, meet the assumptions of bivariate normality. Although there is little evidence of correlation among most of the variates, the three largest variates for each variable are correlated, and this induces the misleadingly high product-moment correlation coefficient.

The significance of τ for sample sizes greater than 40 can easily be tested by a standard error shown in Box 12.3. For sample sizes up to 40, look up critical values of τ in Table **XIV**.

Exercises

12.1   Graph the following data in the form of a bivariate scatter diagram. Compute the correlation coefficient and set 95% confidence intervals to ρ. The data were collected for a study of geographic variation in the aphid *Pemphigus populitransversus*. The values in the table represent locality means based on equal sample sizes for 23 localities in eastern North America. The variables, extracted from Sokal and Thomas (1965), are expressed in millimeters. Y_1 = tibia length, Y_2 = tarsus length. The correlation coefficient will estimate correlation of these two variables over localities. ANS. $r = 0.910$, $P < 0.01$.

Locality code number	Y_1	Y_2
1	0.631	0.140
2	0.644	0.139
3	0.612	0.140
4	0.632	0.141
5	0.675	0.155
6	0.653	0.148
7	0.655	0.146
8	0.615	0.136
9	0.712	0.159
10	0.626	0.140
11	0.597	0.133
12	0.625	0.144
13	0.657	0.147
14	0.586	0.134
15	0.574	0.134
16	0.551	0.127
17	0.556	0.130
18	0.665	0.147
19	0.585	0.138
20	0.629	0.150
21	0.671	0.148
22	0.703	0.151
23	0.662	0.142

12.2   The following data were extracted from a larger study by Brower (1959) on speciation in a group of swallowtail butterflies. Morphological measurements are in millimeters coded × 8.

Species	Specimen number	Y_1 Length of 8th tergile	Y_2 Length of superuncus
Papilio	1	24.0	14.0
multicaudatus	2	21.0	15.0
	3	20.0	17.5
	4	21.5	16.5
	5	21.5	16.0
	6	25.5	16.0
	7	25.5	17.5
	8	28.5	16.5
	9	23.5	15.0
	10	22.0	15.5
	11	22.5	17.5
	12	20.5	19.0
	13	21.0	13.5
	14	19.5	19.0
	15	26.0	18.0
	16	23.0	17.0
	17	21.0	18.0
	18	21.0	17.0
	19	20.5	16.0
	20	22.5	15.5
Papilio	21	20.0	11.5
rutulus	22	21.5	11.0
	23	18.5	10.0
	24	20.0	11.0
	25	19.0	11.0
	26	20.5	11.0
	27	19.5	11.0
	28	19.0	10.5
	29	21.5	11.0
	30	20.0	11.5
	31	21.5	10.0
	32	20.5	12.0
	33	20.0	10.5
	34	21.5	12.5
	35	17.5	12.0
	36	21.0	12.5
	37	21.0	11.5
	38	21.0	12.0
	39	19.5	10.5
	40	19.0	11.0
	41	18.0	11.5
	42	21.5	10.5
	43	23.0	11.0
	44	22.5	11.5
	45	19.0	13.0
	46	22.5	14.0
	47	21.0	12.5

Compute the correlation coefficient separately for each species and test significance of each. Test whether the two correlation coefficients differ significantly.

12.3 A pathologist measured the concentration of a toxic substance in the liver and in the peripheral blood (in $\mu g/kg$) in order to ascertain if the liver concentration is related to the blood concentration. Calculate τ and test its significance.

Liver	Blood
0.296	0.283
0.315	0.323
0.022	0.159
0.361	0.381
0.202	0.208
0.444	0.411
0.252	0.254
0.371	0.352
0.329	0.319
0.183	0.177
0.369	0.315
0.199	0.259
0.353	0.353
0.251	0.303
0.346	0.293

ANS. $\tau = 0.733$.

12.4 The following table of data is from an unpublished morphometric study of the cottonwood *Populus deltoides* by T. J. Crovello. Twenty-six leaves from one tree were measured when fresh and again after drying. The variables shown are fresh-leaf width (Y_1) and dry-leaf width (Y_2), both in millimeters. Calculate r and test its significance.

Y_1	Y_2	Y_1	Y_2
90	88	100	97
88	87	110	105
55	52	95	90
100	95	99	98
86	83	92	92
90	88	80	82
82	77	110	106
78	75	105	97
115	109	101	98
100	95	95	91
110	105	80	76
84	78	103	97
76	71		

12.5   Brown and Comstock (1952) found the following correlations between the length
of the wing and the width of a band on the wing of females of two samples
of the butterfly *Heliconius charitonius*:

Sample	n	r
1	100	0.29
2	46	0.70

Test whether the samples were drawn from populations with the same value of
ρ. ANS. No, $t_s = -3.104$, $P < 0.01$.

12.6   Test for the presence of association between tibia length and tarsus length in
the data of Exercise 12.1 using Kendall's coefficient of rank correlation.

CHAPTER 13

Analysis of Frequencies

Almost all our work so far has dealt with estimation of parameters and tests of hypotheses for continuous variables. The present chapter treats an important class of cases, tests of hypotheses about frequencies. Biological variables may be distributed into two or more classes, depending on some criterion such as arbitrary class limits in a continuous variable or a set of mutually exclusive attributes. An example of the former would be a frequency distribution of birth weights (a continuous variable arbitrarily divided into a number of contiguous classes); one of the latter would be a qualitative frequency distribution such as the frequency of individuals of ten different species obtained from a soil sample. For any such distribution we may hypothesize that it has been sampled from a population in which the frequencies of the various classes represent certain parametric proportions of the total frequency. We need a test of *goodness of fit* for our observed frequency distribution to the expected frequency distribution representing our hypothesis. You may recall that we first realized the need for such a test in Chapters 4 and 5, where we calculated expected binomial, Poisson, and normal frequency distributions but were unable to decide whether an observed sample distribution departed significantly from the theoretical one.

In Section 13.1 we introduce the idea of goodness of fit, discuss the types of significance tests that are appropriate, explain the basic rationale behind such tests, and develop general computational formulas for these tests.

Section 13.2 illustrates the actual computations for goodness of fit when the data are arranged by a single criterion of classification, as in a one-way quantitative or qualitative frequency distribution. This design applies to cases expected to follow one of the well-known frequency distributions such as the binomial, Poisson, or normal distribution. It applies as well to expected distributions following some other law suggested by the scientific subject matter under investigation, such as, for example, tests of goodness of fit of observed genetic ratios against expected Mendelian frequencies.

In Section 13.3 we proceed to significance tests of frequencies in two-way classifications—called *tests of independence*. We shall discuss the common tests of 2×2 tables in which each of two criteria of classification divides the frequencies into two classes, yielding a four-cell table, as well as $R \times C$ tables with more rows and columns.

Throughout this chapter we carry out goodness of fit tests by the G statistic. We briefly mention chi-square tests, which are the traditional way of analyzing such cases. But as is explained at various places throughout the text, G tests have general theoretical advantages over chi-square tests, as well as being computationally simpler, not only by computer, but also on most pocket or tabletop calculators.

13.1 Tests for goodness of fit: Introduction

The basic idea of a goodness of fit test is easily understood, given the extensive experience you now have with statistical hypothesis testing. Let us assume that a geneticist has carried out a crossing experiment between two F_1 hybrids and obtains an F_2 progeny of 90 offspring, 80 of which appear to be wild type and 10 of which are the mutant phenotype. The geneticist assumes dominance and expects a 3:1 ratio of the phenotypes. When we calculate the actual ratios, however, we observe that the data are in a ratio $80/10 = 8:1$. Expected values for p and q are $\hat{p} = 0.75$ and $\hat{q} = 0.25$ for the wild type and mutant, respectively. Note that we use the caret (generally called "hat" in statistics) to indicate hypothetical or expected values of the binomial proportions. However, the observed proportions of these two classes are $p = 0.89$ and $q = 0.11$, respectively. Yet another way of noting the contrast between observation and expectation is to state it in frequencies: the observed frequencies are $f_1 = 80$ and $f_2 = 10$ for the two phenotypes. Expected frequencies should be $\hat{f}_1 = \hat{p}n = 0.75(90) = 67.5$ and $\hat{f}_2 = \hat{q}n = 0.25(90) = 22.5$, respectively, where n refers to the sample size of offspring from the cross. Note that when we sum the expected frequencies they yield $67.5 + 22.5 = n = 90$, as they should.

The obvious question that comes to mind is whether the deviation from the 3:1 hypothesis observed in our sample is of such a magnitude as to be improbable. In other words, do the observed data differ enough from the expected values to cause us to reject the null hypothesis? For the case just considered, you already know two methods for coming to a decision about the null hypothesis. Clearly, this is a binomial distribution in which p is the probability of being a wild type and q is the probability of being a mutant. It is possible to work out the probability of obtaining an outcome of 80 wild type and 10 mutants as well as all "worse" cases for $\hat{p} = 0.75$ and $\hat{q} = 0.25$, and a sample of $n = 90$ offspring. We use the conventional binomial expression here $(\hat{p} + \hat{q})^n$ except that p and q are hypothesized, and we replace the symbol k by n, which we adopted in Chapter 4 as the appropriate symbol for the sum of all the frequencies in a frequency distribution. In this example, we have only one sample, so what would ordinarily be labeled k in the binomial is, at the same time, n. Such an example was illustrated in Table 4.3 and Section 4.2, and we can compute the cumulative probability of the tail of the binomial distribution. When this is done, we obtain a probability of 0.000,849 for all outcomes as deviant or more deviant from the hypothesis. Note that this is a one-tailed test, the alternative hypothesis being that there are, in fact, more wild-type offspring than the Mendelian hypothesis would postulate. Assuming $\hat{p} = 0.75$ and $\hat{q} = 0.25$, the observed sample is, consequently, a very unusual outcome, and we conclude that there is a significant deviation from expectation.

A less time-consuming approach based on the same principle is to look up confidence limits for the binomial proportions, as was done for the sign test in Section 10.3. Interpolation in Table **IX** shows that for a sample of $n = 90$, an observed percentage of 89% would yield approximate 99% confidence limits of 78 and 96 for the true percentage of wild-type individuals. Clearly, the hypothesized value for $\hat{p} = 0.75$ is beyond the 99% confidence bounds.

Now, let us develop a third approach by a goodness of fit test. Table 13.1 illustrates how we might proceed. The first column gives the observed frequencies f representing the outcome of the experiment. Column (2) shows the observed frequencies as (observed) proportions p and q computed as f_1/n and f_2/n, respectively. Column (3) lists the expected proportions for the particular null hypothesis being tested. In this case, the hypothesis is a $3:1$ ratio, corresponding to expected proportions $\hat{p} = 0.75$ and $\hat{q} = 0.25$, as we have seen. In column (4) we show the expected frequencies, which we have already calculated for these proportions as $\hat{f}_1 = \hat{p}n = 0.75(90) = 67.5$ and $f_2 = \hat{q}n = 0.25(90) = 22.5$.

The log likelihood ratio test for goodness of fit may be developed as follows. Using Expression (4.1) for the expected relative frequencies in a binomial distribution, we compute two quantities of interest to us here:

$$C(90, 80)(\tfrac{80}{90})^{80}(\tfrac{10}{90})^{10} = 0.132,683,8$$

$$C(90, 80)(\tfrac{3}{4})^{80}(\tfrac{1}{4})^{10} = 0.000,551,754,9$$

The first quantity is the probability of observing the sampled results (80 wild type and 10 mutants) on the hypothesis that $\hat{p} = p$—that is, that the population parameter equals the observed sample proportion. The second is the probability of observing the sampled results assuming that $\hat{p} = \tfrac{3}{4}$, as per the Mendelian null

TABLE 13.1

Developing the *G* test (likelihood ratio test) and the chi-square test for goodness of fit. Observed and expected frequencies from the outcome of a genetic cross, assuming a 3:1 ratio of phenotypes among the offspring.

Phenotypes	(1) Observed frequencies f	(2) Observed proportions f/n	(3) Expected proportions $\hat{p}$ and $\hat{q}$	(4) Expected frequencies $\hat{f}$	(5) Ratio $f/\hat{f}$	(6) $f \ln(f/\hat{f})$	(7) Deviations from expectation $f - \hat{f}$	(8) Deviations squared $(f - \hat{f})^2$	(9) $(f - \hat{f})^2/\hat{f}$
Wild type	80	$p = \frac{8}{9}$	$\hat{p} = 0.75$	$\hat{p}n = 67.5$	1.185,185	13.591,92	12.5	156.25	2.314,81
Mutant	$\underline{10}$	$q = \frac{1}{9}$	$\hat{q} = 0.25$	$\hat{q}n = \underline{22.5}$	0.444,444	$\underline{-8.109,30}$	$\underline{-12.5}$	156.25	$\underline{6.944,44}$
Sum	90	1.0	1.0	90.0		$\ln L = \ 5.482,62$	0		$X^2 = 9.259,26$

hypothesis. Note that these expressions yield the probabilities for the observed outcomes only, *not for observed and all worse outcomes*. Thus, $P = 0.000,551,8$ is less than the earlier computed $P = 0.000,849$, which is the probability of 10 *and fewer* mutants, assuming $\hat{p} = \frac{3}{4}$, $\hat{q} = \frac{1}{4}$.

The first probability (0.132,683,8) is greater than the second (0.000,551,754,9), since the hypothesis is based on the observed data. If the observed proportion p is in fact equal to the proportion $\hat{p}$ postulated under the null hypothesis, then the two computed probabilities will be equal and their ratio, L, will equal 1.0. The greater the difference between p and $\hat{p}$ (the expected proportion under the null hypothesis), the higher the ratio will be (the probability based on p is divided by the probability based on $\hat{p}$ or defined by the null hypothesis). This indicates that the ratio of these two probabilities or *likelihoods* can be used as a statistic to measure the degree of agreement between sampled and expected frequencies. A test based on such a ratio is called a *likelihood ratio test*. In our case, $L = 0.132,683,8/0.000,551,754,9 = 240.4761$.

It has been shown that the distribution of

$$G = 2 \ln L \tag{13.1}$$

can be approximated by the χ^2 distribution when sample sizes are large (for a definition of "large" in this case, see Section 13.2). The appropriate number of degrees of freedom in Table 13.1 is 1 because the frequencies in the two cells for these data add to a constant sample size, 90. The outcome of the sampling experiment could have been any number of mutants from 0 to 90, but the number of wild type consequently would have to be constrained so that the total would add up to 90. One of the cells in the table is free to vary, the other is constrained. Hence, there is one degree of freedom.

In our case,

$$G = 2 \ln L = 2(5.482,62) = 10.9652$$

If we compare this observed value with a χ^2 distribution with one degree of freedom, we find that the result is significant ($P < 0.001$). Clearly, we reject the 3:1 hypothesis and conclude that the proportion of wild type is greater than 0.75. The geneticist must, consequently, look for a mechanism explaining this departure from expectation.

We shall now develop a simple computational formula for G. Referring back to Expression (4.1), we can rewrite the two probabilities computed earlier as

$$C(n, f_1)p^{f_1}q^{f_2} \tag{13.2}$$

and

$$C(n, f_1)\hat{p}^{f_1}\hat{q}^{f_2} \tag{13.2a}$$

But

$$L = \frac{C(n, f_1)p^{f_1}q^{f_2}}{C(n, f_1)\hat{p}^{f_1}\hat{q}^{f_2}} = \left(\frac{p}{\hat{p}}\right)^{f_1}\left(\frac{q}{\hat{q}}\right)^{f_2}$$

Since $f_1 = np$ and $\hat{f}_1 = n\hat{p}$ and similarly $f_2 = nq$ and $\hat{f}_2 = n\hat{q}$,

$$L = \left(\frac{f_1}{\hat{f}_1}\right)^{f_1} \left(\frac{f_2}{\hat{f}_2}\right)^{f_2}$$

and

$$\ln L = f_1 \ln\left(\frac{f_1}{\hat{f}_1}\right) + f_2 \ln\left(\frac{f_2}{\hat{f}_2}\right) \tag{13.3}$$

The computational steps implied by Expression (13.3) are shown in columns (5) and (6) of Table 13.1. In column (5) are given the ratios of observed over expected frequencies. These ratios would be 1 in the unlikely case of a perfect fit of observations to the hypothesis. In such a case, the logarithms of these ratios entered in column (6) would be 0, as would their sum. Consequently, G, which is twice the natural logarithm of L, would be 0, indicating a perfect fit of the observations to the expectations.

It has been shown that the distribution of G follows a χ^2 distribution. In the particular case we have been studying—the two phenotype classes—the appropriate χ^2 distribution would be the one for one degree of freedom. We can appreciate the reason for the single degree of freedom when we consider the frequencies in the two classes of Table 13.1 and their sum: $80 + 10 = 90$. In such an example, the total frequency is fixed. Therefore, if we were to vary the frequency of any one class, the other class would have to compensate for changes in the first class to retain a correct total. Here the meaning of *one degree of freedom* becomes quite clear. One of the classes is free to vary; the other is not.

The test for goodness of fit can be applied to a distribution with more than two classes. If we designate the number of frequency classes in the Table as a, the operation can be expressed by the following general computational formula, whose derivation, based on the multinominal expectations (for more than two classes), is shown in Appendix A1.9:

$$G = 2 \sum_{i}^{a} f_i \ln\left(\frac{f_i}{\hat{f}_i}\right) \tag{13.4}$$

Thus the formula can be seen as the sum of the independent contributions of departures from expectation $(\ln (f_i/\hat{f}_i))$ weighted by the frequency of the particular class (f_i). If the expected values are given as a proportion, a convenient computational formula for G, also derived in Appendix A1.9, is

$$G = 2\left[\sum_{i}^{a} f_i \ln\left(\frac{f_i}{\hat{p}_i}\right) - n \ln n\right] \tag{13.5}$$

To evaluate the outcome of our test of goodness of fit, we need to know the appropriate number of degrees of freedom to be applied to the χ^2 distribution. For a classes, the number of degrees of freedom is $a - 1$. Since the sum of frequencies in any problem is fixed, this means that $a - 1$ classes are free to vary, whereas the ath class must constitute the difference between the total sum and the sum of the previous $a - 1$ classes.

In some goodness of fit tests involving more than two classes, we subtract more than one degree of freedom from the number of classes, a. These are instances where the parameters for the null hypothesis have been extracted from the sample data themselves, in contrast with the null hypotheses encountered in Table 13.1. In the latter case, the hypothesis to be tested was generated on the basis of the investigator's general knowledge of the specific problem and of Mendelian genetics. The values of $\hat{p} = 0.75$ and $\hat{q} = 0.25$ were dictated by the 3:1 hypothesis and were not estimated from the sampled data. For this reason, the expected frequencies are said to have been based on an *extrinsic hypothesis*, a hypothesis external to the data. By contrast, consider the expected Poisson frequencies of yeast cells in a hemacytometer (Box 4.1). You will recall that to compute these frequencies, you needed values for μ, which you estimated from the sample mean $\bar{Y}$. Therefore, the parameter of the computed Poisson distribution came from the sampled observations themselves. The expected Poisson frequencies represent an *intrinsic hypothesis*. In such a case, to obtain the correct number of degrees of freedom for the test of goodness of fit, we would subtract from a, the number of classes into which the data had been grouped, not only one degree of freedom for n, the sum of the frequencies, but also one further degree of freedom for the estimate of the mean. Thus, in such a case, a sample statistic G would be compared with chi-square for $a - 2$ degrees of freedom.

Now let us introduce you to an alternative technique. This is the traditional approach with which we must acquaint you because you will see it applied in the earlier literature and in a substantial proportion of current research publications. We turn once more to the genetic cross with 80 wild-type and 10 mutant individuals. The computations are laid out in columns (7), (8), and (9) in Table 13.1.

We first measure $f - \hat{f}$, the deviation of observed from expected frequencies. Note that the sum of these deviations equals zero, for reasons very similar to those causing the sum of deviations from a mean to add to zero. Following our previous approach of making all deviations positive by squaring them, we square $(f - \hat{f})$ in column (8) to yield a measure of the magnitude of the deviation from expectation. This quantity must be expressed as a proportion of the expected frequency. After all, if the expected frequency were 13.0, a deviation of 12.5 would be an extremely large one, comprising almost 100% of $\hat{f}$, but such a deviation would represent only 10% of an expected frequency of 125.0. Thus, we obtain column (9) as the quotient of division of the quantity in column (8) by that in column (4). Note that the magnitude of the quotient is greater for the second line, in which the $\hat{f}$ is smaller. Our next step in developing our test statistic is to sum the quotients, which is done at the foot of column (9), yielding a value of 9.259,26.

This test is called the *chi-square test* because the resultant statistic, X^2, is distributed as chi-square with $a - 1$ degrees of freedom. Many persons inappropriately call the statistic obtained as the sum of column (9) a chi-square. However, since the sample statistic is not a chi-square, we have followed the increasingly prevalent convention of labeling the sample statistic X^2 rather than χ^2. The value of $X^2 = 9.259,26$ from Table 13.1, when compared with the critical value of χ^2 (Table **IV**), is highly significant ($P < 0.005$). The chi-square test is always one-tailed. Since the deviations are squared, negative and positive deviations both result in positive values of X^2. Clearly, we reject the 3:1 hypothesis and conclude that the proportion of wild type is greater than 0.75. The geneticist must, consequently, look for a mechanism explaining this departure from expectation. Our conclusions are the same as with the G test. In general, X^2 will be numerically similar to G.

We can apply the chi-square test for goodness of fit to a distribution with more than two classes as well. The operation can be described by the formula

$$X^2 = \sum^a \frac{(f_i - \hat{f}_i)^2}{\hat{f}_i} \tag{13.6}$$

which is a generalization of the computations carried out in columns (7), (8), and (9) of Table 13.1. The pertinent degrees of freedom are again $a - 1$ in the case of an extrinsic hypothesis and vary in the case of an intrinsic hypothesis. The formula is straightforward and can be applied to any of the examples we show in the next section, although we carry these out by means of the G test.

13.2 Single-classification goodness of fit tests

Before we discuss in detail the computational steps involved in tests of goodness of fit of single-classification frequency distributions, some remarks on the choice of a test statistic are in order. We have already stated that the traditional method for such a test is the chi-square test for goodness of fit. However, the newer approach by the G test has been recommended on theoretical grounds. The major advantage of the G test is that it is computationally simpler, especially in more complicated designs. Earlier reservations regarding G when desk calculators are used no longer apply. The common presence of natural logarithm keys on pocket and tabletop calculators makes G as easy to compute as X^2.

The G tests of goodness of fit for single-classification frequency distributions are given in Box 13.1. Expected frequencies in three or more classes can be based on either extrinsic or intrinsic hypotheses, as discussed in the previous section. Examples of goodness of fit tests with more than two classes might be as follows: A genetic cross with four phenotypic classes might be tested against an expected ratio of 9:3:3:1 for these classes. A phenomenon that occurs over various time periods could be tested for uniform frequency of occurrence—for example, number of births in a city over 12 months: Is the frequency of births equal in each month? In such a case the expected frequencies are computed as being equally likely in each class. Thus, for a classes, the expected frequency for any one class would be n/a.

BOX 13.1
G Test for Goodness of Fit. Single Classification.

1. *Frequencies divided into a ≥ 2 classes:* Sex ratio in 6115 sibships of 12 in Saxony. The fourth column gives the expected frequencies, assuming a binomial distribution. These were first computed in Table 4.4 but are here given to five-decimal-place precision to give sufficient accuracy to the computation of *G*.

(1) ♂♂	(2) ♀♀	(3) f	(4) $\hat{f}$	(5) Deviation from expectation
12	0	7 ⎫ 52	2.347,27 ⎫ 28.429,73	+
11	1	45 ⎭	26.082,46 ⎭	
10	2	181	132.835,70	+
9	3	478	410.012,56	+
8	4	829	854.246,65	−
7	5	1112	1265.630,31	−
6	6	1343	1367.279,36	−
5	7	1033	1085.210,70	−
4	8	670	628.055,01	+
3	9	286	258.475,13	+
2	10	104	71.803,17	+
1	11	24 ⎫ 27	12.088,84 ⎫ 13.021,68	+
0	12	3 ⎭	0.932,84 ⎭	
		6115 = n	6115.000,00	

Since expected frequencies $\hat{f}_i < 3$ for $a = 13$ classes should be avoided, we lump the classes at both tails with the adjacent classes to create classes of adequate size. Corresponding classes of observed frequencies f_i should be lumped to match. The number of classes after lumping is $a = 11$.

Compute G by Expression (13.4):

$$G = 2\sum^a f_i \ln\left(\frac{f_i}{\hat{f}_i}\right)$$

$$= 2\left(52 \ln\left(\frac{52}{28.429,73}\right) + 181 \ln\left(\frac{181}{132.835,70}\right) + \cdots + 27 \ln\left(\frac{27}{13.021,68}\right)\right)$$

$$= 94.871,55$$

Since there are $a = 11$ classes remaining, the degrees of freedom would be $a - 1 = 10$, if this were an example tested against expected frequencies based on an extrinsic hypothesis. However, because the expected frequencies are based on a binomial distribution with mean $\hat{p}_\delta$ estimated from the p_δ of the sample, a further degree of freedom is removed, and the sample value of G is compared with a χ^2 distribution with $a - 2 = 11 - 2 = 9$ degrees of freedom. We applied Williams' correction to G, to obtain a better approximation to χ^2. In the formula computed below, v symbolizes the pertinent degrees of freedom of the

BOX 13.1
Continued

problem. We obtain

$$q = 1 + \frac{a^2 - 1}{6nv}$$

$$= 1 + \frac{11^2 - 1}{6(6115)(9)} = 1.000,363,4$$

$$G_{adj} = \frac{G}{q} = \frac{94.871,55}{1.000,363,4} = 94.837,09$$

$$G_{adj} = 94.837,09 > \chi^2_{0.001[9]} = 27.877$$

The null hypothesis—that the sample data follow a binomial distribution—is therefore rejected decisively.

Typically, the following degrees of freedom will pertain to G tests for goodness of fit with expected frequencies based on a hypothesis *intrinsic* to the sample data (a is the number of classes after lumping, if any):

Distribution	Parameters estimated from sample	df
Binomial	$\hat{p}$	$a - 2$
Normal	μ, σ	$a - 3$
Poisson	μ	$a - 2$

When the parameters for such distributions are estimated from hypotheses *extrinsic* to the sampled data, the degrees of freedom are uniformly $a - 1$.

2. *Special case of frequencies divided in $a = 2$ classes:* In an F_2 cross in drosophila, the following 176 progeny were obtained, of which 130 were wild-type flies and 46 ebony mutants. Assuming that the mutant is an autosomal recessive, one would expect a ratio of 3 wild-type flies to each mutant fly. To test whether the observed results are consistent with this 3:1 hypothesis, we set up the data as follows.

Flies	f	Hypothesis	$\hat{f}$
Wild type	$f_1 = 130$	$\hat{p} = 0.75$	$\hat{p}n = 132.0$
Ebony mutant	$f_2 = 46$	$\hat{q} = 0.25$	$\hat{q}n = 44.0$
	$n = 176$		176.0

Computing G from Expression (13.4), we obtain

$$G = 2 \sum^a f_i \ln\left(\frac{f_i}{\hat{f}_i}\right)$$

$$= 2\left[130 \ln\left(\tfrac{130}{132}\right) + 46 \ln\left(\tfrac{46}{44}\right)\right]$$

$$= 0.120,02$$

BOX 13.1
Continued

Williams' correction for the two-cell case is $q = 1 + 1/2n$, which is

$$1 + \frac{1}{2(176)} = 1.002,84$$

in this example.

$$G_{adj} = \frac{G}{q} = \frac{0.120,02}{1.002,84} = 0.1197$$

Since $G_{adj} \ll \chi^2_{0.05[1]} = 3.841$, we clearly do not have sufficient evidence to reject our null hypothesis.

The case presented in Box 13.1, however, is one in which the expected frequencies are based on an intrinsic hypothesis. We use the sex ratio data in sibships of 12, first introduced in Table 4.4, Section 4.2. As you will recall, the expected frequencies in these data are based on the binomial distribution, with the parametric proportion of males $\hat{p}_\delta$ estimated from the observed frequencies of the sample ($p_\delta = 0.519,215$). The computation of this case is outlined fully in Box 13.1.

The G test does not yield very accurate probabilities for small $\hat{f}_i$. The cells with $\hat{f}_i < 3$ (when $a \geq 5$) or $\hat{f}_i < 5$ (when $a < 5$) are generally lumped with adjacent classes so that the new $\hat{f}_i$ are large enough. The lumping of classes results in a less powerful test with respect to alternative hypotheses. By these criteria the classes of $\hat{f}_i$ at both tails of the distribution are too small. We lump them by adding their frequencies to those in contiguous classes, as shown in Box 13.1. Clearly, the observed frequencies must be lumped to match. The number of classes a is the number *after* lumping has taken place. In our case, $a = 11$.

Because the actual type I error of G tests tends to be higher than the intended level, a correction for G to obtain a better approximation to the chi-square distribution has been suggested by Williams (1976). He divides G by a correction factor q (not to be confused with a proportion) to be computed as $q = 1 + (a^2 - 1)/6nv$. In this formula, v is the number of degrees of freedom appropriate to the G test. The effect of this correction is to reduce the observed value of G slightly.

Since this is an example with expected frequencies based on an intrinsic hypothesis, we have to subtract more than one degree of freedom from a for the significance test. In this case, we estimated $\hat{p}_\delta$ from the sample, and therefore a second degree of freedom is subtracted from a, making the final number of degrees of freedom $a - 2 = 11 - 2 = 9$. Comparing the corrected sample value of $G_{adj} = 94.837,09$ with the critical value of χ^2 at 9 degrees of freedom, we find it highly significant ($P \ll 0.001$, assuming that the null hypothesis is correct). We therefore reject this hypothesis and conclude that the sex ratios are not binomially distributed. As is evident from the pattern of deviations, there is an excess of sibships in which one sex or the other predominates. Had we applied the chi-square test to these data, the critical value would have been the same ($\chi^2_{\alpha[9]}$).

Next we consider the case for $a = 2$ cells. The computation is carried out by means of Expression (13.4), as before. In tests of goodness of fit involving only two classes, the value of G as computed from this expression will typically result in type I errors at a level higher than the intended one. Williams' correction reduces the value of G and results in a more conservative test. An alternative correction that has been widely applied is the *correction for continuity*, usually applied in order to make the value of G or X^2 approximate the χ^2 distribution more closely. We have found the continuity correction too conservative and therefore recommend that Williams' correction be applied routinely, although it will have little effect when sample sizes are large. For sample sizes of 25 or less, work out the exact probabilities as shown in Table 4.3, Section 4.2.

The example of the two cell case in Box 13.1 is a genetic cross with an expected 3:1 ratio. The G test is adjusted by Williams' correction. The expected frequencies differ very little from the observed frequencies, and it is no surprise, therefore, that the resulting value of G_{adj} is far less than the critical value of χ^2 at one degree of freedom. Inspection of the chi-square table reveals that roughly 80% of all samples from a population with the expected ratio would show greater deviations than the sample at hand.

13.3 Tests of independence: Two-way tables

The notion of statistical or probabilistic independence was first introduced in Section 4.1, where it was shown that if two events were independent, the probability of their occurring together could be computed as the product of their separate probabilities. Thus, if among the progeny of a certain genetic cross the probability that a kernel of corn will be red is $\frac{1}{2}$ and the probability that the kernel will be dented is $\frac{1}{3}$, the probability of obtaining a kernel both dented and red will be $\frac{1}{2} \times \frac{1}{3} = \frac{1}{6}$, if the joint occurrences of these two characteristics are statistically independent.

The appropriate statistical test for this genetic problem would be to test the frequencies for goodness of fit to the expected ratios of 2 (red, not dented):2 (not red, not dented):1 (red, dented):1 (not red, dented). This would be a simultaneous test of two null hypotheses: that the expected proportions are $\frac{1}{2}$ and $\frac{1}{3}$ for red and dented, respectively, and that these two properties are independent. The first null hypothesis tests the Mendelian model in general. The second tests whether these characters assort independently—that is, whether they are determined by genes located in different linkage groups. If the second hypothesis must be rejected, this is taken as evidence that the characters are linked—that is, located on the same chromosome.

There are numerous instances in biology in which the second hypothesis, concerning the independence of two properties, is of great interest and the first hypothesis, regarding the true proportion of one or both properties, is of little interest. In fact, often no hypothesis regarding the parametric values $\hat{p}_i$ can be formulated by the investigator. We shall cite several examples of such situations, which lead to the test of independence to be learned in this section. We employ this test whenever we wish to test whether two different properties, each occurring in two states, are dependent on each other. For instance, specimens of a certain moth may occur in two color phases—light and dark. Fifty specimens of each phase may be exposed in the open, subject to predation by birds. The number of surviving moths is counted after a fixed interval of time. The proportion predated may differ in the two color phases. The two properties in this example are color and survival. We can divide our sample into four classes: light-colored survivors, light-colored prey, dark survivors, and dark prey. If the probability of being preyed upon is independent of the color of the moth, the expected frequencies of these four classes can be simply computed as independent products of the proportion of each color (in our experiment, $\frac{1}{2}$) and the overall proportion preyed upon in the entire sample. Should the statistical test of independence explained below show that the two properties are not independent, we are led to conclude that one of the color phases is more susceptible to predation than the other. In this example, this is the issue of biological importance; the exact proportions of the two properties are of little interest here. The proportion of the color phases is arbitrary, and the proportion of survivors is of interest only insofar as it differs for the two phases.

A second example might relate to a sampling experiment carried out by a plant ecologist. A random sample is obtained of 100 individuals of a fairly rare species of tree distributed over an area of 400 square miles. For each tree the ecologist notes whether it is rooted in a serpentine soil or not, and whether the leaves are pubescent or smooth. Thus the sample of $n = 100$ trees can be divided into four groups: serpentine-pubescent, serpentine-smooth, nonserpentine-pubescent, and nonserpentine-smooth. If the probability that a tree is or is not pubescent is independent of its location, our null hypothesis of the independence of these properties will be upheld. If, on the other hand, the proportion of pubescence differs for the two types of soils, our statistical test will most probably result in rejection of the null hypothesis of independence. Again, the expected frequencies will simply be products of the independent proportions of the two properties—serpentine versus nonserpentine, and pubescent versus smooth. In this instance the proportions may themselves be of interest to the investigator.

An analogous example may occur in medicine. Among 10,000 patients admitted to a hospital, a certain proportion may be diagnosed as exhibiting disease X. At the same time, all patients admitted are tested for several blood groups. A certain proportion of these are members of blood group Y. Is there some association between membership in blood group Y and susceptibility to the disease X?

The example we shall work out in detail is from immunology. A sample of 111 mice was divided into two groups: 57 that received a standard dose of pathogenic bacteria followed by an antiserum, and a control group of 54 that received the bacteria but no antiserum. After sufficient time had elapsed for an incubation period and for the disease to run its course, 38 dead mice and 73 survivors were counted. Of those that died, 13 had received bacteria *and* antiserum while 25 had received bacteria only. A question of interest is whether the antiserum had in any way protected the mice so that there were proportionally more survivors in that group. Here again the proportions of these properties are of no more interest than in the first example (predation on moths).

Such data are conveniently displayed in the form of a *two-way table* as shown below. Two-way and multiway tables (more than two criteria) are often known as *contingency tables*. This type of two-way table, in which each of the two criteria is divided into two classes, is known as a 2×2 *table*.

	Dead	Alive	$\sum$
Bacteria and antiserum	13	44	57
Bacteria only	25	29	54
$\sum$	38	73	111

Thus 13 mice received bacteria and antiserum but died, as seen in the table. The marginal totals give the number of mice exhibiting any one property: 57 mice received bacteria and antiserum; 73 mice survived the experiment. Altogether 111 mice were involved in the experiment and constitute the total sample.

In discussing such a table it is convenient to label the cells of the table and the row and column sums as follows:

a	b	$a + b$
c	d	$c + d$
$a + c$	$b + d$	n

From a two-way table one can systematically compute the expected frequencies (based on the null hypothesis of independence) and compare them with the observed frequencies. For example, the expected frequency for cell d (bacteria, alive) would be

$$\hat{f}_{\text{bact,alv}} = n\hat{p}_{\text{bact,alv}} = n\hat{p}_{\text{bact}} \times \hat{p}_{\text{alv}} = n\left(\frac{c + d}{n}\right)\left(\frac{b + d}{n}\right) = \frac{(c + d)(b + d)}{n}$$

which in our case would be $(54)(73)/111 = 35.514$, a higher value than the observed frequency of 29. We can proceed similarly to *compute the expected frequencies for each cell in the table by multiplying a row total by a column total, and dividing the product by the grand total.* The expected frequencies can be conveniently displayed in the form of a two-way table:

	Dead	Alive	Σ
Bacteria and antiserum	19.514	37.486	57.000
Bacteria only	18.486	35.514	54.000
Σ	38.000	73.000	111.000

You will note that the row and column sums of this table are identical to those in the table of observed frequencies, which should not surprise you, since the expected frequencies were computed on the basis of these row and column totals. It should therefore be clear that a test of independence will not test whether any property occurs at a given proportion but can only test whether or not the two properties are manifested independently.

The statistical test appropriate to a given 2×2 table depends on the underlying model that it represents. There has been considerable confusion on this subject in the statistical literature. For our purposes here it is not necessary to distinguish among the three models of contingency tables. The G test illustrated in Box 13.2 will give at least approximately correct results with moderate- to large-sized samples regardless of the underlying model. When the test is applied to the above immunology example, using the formulas given in Box 13.2, one obtains $G_{adj} = 6.7732$. One could also carry out a chi-square test on the deviations of the observed from the expected frequencies using Expression (13.2). This would yield $\chi^2 = 6.7966$, using the expected frequencies in the table above. Let us state without explanation that the observed G or X^2 should be compared with χ^2 for one degree of freedom. We shall examine the reasons for this at the end of this section. The probability of finding a fit as bad, or worse, to these data is $0.005 < P < 0.01$. We conclude, therefore, that mortality in these mice is not independent of the presence of antiserum. We note that the percentage mortality among those animals given bacteria *and* antiserum is $(13)(100)/57 = 22.8\%$, considerably lower than the mortality of $(25)(100)/54 = 46.3\%$ among the mice to whom only bacteria had been administered. Clearly, the antiserum has been effective in reducing mortality.

In Box 13.2 we illustrate the G test applied to the sampling experiment in plant ecology, dealing with trees rooted in two different soils and possessing two types of leaves. With small sample sizes ($n < 200$), it is desirable to apply *Williams' correction*, the application of which is shown in the box. The result of the analysis shows clearly that we cannot reject the null hypothesis of independence between soil type and leaf type. The presence of pubescent leaves is independent of whether the tree is rooted in serpentine soils or not.

Tests of independence need not be restricted to 2×2 tables. In the two-way cases considered in this section, we are concerned with only two properties, but each of these properties may be divided into any number of classes. Thus organisms may occur in four color classes and be sampled at five different times during the year, yielding a 4×5 test of independence. Such a test would examine whether the color proportions exhibited by the marginal totals are independent of the times at which the individuals have been sampled. Such tests

BOX 13.2

2 × 2 test of independence.

A plant ecologist samples 100 trees of a rare species from a 400-square-mile area. He records for each tree whether it is rooted in serpentine soils or not, and whether its leaves are pubescent or smooth.

Soil	Pubescent	Smooth	Totals
Serpentine	12	22	34
Not Serpentine	16	50	66
Totals	28	72	$100 = n$

The conventional algebraic representation of this table is as follows:

$$\begin{array}{ccc} & & \Sigma \\ a & b & a + b \\ c & d & c + d \\ \Sigma \quad a + c & b + d & a + b + c + d = n \end{array}$$

Compute the following quantities.

1. $\sum f \ln f$ for the cell frequencies $= 12 \ln 12 + 22 \ln 22 + 16 \ln 16 + 50 \ln 50$
$$= 337.784,38$$

2. $\sum f$ for the row and column totals $= 34 \ln 34 + 66 \ln 66 + 28 \ln 28 + 72 \ln 72$
$$= 797.635,16$$

3. $n \ln n = 100 \ln 100 = 460.517,02$

4. Compute G as follows:

$$G = 2(\text{quantity } \mathbf{1} - \text{quantity } \mathbf{2} + \text{quantity } \mathbf{3})$$
$$= 2(337.784,38 - 797.635,16 + 460.517,02)$$
$$= 2(0.666,24) = 1.332,49$$

Williams' correction for a 2×2 table is

$$q = 1 + \frac{\left(\dfrac{n}{a+b} + \dfrac{n}{c+d} - 1\right)\left(\dfrac{n}{a+c} + \dfrac{n}{b+d} - 1\right)}{6n}$$

For these data we obtain

$$q = 1 + \frac{\left(\frac{100}{34} + \frac{100}{66} - 1\right)\left(\frac{100}{28} + \frac{100}{72} - 1\right)}{6(100)}$$

$$= 1.022,81$$

$$G_{adj} = \frac{G}{q} = \frac{1.332,49}{1.022,81} = 1.3028$$

Compare G_{adj} with critical value of χ^2 for one degree of freedom. Since our observed G_{adj} is much less than $\chi^2_{0.05[1]} = 3.841$, we accept the null hypothesis that the leaf type is independent of the type of soil in which the tree is rooted.

■

BOX 13.3
$R \times C$ test of independence using the G test.

Frequencies for the M and N blood groups in six populations from Lebanon.

Populations ($b = 6$)	Genotypes ($a = 3$)			Totals	%MM	%MN	%NN
	MM	MN	NN				
Druse	59	100	44	203	29.06	49.26	21.67
Greek Catholic	64	98	41	203	31.53	48.28	20.20
Greek Orthodox	44	94	49	187	23.53	50.27	26.20
Maronites	342	435	165	942	36.31	46.18	17.52
Shiites	140	259	104	503	27.83	51.49	20.68
Sunni Moslems	169	168	91	428	39.49	39.25	21.03
Totals	818	1154	494	2466			

Source: Ruffie and Taleb (1965).

Compute the following quantities.

1. Sum of transforms of the frequencies in the body of the contingency table

$$= \sum^b \sum^a f_{ij} \ln f_{ij} = 59 \ln 59 + 100 \ln 100 + \cdots + 91 \ln 91$$
$$= 240.575 + 460.517 + \cdots + 40.488 = 12{,}752.715$$

2. Sum of transforms of the row totals

$$= \sum^b \left(\sum^a f_{ij} \right) \ln \left(\sum^a f_{ij} \right)$$
$$= 203 \ln 203 + \cdots + 428 \ln 428 = 1078.581 + \cdots + 2593.305$$
$$= 15{,}308.461$$

3. Sum of the transforms of the column totals

$$= \sum^a \left(\sum^b f_{ij} \right) \ln \left(\sum^b f_{ij} \right)$$
$$= 818 \ln 818 + \cdots + 494 \ln 494 = 5486.213 + \cdots + 3064.053 =$$

4. Transform of the grand total $= n \ln n = 2466 \ln 2466 = 19{,}260.330$

5. $G = 2($quantity **1** $-$ quantity **2** $-$ quantity **3** $+$ quantity **4**$)$

$$= 2(12{,}752.715 - 15{,}308.46 - 16{,}687.108 + 19{,}260.330) = 2(17.475) = 34.951$$

6. The lower bound estimate of q using Williams' correction for an $a \times b$ table is

BOX 13.3
Continued

$$q_{min} = 1 + \frac{(a + 1)(b + 1)}{6n}$$

$$= 1 + \frac{(3 + 1)(6 + 1)}{6(2466)}$$

$$= 1.001,892$$

Thus $G_{adj} = G/q_{min} = 34.951/1.001,892 = 34.885$.

This value is to be compared with a χ^2 distribution with $(a - 1)(b - 1)$ degrees of freedom, where a is the number of columns and b the number of rows in the table. In our case, $df = (3 - 1)(6 - 1) = 10$.

Since $\chi^2_{0.001[10]} = 29.588$, our G value is significant at $P < 0.001$, and we must reject our null hypothesis that genotype frequency is independent of the population sampled.

are often called $R \times C$ *tests of independence*, R and C standing for the number of rows and columns in the frequency table. Another case, examined in detail in Box 13.3, concerns the MN blood groups which occur in human populations in three genotypes—MM, MN, and NN. Frequencies of these blood groups can be obtained in samples of human populations and the samples compared for differences in these frequencies. In Box 13.3 we feature frequencies from six Lebanese populations and test whether the proportions of the three groups are independent of the populations sampled, or in other words, whether the frequencies of the three genotypes differ among these six populations.

As shown in Box 13.3, the following is a simple general rule for computation of the G test of independence:

$$G = 2[(\sum f \ln f \text{ for the cell frequencies})$$
$$- (\sum f \ln f \text{ for the row and column totals}) + n \ln n]$$

The transformations can be computed using the natural logarithm function found on most calculators. In the formulas in Box 13.3 we employ a double subscript to refer to entries in a two-way table, as in the structurally similar case of two-way anova. The quantity f_{ij} in Box 13.3 refers to the observed frequency in row i and column j of the table. Williams' correction is now more complicated. We feature a lower bound estimate of its correct value. The adjustment will be minor when sample size is large, as in this example, and need be carried out only when the sample size is small and the observed G value is of marginal significance.

The results in Box 13.3 show clearly that the frequency of the three genotypes is dependent upon the population sampled. We note the lower frequency of the

MM genotypes in the third population (Greek Orthodox) and the much lower frequency of the MN heterozygotes in the last population (Sunni Moslems).

The *degrees of freedom for tests of independence* are always the same and can be computed using the rules given earlier (Section 13.2). There are k cells in the table but we must subtract one degree of freedom for each independent parameter we have estimated from the data. We must, of course, subtract one degree of freedom for the observed total sample size, n. We have also estimated $a - 1$ row probabilities and $b - 1$ column probabilities, where a and b are the number of rows and columns in the table, respectively. Thus, there are $k - (a - 1) - (b - 1) - 1 = k - a - b + 1$ degrees of freedom for the test. But since $k = a \times b$, this expression becomes $(a \times b) - a - b + 1 = (a - 1) \times (b - 1)$, the conventional expression for the degrees of freedom in a two-way test of independence. Thus, the degrees of freedom in the example of Box 13.3, a 6×3 case, was $(6 - 1) \times (3 - 1) = 10$. In all 2×2 cases there is clearly only $(2 - 1) \times (2 - 1) = 1$ degree of freedom.

Another name for test of independence is *test of association*. If two properties are not independent of each other they are *associated*. Thus, in the example testing relative frequency of two leaf types on two different soils, we can speak of an association between leaf types and soils. In the immunology experiment there is a negative association between presence of antiserum and mortality. *Association* is thus similar to correlation, but it is a more general term, applying to attributes as well as continuous variables. In the 2×2 tests of independence of this section, one way of looking for suspected lack of independence was to examine the percentage occurrence of one of the properties in the two classes based on the other property. Thus we compared the percentage of smooth leaves on the two types of soils, or we studied the percentage mortality with or without antiserum. This way of looking at a test of independence suggests another interpretation of these tests as tests for the significance of differences between two percentages.

Exercises

13.1 In an experiment to determine the mode of inheritance of a *green* mutant, 146 wild-type and 30 mutant offspring were obtained when F_1 generation houseflies were crossed. Test whether the data agree with the hypothesis that the ratio of wild type of mutants is 3:1. ANS. $G = 6.4624$, $G_{adj} = 6.441$, 1 df, $\chi^2_{0.05[1]} = 3.841$.

13.2 Locality A has been exhaustively collected for snakes of species S. An examination of the 167 adult males that have been collected reveals that 35 of these have pale-colored bands around their necks. From locality B, 90 miles away, we obtain a sample of 27 adult males of the same species, 6 of which show the bands. What is the chance that both samples are from the same statistical population with respect to frequency of bands?

13.3 Of 445 specimens of the butterfly *Erebia epipsodea* from mountainous areas, 2.5% have light color patches on their wings. Of 65 specimens from the prairie, 70.8% have such patches (unpublished data by P. R. Ehrlich). Is this difference significant? *Hint*: First work backwards to obtain original frequencies. ANS. $G = 175.5163$, 1 df, $G_{adj} = 171.4533$.

13.4   Test whether the percentage of nymphs of the aphid *Myzus persicae* that developed into winged forms depends on the type of diet provided. Stem mothers had been placed on the diets one day before the birth of the nymphs (data by Mittler and Dadd, 1966).

Type of diet	% winged forms	n
Synthetic diet	100	216
Cotyledon "sandwich"	92	230
Free cotyledon	36	75

13.5   In a study of polymorphism of chromosomal inversions in the grasshopper *Moraba scurra*, Lewontin and White (1960) gave the following results for the composition of a population at Royalla "B" in 1958.

			Chromosome CD	
		St/St	St/B1	B1/B1
Chromosome EF	Td/Td	22	96	75
	St/Td	8	56	64
	St/St	0	6	6

Are the frequencies of the three different combinations of chromosome EF independent of those of the frequencies of the three combinations of chromosome CD? ANS. $G = 7.396$.

13.6   Test agreement of observed frequencies with those expected on the basis of a binomial distribution for the data given in Tables 4.1 and 4.2.

13.7   Test agreement of observed frequencies with those expected on the basis of a Poisson distribution for the data given in Table 4.5 and Table 4.6. ANS. For Table 4.5: $G = 49.9557$, 3 *df*, $G_{adj} = 49.8914$. For Table 4.6: $G = 20.6077$, 2 *df*, $G_{adj} = 20.4858$.

13.8   In clinical tests of the drug Nimesulide, Pfändner (1984) reports the following results. The drug was given, together with an antibiotic, to 20 persons. A control group of 20 persons with urinary infections were given the antibiotic and a placebo. The results, edited for purposes of this exercise, are as follows:

	Antibiotic + Nimesulide	Antibiotic + placebo
Negative opinion	1	16
Positive opinion	19	4

Analyze and interpret the results.

13.9   Refer to the distributions of melanoma over body regions shown in Table 2.1. Is there evidence for differential susceptibility to melanoma of differing body regions in males and females? ANS. $G = 160.2366$, 5 *df*, $G_{adj} = 158.6083$.

Mathematical Appendix

A1.1 Demonstration that the sum of the deviations from the mean is equal to zero.

We have to learn two common rules of statistical algebra. We can open a pair of parentheses with a Σ sign in front of them by treating the Σ as though it were a common factor. We have

$$\sum_{i=1}^{n} (A_i + B_i) = (A_1 + B_1) + (A_2 + B_2) + \cdots + (A_n + B_n)$$

$$= (A_1 + A_2 + \cdots + A_n) + (B_1 + B_2 + \cdots + B_n)$$

Therefore,

$$\sum_{i=1}^{n} (A_i + B_i) = \sum_{i=1}^{n} A_i + \sum_{i=1}^{n} B_i$$

Also, when $\Sigma_{i=1}^{n} C$ is developed during an algebraic operation, where C is a constant, this can be computed as follows:

$$\sum_{i=1}^{n} C = C + C + \cdots + C \qquad (n \text{ terms})$$

$$= nC$$

Since in a given problem a mean is a constant value, $\Sigma^n \bar{Y} = n\bar{Y}$. If you wish, you may check these rules, using simple numbers. In the subsequent demonstration and others to follow, whenever all summations are over n items, we have simplified the notation by dropping subscripts for variables and superscripts above summation signs.

We wish to prove that $\Sigma y = 0$. By definition,

$$\Sigma y = \Sigma(Y - \bar{Y})$$
$$= \Sigma Y - n\bar{Y}$$
$$= \Sigma Y - \frac{n\Sigma Y}{n} \qquad \left(\text{since } \bar{Y} = \frac{\Sigma Y}{n}\right)$$
$$= \Sigma Y - \Sigma Y$$

Therefore, $\Sigma y = 0$.

A1.2   Demonstration that Expression (3.8), the computational formula for the sum of squares, equals Expression (3.7), the expression originally developed for this statistic.

We wish to prove that $\Sigma(Y - \bar{Y})^2 = \Sigma Y^2 - ((\Sigma Y)^2/n)$. We have

$$\Sigma(Y - \bar{Y})^2 = \Sigma(Y^2 - 2Y\bar{Y} + \bar{Y}^2)$$
$$= \Sigma Y^2 - 2\bar{Y}\Sigma Y + n\bar{Y}^2$$
$$= \Sigma Y^2 - \frac{2(\Sigma Y)^2}{n} + \frac{n(\Sigma Y)^2}{n^2} \qquad \left(\text{since } \bar{Y} = \frac{\Sigma Y}{n}\right)$$
$$= \Sigma Y^2 - \frac{2(\Sigma Y)^2}{n} + \frac{(\Sigma Y)^2}{n}$$

Hence,

$$\Sigma(Y - \bar{Y})^2 = \Sigma Y^2 - \frac{(\Sigma Y)^2}{n}$$

A1.3   Simplified formulas for standard error of the difference between two means.

The standard error squared from Expression (8.2) is

$$\left[\frac{(n_1 - 1)s_1^2 + (n_2 - 1)s_2^2}{n_1 + n_2 - 2}\right]\left(\frac{n_1 + n_2}{n_1 n_2}\right)$$

When $n_1 = n_2 = n$, this simplifies to

$$\left[\frac{(n - 1)s_1^2 + (n - 1)s_2^2}{2n - 2}\right]\left(\frac{2n}{n^2}\right) = \left[\frac{(n - 1)(s_1^2 + s_2^2)(2)}{2(n - 1)(n)}\right] = \frac{1}{n}(s_1^2 + s_2^2)$$

which is the standard error squared of Expression (8.3).

When $n_1 \neq n_2$ but each is large, so that $(n_1 - 1) \approx n_1$ and $(n_2 - 1) \approx n_2$, the standard error squared of Expression (8.2) simplifies to

$$\left[\frac{n_1 s_1^2 + n_2 s_2^2}{n_1 + n_2}\right]\left(\frac{n_1 + n_2}{n_1 n_2}\right) = \frac{n_1 s_1^2}{n_1 n_2} + \frac{n_2 s_2^2}{n_1 n_2} = \frac{s_1^2}{n_2} + \frac{s_2^2}{n_1}$$

which is the standard error squared of Expression (8.4).

A1.4   Demonstration that t_s^2 obtained from a test of significance of the difference between two means (as in Box 8.2) is identical to the F_s value obtained in a single-classification anova of two equal-sized groups (in the same box).

$$t_s \text{ (from Box 8.2)} = \frac{\bar{Y}_1 - \bar{Y}_2}{\sqrt{\frac{1}{n}(s_1^2 + s_2^2)}} = \frac{\bar{Y}_1 - \bar{Y}_2}{\sqrt{\frac{1}{n(n-1)}\left(\sum^n y_1^2 + \sum^n y_2^2\right)}}$$

$$t_s^2 = \frac{(\bar{Y}_1 - \bar{Y}_2)^2}{\frac{1}{n(n-1)}\left(\sum^n y_1^2 + \sum^n y_2^2\right)} = \frac{n(n-1)(\bar{Y}_1 - \bar{Y}_2)^2}{\sum^n y_1^2 + \sum^n y_2^2}$$

In the two-sample anova,

$$MS_{\text{means}} = \frac{1}{2-1}\sum^2(\bar{Y}_i - \bar{\bar{Y}})^2$$

$$= (\bar{Y}_1 - \bar{\bar{Y}})^2 + (\bar{Y}_2 - \bar{\bar{Y}})^2$$

$$= \left(\bar{Y}_1 - \frac{\bar{Y}_1 + \bar{Y}_2}{2}\right)^2 + \left(\bar{Y}_2 - \frac{\bar{Y}_1 + \bar{Y}_2}{2}\right)^2 \qquad \text{(since } \bar{\bar{Y}} = (\bar{Y}_1 + \bar{Y}_2)/2\text{)}$$

$$= \left(\frac{\bar{Y}_1 - \bar{Y}_2}{2}\right)^2 + \left(\frac{\bar{Y}_2 - \bar{Y}_1}{2}\right)^2$$

$$= \tfrac{1}{2}(\bar{Y}_1 - \bar{Y}_2)^2$$

since the squares of the numerators are identical. Then

$$MS_{\text{groups}} = n \times MS_{\text{means}} = n[\tfrac{1}{2}(\bar{Y}_1 - \bar{Y}_2)^2]$$

$$= \frac{n}{2}(\bar{Y}_1 - \bar{Y}_2)^2$$

$$MS_{\text{within}} = \frac{\sum^n y_1^2 + \sum^n y_2^2}{2(n-1)}$$

$$F_s = \frac{MS_{\text{groups}}}{MS_{\text{within}}}$$

$$F_s = \frac{\frac{n}{2}(\bar{Y}_1 - \bar{Y}_2)^2}{\left(\sum^n y_1^2 + \sum^n y_2^2\right)\bigg/[2(n-1)]}$$

$$= \frac{n(n-1)(\bar{Y}_1 - \bar{Y}_2)^2}{\sum^n y_1^2 + \sum^n y_2^2}$$

$$= t_s^2$$

A1.5 Demonstration that Expression (11.5), the computational formula for the sum of products, equals $\Sigma(X - \bar{X})(Y - \bar{Y})$, the expression originally developed for this quantity.

All summations are over n items. We have

$$\sum xy = \sum(X - \bar{X})(Y - \bar{Y})$$
$$= \sum XY - \bar{X}\sum Y - \bar{Y}\sum X + n\bar{X}\bar{Y} \qquad \text{(since } \sum \bar{X}\bar{Y} = n\bar{X}\bar{Y}\text{)}$$
$$= \sum XY - \bar{X}n\bar{Y} - \bar{Y}n\bar{X} + n\bar{X}\bar{Y} \qquad \begin{array}{l}\text{(since } \sum Y/n = \bar{Y}, \\ \sum Y = n\bar{Y}; \text{ similarly, } \sum X = n\bar{X}\text{)}\end{array}$$
$$= \sum XY - n\bar{X}\bar{Y}$$
$$= \sum XY - n\bar{X}\frac{\sum Y}{\cdot}$$
$$= \sum XY - \bar{X}\sum Y$$

Similarly,

$$\sum xy = \sum XY - \bar{Y}\sum X$$

and

$$\sum xy = \sum XY - \frac{(\sum X)(\sum Y)}{n} \qquad\qquad (11.5)$$

A1.6 Derivation of computational formula for $\sum d_{Y\cdot X}^2 = \sum y^2 - ((\sum xy)^2/\sum x^2)$.

By definition, $d_{Y\cdot X} = Y - \hat{Y}$. Since $\bar{Y} = \bar{\hat{Y}}$, we can subtract $\bar{Y}$ from both Y and $\hat{Y}$ to obtain

$$d_{Y\cdot X} = y - \hat{y} = y - bx \qquad \text{(since } \hat{y} = bx\text{)}$$

Therefore,

$$\sum d_{Y\cdot X}^2 = \sum(y - bx)^2 = \sum y^2 - 2b\sum xy + b^2\sum x^2$$
$$= \sum y^2 - 2\frac{\sum xy}{\sum x^2}\sum xy + \frac{(\sum xy)^2}{(\sum x^2)^2}\sum x^2 = \sum y^2 - 2\frac{(\sum xy)^2}{\sum x^2} + \frac{(\sum xy)^2}{\sum x^2}$$

or

$$\sum d_{Y \cdot X}^2 = \sum y^2 - \frac{(\sum xy)^2}{\sum x^2} \qquad (11.6)$$

A1.7 Demonstration that the sum of squares of the dependent variable in regression can be partitioned exactly into explained and unexplained sums of squares, the cross products canceling out.

By definition (Section 11.5),

$$y = \hat{y} + d_{Y \cdot X}$$

$$\sum y^2 = \sum (\hat{y} + d_{Y \cdot X})^2 = \sum \hat{y}^2 + \sum d_{Y \cdot X}^2 + 2 \sum \hat{y} d_{Y \cdot X}$$

If we can show that $\sum \hat{y} d_{Y \cdot X} = 0$, then we have demonstrated the required identity. We have

$$\sum \hat{y} d_{Y \cdot X} = \sum bx(y - bx) \qquad \text{[since } \hat{y} = bx \text{ from Expression (11.3) and}$$
$$d_{Y \cdot X} = y - bx \text{ from Appendix A1.6]}$$

$$= b \sum xy - b^2 \sum x^2$$

$$= b \sum xy - b \frac{\sum xy}{\sum x^2} \sum x^2 \qquad \left(\text{since } b = \frac{\sum xy}{\sum x^2} \right)$$

$$= b \sum xy - b \sum xy$$

$$= 0$$

Therefore, $\sum y^2 = \sum \hat{y}^2 + \sum d_{Y \cdot X}^2$, or, written out in terms of variates,

$$\sum (Y - \bar{Y})^2 = \sum (\hat{Y} - \bar{Y})^2 + \sum (Y - \hat{Y})^2$$

A1.8 Proof that the variance of the sum of two variables is

$$\sigma_{(Y_1 + Y_2)}^2 = \sigma_1^2 + \sigma_2^2 + 2\rho_{12}\sigma_1\sigma_2$$

where σ_1 and σ_2 are standard deviations of Y_1 and Y_2, respectively, and ρ_{12} is the parametric correlation coefficient between Y_1 and Y_2.

If $Z = Y_1 + Y_2$, then

$$\sigma_z^2 = \frac{1}{n} \sum (Z - \bar{Z})^2 = \frac{1}{n} \sum \left[(Y_1 + Y_2) - \frac{1}{n} \sum (Y_1 + Y_2) \right]^2$$

$$= \frac{1}{n} \sum \left[(Y_1 + Y_2) - \frac{1}{n} \sum Y_1 - \frac{1}{n} \sum Y_2 \right]^2 = \frac{1}{n} \sum [(Y_1 + Y_2) - \bar{Y}_1 - \bar{Y}_2]^2$$

$$= \frac{1}{n} \sum \left[(Y_1 - \bar{Y}_1) + (Y_2 - \bar{Y}_2) \right]^2 = \frac{1}{n} \sum (y_1 + y_2)^2$$

$$= \frac{1}{n} \sum (y_1^2 + y_2^2 + 2y_1 y_2) = \frac{1}{n} \sum y_1^2 + \frac{1}{n} \sum y_2^2 + \frac{2}{n} \sum y_1 y_2$$

$$= \sigma_1^2 + \sigma_2^2 + 2\sigma_{12}$$

But, since $\rho_{12} = \sigma_{12}/\sigma_1\sigma_2$, we have

$$\sigma_{12} = \rho_{12}\sigma_1\sigma_2$$

Therefore

$$\sigma_Z^2 = \sigma_{(Y_1 + Y_2)}^2 = \sigma_1^2 + \sigma_2^2 + 2\rho_{12}\sigma_1\sigma_2$$

Similarly,

$$\sigma_D^2 = \sigma_{(Y_1 - Y_2)}^2 = \sigma_1^2 + \sigma_2^2 - 2\rho_{12}\sigma_1\sigma_2$$

The analogous expressions apply to sample statistics. Thus

$$s_{(Y_1 + Y_2)}^2 = s_1^2 + s_2^2 + 2r_{12}s_1s_2 \tag{12.8}$$

$$s_{(Y_1 - Y_2)}^2 = s_1^2 + s_2^2 - 2r_{12}s_1s_2 \tag{12.9}$$

A1.9 Proof that the general expression for the G test can be simplified to Expressions (13.4) and (13.5).

In general, G is twice the natural logarithm of the ratio of the probability of the sample with all parameters estimated from the data and the probability of the sample assuming the null hypothesis is true. Assuming a multinomial distribution, this ratio is

$$L = \dfrac{\dfrac{n!}{f_1!f_2!\cdots f_a!}\,p_1^{f_1}p_2^{f_2}\cdots p_a^{f_a}}{\dfrac{n!}{f_1!f_2!\cdots f_a!}\,\hat{p}_1^{f_1}\hat{p}_2^{f_2}\cdots \hat{p}_a^{f_a}}$$

$$= \prod_{i=1}^{a}\left(\frac{p_i}{\hat{p}_i}\right)^{f_i}$$

where f_i is the observed frequency, p_i is the observed proportion, and $\hat{p}_i$ the expected proportion of class i, while n is sample size, the sum of the observed frequencies over the a classes.

$$G = 2\ln L$$

$$= 2\sum^{a} f_i \ln\left(\frac{p_i}{\hat{p}_i}\right)$$

Since $f_i = np_i$ and $\hat{f}_i = n\hat{p}_i$,

$$G = 2\sum^{a} f_i \ln\left(\frac{f_i}{\hat{f}_i}\right) \tag{13.4}$$

If we now replace $\hat{f}_i$ by $n\hat{p}_i$,

$$G = 2\sum^{a} f_i \ln\left(\frac{f_i}{n\hat{p}_i}\right) = 2\left[\sum^{a} f_i \ln\left(\frac{f_i}{\hat{p}_i}\right) - \sum^{a} f_i \ln n\right]$$

$$= 2\left[\sum^{a} f_i \ln\left(\frac{f_i}{\hat{p}_i}\right) - n\ln n\right] \tag{13.5}$$

Statistical Tables

TABLE I
Twenty-five hundred random digits.

	1	2	3	4	5	6	7	8	9	10	
1	48461	14952	72619	73689	52059	37086	60050	86192	67049	64739	1
2	76534	38149	49692	31366	52093	15422	20498	33901	10319	43397	2
3	70437	25861	38504	14752	23757	59660	67844	78815	23758	86814	3
4	59584	03370	42806	11393	71722	93804	09095	07856	55589	46020	4
5	04285	58554	16085	51555	27501	73883	33427	33343	45507	50063	5
6	77340	10412	69189	85171	29082	44785	83638	02583	96483	76553	6
7	59183	62687	91778	80354	23512	97219	65921	02035	59847	91403	7
8	91800	04281	39979	03927	82564	28777	59049	97532	54540	79472	8
9	12066	24817	81099	48940	69554	55925	48379	12866	51232	21580	9
10	69907	91751	53512	23748	65906	91385	84983	27915	48491	91068	10
11	80467	04873	54053	25955	48518	13815	37707	68687	15570	08890	11
12	78057	67835	28302	45048	56761	97725	58438	91528	24645	18544	12
13	05648	39387	78191	88415	60269	94880	58812	42931	71898	61534	13
14	22304	39246	01350	99451	61862	78688	30339	60222	74052	25740	14
15	61346	50269	67005	40442	33100	16742	61640	21046	31909	72641	15
16	66793	37696	27965	30459	91011	51426	31006	77468	61029	57108	16
17	86411	48809	36698	42453	83061	43769	39948	87031	30767	13953	17
18	62098	12825	81744	28882	27369	88183	65846	92545	09065	22655	18
19	68775	06261	54265	16203	23340	84750	16317	88686	86842	00879	19
20	52679	19595	13687	74872	89181	01939	18447	10787	76246	80072	20
21	84096	87152	20719	25215	04349	54434	72344	93008	83282	31670	21
22	63964	55937	21417	49944	38356	98404	14850	17994	17161	98981	22
23	31191	75131	72386	11689	95727	05414	88727	45583	22568	77700	23
24	30545	68523	29850	67833	05622	89975	79042	27142	99257	32349	24
25	52573	91001	52315	26430	54175	30122	31796	98842	37600	26025	25
26	16586	81842	01076	99414	31574	94719	34656	80018	86988	79234	26
27	81841	88481	61191	25013	30272	23388	22463	65774	10029	58376	27
28	43563	66829	72838	08074	57080	15446	11034	98143	74989	26885	28
29	19945	84193	57581	77252	85604	45412	43556	27518	90572	00563	29
30	79374	23796	16919	99691	80276	32818	62953	78831	54395	30705	30
31	48503	26615	43980	09810	38289	66679	73799	48418	12647	40044	31
32	32049	65541	37937	41105	70106	89706	40829	40789	59547	00783	32
33	18547	71562	95493	34112	76895	46766	96395	31718	48302	45893	33
34	03180	96742	61486	43305	34183	99605	67803	13491	09243	29557	34
35	94822	24738	67749	83748	59799	25210	31093	62925	72061	69991	35
36	34330	60599	85828	19152	68499	27977	35611	96240	62747	89529	36
37	43770	81537	59527	95674	76692	86420	69930	10020	72881	12532	37
38	56908	77192	50623	41215	14311	42834	80651	93750	59957	31211	38
39	32787	07189	80539	75927	75475	73965	11796	72140	48944	74156	39
40	52441	78392	11733	57703	29133	71164	55355	31006	25526	55790	40
41	22377	54723	18227	28449	04570	18882	00023	67101	06895	08915	41
42	18376	73460	88841	39602	34049	20589	05701	08249	74213	25220	42
43	53201	28610	87957	21497	64729	64983	71551	99016	87903	63875	43
44	34919	78901	59710	27396	02593	05665	11964	44134	00273	76358	44
45	33617	92159	21971	16901	57383	34262	41744	60891	57624	06962	45
46	70010	40964	98780	72418	52571	18415	64362	90636	38034	04909	46
47	19282	68447	35665	31530	59832	49181	21914	65742	89815	39231	47
48	91429	73328	13266	54898	68795	40948	80808	63887	89939	47938	48
49	97637	78393	33021	05867	86520	45363	43066	00988	64040	09803	49
50	95150	07625	05255	83254	93943	52325	93230	62668	79529	65964	50

TABLE **II**
Areas of the normal curve

y/σ	0.00	0.01	0.02	0.03	0.04	0.05	0.06	0.07	0.08	0.09	y/σ
0.0	.0000	.0040	.0080	.0120	.0160	.0199	.0239	.0279	.0319	.0359	0.0
0.1	.0398	.0438	.0478	.0517	.0557	.0596	.0636	.0675	.0714	.0753	0.1
0.2	.0793	.0832	.0871	.0910	.0948	.0987	.1026	.1064	.1103	.1141	0.2
0.3	.1179	.1217	.1255	.1293	.1331	.1368	.1406	.1443	.1480	.1517	0.3
0.4	.1554	.1591	.1628	.1664	.1700	.1736	.1772	.1808	.1844	.1879	0.4
0.5	.1915	.1950	.1985	.2019	.2054	.2088	.2123	.2157	.2190	.2224	0.5
0.6	.2257	.2291	.2324	.2357	.2389	.2422	.2454	.2486	.2517	.2549	0.6
0.7	.2580	.2611	.2642	.2673	.2704	.2734	.2764	.2794	.2823	.2852	0.7
0.8	.2881	.2910	.2939	.2967	.2995	.3023	.3051	.3078	.3106	.3133	0.8
0.9	.3159	.3186	.3212	.3238	.3264	.3289	.3315	.3340	.3365	.3389	0.9
1.0	.3413	.3438	.3461	.3485	.3508	.3531	.3554	.3577	.3599	.3621	1.0
1.1	.3643	.3665	.3686	.3708	.3729	.3749	.3770	.3790	.3810	.3830	1.1
1.2	.3849	.3869	.3888	.3907	.3925	.3944	.3962	.3980	.3997	.4015	1.2
1.3	.4032	.4049	.4066	.4082	.4099	.4115	.4131	.4147	.4162	.4177	1.3
1.4	.4192	.4207	.4222	.4236	.4251	.4265	.4279	.4292	.4306	.4319	1.4
1.5	.4332	.4345	.4357	.4370	.4382	.4394	.4406	.4418	.4429	.4441	1.5
1.6	.4452	.4463	.4474	.4484	.4495	.4505	.4515	.4525	.4535	.4545	1.6
1.7	.4554	.4564	.4573	.4582	.4591	.4599	.4608	.4616	.4625	.4633	1.7
1.8	.4641	.4649	.4656	.4664	.4671	.4678	.4686	.4693	.4699	.4706	1.8
1.9	.4713	.4719	.4726	.4732	.4738	.4744	.4750	.4756	.4761	.4767	1.9
2.0	.4772	.4778	.4783	.4788	.4793	.4798	.4803	.4808	.4812	.4817	2.0
2.1	.4821	.4826	.4830	.4834	.4838	.4842	.4846	.4850	.4854	.4857	2.1
2.2	.4861	.4864	.4868	.4871	.4875	.4878	.4881	.4884	.4887	.4890	2.2
2.3	.4893	.4896	.4898	.4901	.4904	.4906	.4909	.4911	.4913	.4916	2.3
2.4	.4918	.4920	.4922	.4925	.4927	.4929	.4931	.4932	.4934	.4936	2.4
2.5	.4938	.4940	.4941	.4943	.4945	.4946	.4948	.4949	.4951	.4952	2.5
2.6	.4953	.4955	.4956	.4957	.4959	.4960	.4961	.4962	.4963	.4964	2.6
2.7	.4965	.4966	.4967	.4968	.4969	.4970	.4971	.4972	.4973	.4974	2.7
2.8	.4974	.4975	.4976	.4977	.4977	.4978	.4979	.4979	.4980	.4981	2.8
2.9	.4981	.4982	.4982	.4983	.4984	.4984	.4985	.4985	.4986	.4986	2.9
3.0	.4987	.4987	.4987	.4988	.4988	.4989	.4989	.4989	.4990	.4990	3.0
3.1	.4990	.4991	.4991	.4991	.4992	.4992	.4992	.4992	.4993	.4993	3.1
3.2	.4993	.4993	.4994	.4994	.4994	.4994	.4994	.4995	.4995	.4995	3.2
3.3	.4995	.4995	.4995	.4996	.4996	.4996	.4996	.4996	.4996	.4997	3.3
3.4	.4997	.4997	.4997	.4997	.4997	.4997	.4997	.4997	.4997	.4998	3.4

y/σ	
3.5	.499767
3.6	.499841
3.7	.499892
3.8	.499928
3.9	.499952
4.0	.499968
4.1	.499979
4.2	.499987
4.3	.499991
4.4	.499995
4.5	.499997
4.6	.499998
4.7	.499999
4.8	.499999
4.9	.500000

Tabled area

$\frac{1}{2}-\alpha$

α

$$\text{Argument} = \frac{Y - \mu_Y}{\sigma}$$

Note: The quantity given is the area under the standard normal density function between the mean and the critical point. The area is generally labeled $\frac{1}{2} - \alpha$ (as shown in the figure). By inverse interpolation one can find the number of standard deviations corresponding to a given area.

TABLE **III**

Critical values of Student's t distribution

α

ν	0.9	0.5	0.4	0.2	0.1	0.05	0.02	0.01	0.001	ν
1	.158	1.000	1.376	3.078	6.314	12.706	31.821	63.657	636.619	1
2	.142	.816	1.061	1.886	2.920	4.303	6.965	9.925	31.598	2
3	.137	.765	.978	1.638	2.353	3.182	4.541	5.841	12.924	3
4	.134	.741	.941	1.533	2.132	2.776	3.747	4.604	8.610	4
5	.132	.727	.920	1.476	2.015	2.571	3.365	4.032	6.869	5
6	.131	.718	.906	1.440	1.943	2.447	3.143	3.707	5.959	6
7	.130	.711	.896	1.415	1.895	2.365	2.998	3.499	5.408	7
8	.130	.706	.889	1.397	1.860	2.306	2.896	3.355	5.041	8
9	.129	.703	.883	1.383	1.833	2.262	2.821	3.250	4.781	9
10	.129	.700	.879	1.372	1.812	2.228	2.764	3.169	4.587	10
11	.129	.697	.876	1.363	1.796	2.201	2.718	3.106	4.437	11
12	.128	.695	.873	1.356	1.782	2.179	2.681	3.055	4.318	12
13	.128	.694	.870	1.350	1.771	2.160	2.650	3.012	4.221	13
14	.128	.692	.868	1.345	1.761	2.145	2.624	2.977	4.140	14
15	.128	.691	.866	1.341	1.753	2.131	2.602	2.947	4.073	15
16	.128	.690	.865	1.337	1.746	2.120	2.583	2.921	4.015	16
17	.128	.689	.863	1.333	1.740	2.110	2.567	2.898	3.965	17
18	.127	.688	.862	1.330	1.734	2.101	2.552	2.878	3.922	18
19	.127	.688	.861	1.328	1.729	2.093	2.539	2.861	3.883	19
20	.127	.687	.860	1.325	1.725	2.086	2.528	2.845	3.850	20
21	.127	.686	.859	1.323	1.721	2.080	2.518	2.831	3.819	21
22	.127	.686	.858	1.321	1.717	2.074	2.508	2.819	3.792	22
23	.127	.685	.858	1.319	1.714	2.069	2.500	2.807	3.767	23
24	.127	.685	.857	1.318	1.711	2.064	2.492	2.797	3.745	24
25	.127	.684	.856	1.316	1.708	2.060	2.485	2.787	3.725	25
26	.127	.684	.856	1.315	1.706	2.056	2.479	2.779	3.707	26
27	.127	.684	.855	1.314	1.703	2.052	2.473	2.771	3.690	27
28	.127	.683	.855	1.313	1.701	2.048	2.467	2.763	3.674	28
29	.127	.683	.854	1.311	1.699	2.045	2.462	2.756	3.659	29
30	.127	.683	.854	1.310	1.697	2.042	2.457	2.750	3.646	30
40	.126	.681	.851	1.303	1.684	2.021	2.423	2.704	3.551	40
60	.126	.679	.848	1.296	1.671	2.000	2.390	2.660	3.460	60
120	.126	.677	.845	1.289	1.658	1.980	2.358	2.617	3.373	120
∞	.126	.674	.842	1.282	1.645	1.960	2.326	2.576	3.291	∞

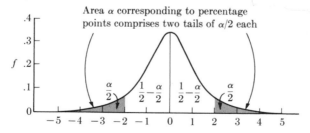

Area α corresponding to percentage points comprises two tails of $\alpha/2$ each

Note: If a one-tailed test is desired, the probabilities at the head of the table must be halved. For degrees of freedom $\nu > 30$, interpolate between the values of the argument ν. The table is designed for harmonic interpolation. Thus, to obtain $t_{0.05[43]}$, interpolate between $t_{0.05[40]} = 2.021$ and $t_{0.05[60]} = 2.000$, which are furnished in the table. Transform the arguments into $120/\nu = 120/43 = 2.791$ and interpolate between $120/60 = 2.000$ and $120/40 = 3.000$ by ordinary linear interpolation:

$$t_{0.05[43]} = (0.791 \times 2.021) + [(1 - 0.791) \times 2.000]$$
$$= 2.017$$

When $\nu > 120$, interpolate between $120/\infty = 0$ and $120/120 = 1$. Values in this table have been taken from a more extensive one (table III) in R. A. Fisher and F. Yates, *Statistical Tables for Biological, Agricultural and Medical Research*, 5th ed. (Oliver & Boyd, Edinburgh, 1958) with permission of the authors and their publishers.

TABLE **IV**
Critical values of the chi-square distribution

α

ν	.995	.975	.9	.5	.1	.05	.025	.01	.005	.001	ν
1	0.000	0.000	0.016	0.455	2.706	3.841	5.024	6.635	7.879	10.828	1
2	0.010	0.051	0.211	1.386	4.605	5.991	7.378	9.210	10.597	13.816	2
3	0.072	0.216	0.584	2.366	6.251	7.815	9.348	11.345	12.838	16.266	3
4	0.207	0.484	1.064	3.357	7.779	9.488	11.143	13.277	14.860	18.467	4
5	0.412	0.831	1.610	4.351	9.236	11.070	12.832	15.086	16.750	20.515	5
6	0.676	1.237	2.204	5.348	10.645	12.592	14.449	16.812	18.548	22.458	6
7	0.989	1.690	2.833	6.346	12.017	14.067	16.013	18.475	20.278	24.322	7
8	1.344	2.180	3.490	7.344	13.362	15.507	17.535	20.090	21.955	26.124	8
9	1.735	2.700	4.168	8.343	14.684	16.919	19.023	21.666	23.589	27.877	9
10	2.156	3.247	4.865	9.342	15.987	18.307	20.483	23.209	25.188	29.588	10
11	2.603	3.816	5.578	10.341	17.275	19.675	21.920	24.725	26.757	31.264	11
12	3.074	4.404	6.304	11.340	18.549	21.026	23.337	26.217	28.300	32.910	12
13	3.565	5.009	7.042	12.340	19.812	22.362	24.736	27.688	29.819	34.528	13
14	4.075	5.629	7.790	13.339	21.064	23.685	26.119	29.141	31.319	36.123	14
15	4.601	6.262	8.547	14.339	22.307	24.996	27.488	30.578	32.801	37.697	15
16	5.142	6.908	9.312	15.338	23.542	26.296	28.845	32.000	34.267	39.252	16
17	5.697	7.564	10.085	16.338	24.769	27.587	30.191	33.409	35.718	40.790	17
18	6.265	8.231	10.865	17.338	25.989	28.869	31.526	34.805	37.156	42.312	18
19	6.844	8.907	11.651	18.338	27.204	30.144	32.852	36.191	38.582	43.820	19
20	7.434	9.591	12.443	19.337	28.412	31.410	34.170	37.566	39.997	45.315	20
21	8.034	10.283	13.240	20.337	29.615	32.670	35.479	38.932	41.401	46.797	21
22	8.643	10.982	14.042	21.337	30.813	33.924	36.781	40.289	42.796	48.268	22
23	9.260	11.688	14.848	22.337	32.007	35.172	38.076	41.638	44.181	49.728	23
24	9.886	12.401	15.659	23.337	33.196	36.415	39.364	42.980	45.558	51.179	24
25	10.520	13.120	16.473	24.337	34.382	37.652	40.646	44.314	46.928	52.620	25
26	11.160	13.844	17.292	25.336	35.563	38.885	41.923	45.642	48.290	54.052	26
27	11.808	14.573	18.114	26.336	36.741	40.113	43.194	46.963	49.645	55.476	27
28	12.461	15.308	18.939	27.336	37.916	41.337	44.461	48.278	50.993	56.892	28
29	13.121	16.047	19.768	28.336	39.088	42.557	45.722	49.588	52.336	58.301	29
30	13.787	16.791	20.599	29.336	40.256	43.773	46.979	50.892	53.672	59.703	30
31	14.458	17.539	21.434	30.336	41.422	44.985	48.232	52.191	55.003	61.098	31
32	15.134	18.291	22.271	31.336	42.585	46.194	49.480	53.486	56.329	62.487	32
33	15.815	19.047	23.110	32.336	43.745	47.400	50.725	54.776	57.649	63.870	33
34	16.501	19.806	23.952	33.336	44.903	48.602	51.966	56.061	58.964	65.247	34
35	17.192	20.569	24.797	34.336	46.059	49.802	53.203	57.342	60.275	66.619	35
36	17.887	21.336	25.643	35.336	47.212	50.998	54.437	58.619	61.582	67.985	36
37	18.586	22.106	26.492	36.335	48.363	52.192	55.668	59.892	62.884	69.346	37
38	19.289	22.878	27.343	37.335	49.513	53.384	56.896	61.162	64.182	70.703	38
39	19.996	23.654	28.196	38.335	50.660	54.572	58.120	62.428	65.476	72.055	39
40	20.707	24.433	29.051	39.335	51.805	55.758	59.342	63.691	66.766	73.402	40
41	21.421	25.215	29.907	40.335	52.949	56.942	60.561	64.950	68.053	74.745	41
42	22.138	25.999	30.765	41.335	54.090	58.124	61.777	66.206	69.336	76.084	42
43	22.859	26.785	31.625	42.335	55.230	59.304	62.990	67.459	70.616	77.419	43
44	23.584	27.575	32.487	43.335	56.369	60.481	64.202	68.710	71.893	78.750	44
45	24.311	28.366	33.350	44.335	57.505	61.656	65.410	69.957	73.166	80.077	45
46	25.042	29.160	34.215	45.335	58.641	62.830	66.617	71.201	74.437	81.400	46
47	25.775	29.956	35.081	46.335	59.774	64.001	67.821	72.443	75.704	82.720	47
48	26.511	30.755	35.949	47.335	60.907	65.171	69.023	73.683	76.969	84.037	48
49	27.249	31.555	36.818	48.335	62.038	66.339	70.222	74.919	78.231	85.351	49
50	27.991	32.357	37.689	49.335	63.167	67.505	71.420	76.154	79.490	86.661	50

Note: For values of $v > 100$, compute approximate critical values of χ^2 by formula as follows: $\chi^2_{\alpha[v]} = \frac{1}{2}(t_{2\alpha[\infty]} + \sqrt{2v-1})^2$, where $t_{2\alpha[\infty]}$ can be looked up in Table **III**. Thus $\chi^2_{0.05[120]}$ is computed as $\frac{1}{2}(t_{0.10[\infty]} + \sqrt{240-1})^2 = \frac{1}{2}(1.645 + \sqrt{239})^2 = \frac{1}{2}(17.10462)^2 = 146.284$. For $\alpha > 0.5$ employ $t_{1-2\alpha[\infty]}$ in the above formula. When $\alpha = 0.5$, $t_{2\alpha} = 0$. Values of chi-square from 1 to 30 degrees of freedom have been taken from a more extensive table by C. M. Thompson (*Biometrika* **32**:188–189, 1941) with permission of the publisher.

TABLE IV
continued

α

ν	.995	.975	.9	.5	.1	.05	.025	.01	.005	.001	ν
51	28.735	33.162	38.560	50.335	64.295	68.669	72.616	77.386	80.747	87.968	51
52	29.481	33.968	39.433	51.335	65.422	69.832	73.810	78.616	82.001	89.272	52
53	30.230	34.776	40.308	52.335	66.548	70.993	75.002	79.843	83.253	90.573	53
54	30.981	35.586	41.183	53.335	67.673	72.153	76.192	81.069	84.502	91.872	54
55	31.735	36.398	42.060	54.335	68.796	73.311	77.380	82.292	85.749	93.168	55
56	32.490	37.212	42.937	55.335	69.918	74.468	78.567	83.513	86.994	94.460	56
57	33.248	38.027	43.816	56.335	71.040	75.624	79.752	84.733	88.237	95.751	57
58	34.008	38.844	44.696	57.335	72.160	76.778	80.936	85.950	89.477	97.039	58
59	34.770	39.662	45.577	58.335	73.279	77.931	82.117	87.166	90.715	98.324	59
60	35.534	40.482	46.459	59.335	74.397	79.082	83.298	88.379	91.952	99.607	60
61	36.300	41.303	47.342	60.335	75.514	80.232	84.476	89.591	93.186	100.888	61
62	37.068	42.126	48.226	61.335	76.630	81.381	85.654	90.802	94.419	102.166	62
63	37.838	42.950	49.111	62.335	77.745	82.529	86.830	92.010	95.649	103.442	63
64	38.610	43.776	49.996	63.335	78.860	83.675	88.004	93.217	96.878	104.716	64
65	39.383	44.603	50.883	64.335	79.973	84.821	89.177	94.422	98.105	105.998	65
66	40.158	45.431	51.770	65.335	81.085	85.965	90.349	95.626	99.331	107.258	66
67	40.935	46.261	52.659	66.335	82.197	87.108	91.519	96.828	100.55	108.526	67
68	41.713	47.092	53.548	67.334	83.308	88.250	92.689	98.028	101.78	109.791	68
69	42.494	47.924	54.438	68.334	84.418	89.391	93.856	99.228	103.00	111.055	69
70	43.275	48.758	55.329	69.334	85.527	90.531	95.023	100.43	104.21	112.317	70
71	44.058	49.592	56.221	70.334	86.635	91.670	96.189	101.62	105.43	113.577	71
72	44.843	50.428	57.113	71.334	87.743	92.808	97.353	102.82	106.65	114.835	72
73	45.629	51.265	58.006	72.334	88.850	93.945	98.516	104.01	107.86	116.092	73
74	46.417	52.103	58.900	73.334	89.956	95.081	99.678	105.20	109.07	117.346	74
75	47.206	52.942	59.795	74.334	91.061	96.217	100.84	106.39	110.29	118.599	75
76	47.997	53.782	60.690	75.334	92.166	97.351	102.00	107.58	111.50	119.850	76
77	48.788	54.623	61.586	76.334	93.270	98.484	103.16	108.77	112.70	121.100	77
78	49.582	55.466	62.483	77.334	94.373	99.617	104.32	109.96	113.91	122.348	78
79	50.376	56.309	63.380	78.334	95.476	100.75	105.47	111.14	115.12	123.594	79
80	51.172	57.153	64.278	79.334	96.578	101.88	106.63	112.33	116.32	124.839	80
81	51.969	57.998	65.176	80.334	97.680	103.01	107.78	113.51	117.52	126.082	81
82	52.767	58.845	66.076	81.334	98.780	104.14	108.94	114.69	118.73	127.324	82
83	53.567	59.692	66.976	82.334	99.880	105.27	110.09	115.88	119.93	128.565	83
84	54.368	60.540	67.876	83.334	100.98	106.39	111.24	117.06	121.13	129.804	84
85	55.170	61.389	68.777	84.334	102.08	107.52	112.39	118.24	122.32	131.041	85
86	55.973	62.239	69.679	85.334	103.18	108.65	113.54	119.41	123.52	132.277	86
87	56.777	63.089	70.581	86.334	104.28	109.77	114.69	120.59	124.72	133.512	87
88	57.582	63.941	71.484	87.334	105.37	110.90	115.84	121.77	125.91	134.745	88
89	58.389	64.793	72.387	88.334	106.47	112.02	116.99	122.94	127.11	135.978	89
90	59.196	65.647	73.291	89.334	107.56	113.15	118.14	124.12	128.30	137.208	90
91	60.005	66.501	74.196	90.334	108.66	114.27	119.28	125.29	129.49	138.438	91
92	60.815	67.356	75.101	91.334	109.76	115.39	120.43	126.46	130.68	139.666	92
93	61.625	68.211	76.006	92.334	110.85	116.51	121.57	127.63	131.87	140.893	93
94	62.437	69.068	76.912	93.334	111.94	117.63	122.72	128.80	133.06	142.119	94
95	63.250	69.925	77.818	94.334	113.04	118.75	123.86	129.97	134.25	143.344	95
96	64.063	70.783	78.725	95.334	114.13	119.87	125.00	131.14	135.43	144.567	96
97	64.878	71.642	79.633	96.334	115.22	120.99	126.14	132.31	136.62	145.789	97
98	65.694	72.501	80.541	97.334	116.32	122.11	127.28	133.48	137.80	147.010	98
99	66.510	73.361	81.449	98.334	117.41	123.23	128.42	134.64	138.99	148.230	99
100	67.328	74.222	82.358	99.334	118.50	124.34	129.56	135.81	140.17	149.449	100

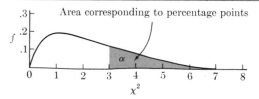

Area corresponding to percentage points

TABLE **V**
Critical values of the F distribution

ν_1 (degrees of freedom of numerator mean squares)

(left margin:) ν_2 (degrees of freedom of denominator mean squares)

α	1	2	3	4	5	6	7	8	9	10	11	12	α
1 .05	161	199	216	225	230	234	237	239	241	241	243	244	.05
.025	648	800	864	900	922	937	948	957	963	969	973	977	.025
.01	4050	5000	5400	5620	5760	5860	5930	5980	6020	6060	6080	6110	.01
2 .05	18.5	19.0	19.2	19.2	19.3	19.3	19.4	19.4	19.4	19.4	19.4	19.4	.05
.025	38.5	39.0	39.2	39.2	39.3	39.3	39.4	39.4	39.4	39.4	39.4	39.4	.025
.01	98.5	99.0	99.2	99.2	99.3	99.3	99.4	99.4	99.4	99.4	99.4	99.4	.01
3 .05	10.1	9.55	9.28	9.12	9.01	8.94	8.89	8.85	8.81	8.79	8.76	8.74	.05
.025	17.4	16.0	15.4	15.1	14.9	14.7	14.6	14.5	14.5	14.4	14.3	14.3	.025
.01	34.1	30.8	29.5	28.7	28.2	27.9	27.7	27.5	27.3	27.2	27.1	27.1	.01
4 .05	7.71	6.94	6.59	6.39	6.26	6.16	6.09	6.04	6.00	5.96	5.93	5.91	.05
.025	12.2	10.6	9.98	9.60	9.36	9.20	9.07	8.98	8.90	8.84	8.79	8.75	.025
.01	21.2	18.0	16.7	16.0	15.5	15.2	15.0	14.8	14.7	14.5	14.4	14.4	.01
5 .05	6.61	5.79	5.41	5.19	5.05	4.95	4.88	4.82	4.77	4.74	4.71	4.68	.05
.025	10.0	8.43	7.76	7.39	7.15	6.98	6.85	6.76	6.68	6.62	6.57	6.52	.025
.01	16.3	13.3	12.1	11.4	11.0	10.7	10.5	10.3	10.2	10.1	9.99	9.89	.01
6 .05	5.99	5.14	4.76	4.53	4.39	4.28	4.21	4.15	4.10	4.06	4.03	4.00	.05
.025	8.81	7.26	6.60	6.23	5.99	5.82	5.70	5.60	5.52	5.46	5.41	5.37	.025
.01	13.7	10.9	9.78	9.15	8.75	8.47	8.26	8.10	7.98	7.87	7.79	7.72	.01
7 .05	5.59	4.74	4.35	4.12	3.97	3.87	3.77	3.73	3.68	3.64	3.60	3.57	.05
.025	8.07	6.54	5.89	5.52	5.29	5.12	4.99	4.89	4.82	4.76	4.71	4.67	.025
.01	12.2	9.55	8.45	7.85	7.46	7.19	6.99	6.84	6.72	6.62	6.54	6.47	.01
8 .05	5.32	4.46	4.07	3.84	3.69	3.58	3.50	3.44	3.39	3.35	3.31	3.28	.05
.025	7.57	6.06	5.42	5.05	4.82	4.65	4.53	4.43	4.36	4.30	4.25	4.20	.025
.01	11.3	8.65	7.59	7.01	6.63	6.37	6.18	6.03	5.91	5.81	5.73	5.67	.01
9 .05	5.12	4.26	3.86	3.63	3.48	3.37	3.29	3.23	3.18	3.14	3.10	3.07	.05
.025	7.21	5.71	5.08	4.72	4.48	4.32	4.20	4.10	4.03	3.96	3.91	3.87	.025
.01	10.6	8.02	6.99	6.42	6.06	5.80	5.61	5.47	5.35	5.26	5.18	5.11	.01
10 .05	4.96	4.10	3.71	3.48	3.33	3.22	3.14	3.07	3.02	2.98	2.94	2.91	.05
.025	6.94	5.46	4.83	4.47	4.24	4.07	3.95	3.85	3.78	3.72	3.67	3.62	.025
.01	10.0	7.56	6.55	5.99	5.64	5.39	5.20	5.06	4.94	4.85	4.77	4.71	.01

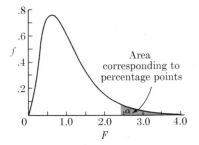

Note: Interpolation for number of degrees of freedom not furnished in the arguments is by means of harmonic interpolation (see footnote for Table **III**). If both ν_1 and ν_2 require interpolation, one needs to interpolate for each of these arguments in turn. Thus to obtain $F_{0.05[55,80]}$, one first interpolates between $F_{0.05[50,60]}$ and $F_{0.05[60,60]}$ and between $F_{0.05[50,120]}$ and $F_{0.05[60,120]}$, to estimate $F_{0.05[55,60]}$ and $F_{0.05[55,120]}$, respectively. One then interpolates between these two values to obtain the desired quantity. Entries for $\alpha = 0.05$, 0.025, 0.01, and 0.005 and for ν_1 and $\nu_2 = 1$ to 10, 12, 15, 20, 24, 30, 40, 60, 120, and ∞ were copied from a table by M. Merrington and C. M. Thompson (*Biometrika* **33**:73–88, 1943) with permission of the publisher.

TABLE V
continued

ν_1 (degrees of freedom of numerator mean squares)

	α	15	20	24	30	40	50	60	120	∞	α
1	.05	246	248	249	250	251	252	252	253	254	.05
	.025	985	993	997	1000	1010	1010	1010	1010	1020	.025
	.01	6160	6210	6230	6260	6290	6300	6310	6340	6370	.01
2	.05	19.4	19.4	19.5	19.5	19.5	19.5	19.5	19.5	19.5	.05
	.025	39.4	39.4	39.5	39.5	39.5	39.5	39.5	39.5	39.5	.025
	.01	99.4	99.4	99.5	99.5	99.5	99.5	99.5	99.5	99.5	.01
3	.05	8.70	8.66	8.64	8.62	8.59	8.58	8.57	8.55	8.53	.05
	.025	14.3	14.2	14.1	14.1	14.0	14.0	14.0	13.9	13.9	.025
	.01	26.9	26.7	26.6	26.5	26.4	26.3	26.3	26.2	26.1	.01
4	.05	5.86	5.80	5.77	5.75	5.72	5.70	5.69	5.66	5.63	.05
	.025	8.66	8.56	8.51	8.46	8.41	8.38	8.36	8.31	8.26	.025
	.01	14.2	14.0	13.9	13.8	13.7	13.7	13.7	13.6	13.5	.01
5	.05	4.62	4.56	4.53	4.50	4.46	4.44	4.43	4.40	4.36	.05
	.025	6.43	6.33	6.28	6.23	6.18	6.14	6.12	6.07	6.02	.025
	.01	9.72	9.55	9.47	9.38	9.29	9.24	9.20	9.11	9.02	.01
6	.05	3.94	3.87	3.84	3.81	3.77	3.75	3.74	3.70	3.67	.05
	.025	5.27	5.17	5.12	5.07	5.01	4.98	4.96	4.90	4.85	.025
	.01	7.56	7.40	7.31	7.23	7.14	7.09	7.06	6.97	6.88	.01
7	.05	3.51	3.44	3.41	3.38	3.34	3.32	3.30	3.27	3.23	.05
	.025	4.57	4.47	4.42	4.36	4.31	4.27	4.25	4.20	4.14	.025
	.01	6.31	6.16	6.07	5.99	5.91	5.86	5.82	5.74	5.65	.01
8	.05	3.22	3.15	3.12	3.08	3.04	3.02	3.01	2.97	2.93	.05
	.025	4.10	4.00	3.95	3.89	3.84	3.80	3.78	3.73	3.67	.025
	.01	5.52	5.36	5.28	5.20	5.12	5.07	5.03	4.95	4.86	.01
9	.05	3.01	2.94	2.90	2.86	2.83	2.81	2.79	2.75	2.71	.05
	.025	3.77	3.67	3.61	3.56	3.51	3.47	3.45	3.39	3.33	.025
	.01	4.96	4.81	4.73	4.65	4.57	4.52	4.48	4.40	4.31	.01
10	.05	2.85	2.77	2.74	2.70	2.66	2.64	2.62	2.58	2.54	.05
	.025	3.52	3.42	3.37	3.31	3.26	3.22	3.20	3.14	3.08	.025
	.01	4.56	4.41	4.33	4.25	4.17	4.12	4.08	4.00	3.91	.01

ν_2 (degrees of freedom of denominator mean squares)

TABLE V
continued

ν_1 (degrees of freedom of numerator mean squares)

	α	1	2	3	4	5	6	7	8	9	10	11	12	α
11	.05	4.84	3.98	3.59	3.36	3.20	3.09	3.01	2.95	2.90	2.85	2.82	2.79	.05
	.025	6.72	5.26	4.63	4.28	4.04	3.88	3.76	3.66	3.59	3.53	3.48	3.43	.025
	.01	9.65	7.21	6.22	5.67	5.32	5.07	4.89	4.74	4.63	4.54	4.46	4.40	.01
12	.05	4.75	3.89	3.49	3.26	3.11	3.00	2.91	2.85	2.80	2.75	2.72	2.69	.05
	.025	6.55	5.10	4.47	4.12	3.89	3.73	3.61	3.51	3.44	3.37	3.32	3.28	.025
	.01	9.33	6.93	5.95	5.41	5.06	4.82	4.64	4.50	4.39	4.30	4.22	4.16	.01
15	.05	4.54	3.68	3.29	3.06	2.90	2.79	2.71	2.64	2.59	2.54	2.51	2.48	.05.
	.025	6.20	4.77	4.15	3.80	3.58	3.41	3.29	3.20	3.12	3.06	3.01	2.96	.025
	.01	8.68	6.36	5.42	4.89	4.56	4.32	4.14	4.00	3.89	3.80	3.73	3.67	.01
20	.05	4.35	3.49	3.10	2.87	2.71	2.60	2.51	2.45	2.39	2.35	2.31	2.28	.05
	.025	5.87	4.46	3.86	3.51	3.29	3.13	3.01	2.91	2.84	2.77	2.72	2.68	.025
	.01	8.10	5.85	4.94	4.43	4.10	3.87	3.70	3.56	3.46	3.37	3.29	3.23	.01
24	.05	4.26	3.40	3.01	2.78	2.62	2.51	2.42	2.36	2.30	2.25	2.22	2.18	.05
	.025	5.72	4.32	3.72	3.38	3.15	2.99	2.87	2.78	2.70	2.64	2.59	2.54	.025
	.01	7.82	5.61	4.72	4.22	3.90	3.67	3.50	3.36	3.26	3.17	3.09	3.03	.01
30	.05	4.17	3.32	2.92	2.69	2.53	2.42	2.33	2.27	2.21	2.16	2.13	2.09	.05
	.025	5.57	4.18	3.59	3.25	3.03	2.87	2.75	2.65	2.57	2.51	2.46	2.41	.025
	.01	7.56	5.39	4.51	4.02	3.70	3.47	3.30	3.17	3.07	2.98	2.90	2.84	.01
40	.05	4.08	3.23	2.84	2.61	2.45	2.34	2.25	2.18	2.12	2.08	2.04	2.04	.05
	.025	5.42	4.05	3.46	3.13	2.90	2.74	2.62	2.53	2.45	2.39	2.33	2.29	.025
	.01	7.31	5.18	4.31	3.83	3.51	3.29	3.12	2.99	2.89	2.80	2.73	2.66	.01
60	.05	4.00	3.15	2.76	2.53	2.37	2.25	2.17	2.10	2.04	1.99	1.95	1.92	.05
	.025	5.29	3.93	3.34	3.01	2.79	2.63	2.51	2.41	2.33	2.27	2.22	2.17	.025
	.01	7.08	4.98	4.13	3.65	3.34	3.12	2.95	2.82	2.72	2.63	2.56	2.50	.01
120	.05	3.92	3.07	2.68	2.45	2.29	2.17	2.09	2.02	1.96	1.91	1.87	1.83	.05
	.025	5.15	3.80	3.23	2.89	2.67	2.52	2.39	2.30	2.22	2.16	2.10	2.05	.025
	.01	6.85	4.79	3.95	3.48	3.17	2.96	2.79	2.66	2.56	2.47	2.40	2.34	.01
∞	.05	3.84	3.00	2.60	2.37	2.21	2.10	2.01	1.94	1.88	1.83	1.79	1.75	.05
	.025	5.02	3.69	3.11	2.79	2.57	2.41	2.29	2.19	2.11	2.05	1.99	1.94	.025
	.01	6.63	4.61	3.78	3.32	3.02	2.80	2.64	2.51	2.41	2.32	2.25	2.18	.01

ν_2 (degrees of freedom of denominator mean squares)

TABLE V
continued

ν_1 (degrees of freedom of numerator mean squares)

	α	15	20	24	30	40	50	60	120	∞	α
11	.05	2.72	2.65	2.61	2.57	2.53	2.51	2.49	2.45	2.40	.05
	.025	3.33	3.23	3.17	3.12	3.06	3.02	3.00	2.94	2.88	.025
	.01	4.25	4.10	4.02	3.94	3.86	3.81	3.78	3.69	3.60	.01
12	.05	2.62	2.54	2.51	2.47	2.43	2.40	2.38	2.34	2.30	.05
	.025	3.18	3.07	3.02	2.96	2.91	2.87	2.85	2.79	2.72	.025
	.01	4.01	3.86	3.78	3.70	3.62	3.57	3.54	3.45	3.36	.01
15	.05	2.40	2.33	2.39	2.25	2.20	2.18	2.16	2.11	2.07	.05
	.025	2.86	2.76	2.70	2.64	2.59	2.55	2.52	2.46	2.40	.025
	.01	3.52	3.37	3.29	3.21	3.13	3.08	3.05	2.96	2.87	.01
20	.05	2.20	2.12	2.08	2.04	1.99	1.97	1.95	1.90	1.84	.05
	.025	2.57	2.46	2.41	2.35	2.29	2.25	2.22	2.16	2.09	.025
	.01	3.09	2.94	2.86	2.78	2.69	2.64	2.61	2.52	2.42	.01
24	.05	2.11	2.03	1.98	1.94	1.89	1.86	1.84	1.79	1.73	.05
	.025	2.44	2.33	2.27	2.21	2.15	2.11	2.08	2.01	1.94	.025
	.01	2.89	2.74	2.66	2.58	2.49	2.44	2.40	2.31	2.21	.01
30	.05	2.01	1.93	1.89	1.84	1.79	1.76	1.74	1.68	1.62	.05
	.025	2.31	2.20	2.14	2.07	2.01	1.97	1.94	1.87	1.79	.025
	.01	2.70	2.55	2.47	2.39	2.30	2.25	2.21	2.11	2.01	.01
40	.05	1.92	1.84	1.79	1.74	1.69	1.66	1.64	1.58	1.51	.05
	.025	2.18	2.07	2.01	1.94	1.88	1.83	1.80	1.72	1.64	.025
	.01	2.52	2.37	2.29	2.20	2.11	2.06	2.02	1.92	1.80	.01
60	.05	1.84	1.75	1.70	1.65	1.59	1.56	1.53	1.47	1.39	.05
	.025	2.06	1.94	1.88	1.82	1.74	1.70	1.67	1.58	1.48	.025
	.01	2.35	2.20	2.12	2.03	1.94	1.88	1.84	1.73	1.60	.01
120	.05	1.75	1.66	1.61	1.55	1.50	1.46	1.43	1.35	1.25	.05
	.025	1.95	1.82	1.76	1.69	1.61	1.56	1.53	1.43	1.31	.025
	.01	2.19	2.03	1.95	1.86	1.76	1.70	1.66	1.53	1.38	.01
∞	.05	1.67	1.57	1.52	1.46	1.39	1.35	1.32	1.22	1.00	.05
	.025	1.83	1.71	1.64	1.57	1.48	1.43	1.39	1.27	1.00	.025
	.01	2.04	1.88	1.79	1.70	1.59	1.52	1.47	1.32	1.00	.01

ν_2 (degrees of freedom of denominator mean squares)

TABLE **VI**
Critical values of F_{max}

α (number of samples)

v	α	2	3	4	5	6	7	8	9	10	11	12
2	.05	39.0	87.5	142.	202.	266.	333.	403.	475.	550.	626.	704.
	.01	199.	448	729.	1036.	1362.	1705.	2063.	2432.	2813.	3204.	3605.
3	.05	15.4	27.8	39.2	50.7	62.0	72.9	83.5	93.9	104.	114.	124.
	.01	47.5	85.	120.	151.	184.	21(6)	24(9)	28(1)	31(0)	33(7)	36(1)
4	.05	9.60	15.5	20.6	25.2	29.5	33.6	37.5	41.1	44.6	48.0	51.4
	.01	23.2	37.	49.	59.	69.	79.	89.	97.	106.	113.	120.
5	.05	7.15	10.8	13.7	16.3	18.7	20.8	22.9	24.7	26.5	28.2	29.9
	.01	14.9	22.	28.	33.	38.	42.	46.	50.	54.	57.	60.
6	.05	5.82	8.38	10.4	12.1	13.7	15.0	16.3	17.5	18.6	19.7	20.7
	.01	11.1	15.5	19.1	22.	25.	27.	30.	32.	34.	36.	37.
7	.05	4.99	6.94	8.44	9.70	10.8	11.8	12.7	13.5	14.3	15.1	15.8
	.01	8.89	12.1	14.5	16.5	18.4	20.	22.	23.	24.	26.	27.
8	.05	4.43	6.00	7.18	8.12	9.03	9.78	10.5	11.1	11.7	12.2	12.7
	.01	7.50	9.9	11.7	13.2	14.5	15.8	16.9	17.9	18.9	19.8	21.
9	.05	4.03	5.34	6.31	7.11	7.80	8.41	8.95	9.45	9.91	10.3	10.7
	.01	6.54	8.5	9.9	11.1	12.1	13.1	13.9	14.7	15.3	16.0	16.6
10	.05	3.72	4.85	5.67	6.34	6.92	7.42	7.87	8.28	8.66	9.01	9.34
	.01	5.85	7.4	8.6	9.6	10.4	11.1	11.8	12.4	12.9	13.4	13.9
12	.05	3.28	4.16	4.79	5.30	5.72	6.09	6.42	6.72	7.00	7.25	7.48
	.01	4.91	6.1	6.9	7.6	8.2	8.7	9.1	9.5	9.9	10.2	10.6
15	.05	2.86	3.54	4.01	4.37	4.68	4.95	5.19	5.40	5.59	5.77	5.93
	.01	4.07	4.9	5.5	6.0	6.4	6.7	7.1	7.3	7.5	7.8	8.0
20	.05	2.46	2.95	3.29	3.54	3.76	3.94	4.10	4.24	4.37	4.49	4.59
	.01	3.32	3.8	4.3	4.6	4.9	5.1	5.3	5.5	5.6	5.8	5.9
30	.05	2.07	2.40	2.61	2.78	2.91	3.02	3.12	3.21	3.29	3.36	3.39
	.01	2.63	3.0	3.3	3.4	3.6	3.7	3.8	3.9	4.0	4.1	4.2
60	.05	1.67	1.85	1.96	2.04	2.11	2.17	2.22	2.26	2.30	2.33	2.36
	.01	1.96	2.2	2.3	2.4	2.4	2.5	2.5	2.6	2.6	2.7	2.7
∞	.05	1.00	1.00	1.00	1.00	1.00	1.00	1.00	1.00	1.00	1.00	1.00
	.01	1.00	1.00	1.00	1.00	1.00	1.00	1.00	1.00	1.00	1.00	1.00

Note: Corresponding to each value of a (number of samples) and v (degrees of freedom) are two critical values of F_{max} representing the upper 5% and 1% percentage points. The corresponding probabilities $\alpha = 0.05$ and 0.01 represent *one tail* of the F_{max} distribution. This table was copied from H. A. David (*Biometrika* **39**:422–424, 1952) with permission of the publisher and author.

TABLE **VII**
Shortest unbiased confidence limits for the variance

ν	Confidence coefficients 0.95	0.99	ν	Confidence coefficients 0.95	0.99	ν	Confidence coefficients 0.95	0.99
2	.2099 23.605	.1505 114.489	14	.5135 2.354	.4289 3.244	26	.6057 1.825	.5261 2.262
3	.2681 10.127	.1983 29.689	15	.5242 2.276	.4399 3.091	27	.6110 1.802	.5319 2.223
4	.3125 6.590	.2367 15.154	16	.5341 2.208	.4502 2.961	28	.6160 1.782	.5374 2.187
5	.3480 5.054	.2685 10.076	17	.5433 2.149	.4598 2.848	29	.6209 1.762	.5427 2.153
6	.3774 4.211	.2956 7.637	18	.5520 2.097	.4689 2.750	30	.6255 1.744	.5478 2.122
7	.4025 3.679	.3192 6.238	19	.5601 2.050	.4774 2.664	40	.6636 1.608	.5900 1.896
8	.4242 3.314	.3400 5.341	20	.5677 2.008	.4855 2.588	50	.6913 1.523	.6213 1.760
9	.4432 3.048	.3585 4.720	21	.5749 1.971	.4931 2.519	60	.7128 1.464	.6458 1.668
10	.4602 2.844	.3752 4.265	22	.5817 1.936	.5004 2.458	70	.7300 1.421	.6657 1.607
11	.4755 2.683	.3904 3.919	23	.5882 1.905	.5073 2.402	80	.7443 1.387	.6824 1.549
12	.4893 2.553	.4043 3.646	24	.5943 1.876	.5139 2.351	90	.7564 1.360	.6966 1.508
13	.5019 2.445	.4171 3.426	25	.6001 1.850	.5201 2.305	100	.7669 1.338	.7090 1.475

Note: The factors in this table have been obtained by dividing the quantity $n - 1$ by the values found in a table prepared by D. V. Lindley, D. A. East, and P. A. Hamilton (*Biometrika* **47**:433–437, 1960).

TABLE **VIII**
Critical values for correlation coefficients

ν	α	r	ν	α	r	ν	α	r
1	.05	.997	16	.05	.468	35	.05	.325
	.01	1.000		.01	.590		.01	.418
2	.05	.950	17	.05	.456	40	.05	.304
	.01	.990		.01	.575		.01	.393
3	.05	.878	18	.05	.444	45	.05	.288
	.01	.959		.01	.561		.01	.372
4	.05	.811	19	.05	.433	50	.05	.273
	.01	.917		.01	.549		.01	.354
5	.05	.754	20	.05	.423	60	.05	.250
	.01	.874		.01	.537		.01	.325
6	.05	.707	21	.05	.413	70	.05	.232
	.01	.834		.01	.526		.01	.302
7	.05	.666	22	.05	.404	80	.05	.217
	.01	.798		.01	.515		.01	.283
8	.05	.632	23	.05	.396	90	.05	.205
	.01	.765		.01	.505		.01	.267
9	.05	.602	24	.05	.388	100	.05	.195
	.01	.735		.01	.496		.01	.254
10	.05	.576	25	.05	.381	120	.05	.174
	.01	.708		.01	.487		.01	.228
11	.05	.553	26	.05	.374	150	.05	.159
	.01	.684		.01	.478		.01	.208
12	.05	.532	27	.05	.367	200	.05	.138
	.01	.661		.01	.470		.01	.181
13	.05	.514	28	.05	.361	300	.05	.113
	.01	.641		.01	.463		.01	.148
14	.05	.497	29	.05	.355	400	.05	.098
	.01	.623		.01	.456		.01	.128
15	.05	.482	30	.05	.349	500	.05	.088
	.01	.606		.01	.449		.01	.115
						1,000	.05	.062
							.01	.081

Note: Upper value is 5% lower value is 1% critical value. This table is reproduced by permission from *Statistical Methods*, 5th edition, by George W. Snedecor, © 1956 by The Iowa State University Press.

TABLE IX
Confidence limits for percentages

This table furnishes confidence limits for percentages based on the binomial distribution.

The first part of the table furnishes limits for samples up to size $n = 30$. The arguments are Y, number of items in the sample that exhibit a given property, and n, sample size. Argument Y is tabled for integral values between 0 and 15, which yield percentages up to 50%. For each sample size n and number of items Y with the given property, three lines of numerical values are shown. The first line of values gives 95% confidence limits for the percentage, the second line lists the observed percentage incidence of the property, and the third line of values furnishes the 99% confidence limits for the percentage. For example, for $Y = 8$ individuals showing the property out of a sample of $n = 20$, the second line indicates that this represents an incidence of the property of 40.00%, the first line yields the 95% confidence limits of this percentage as 19.10% to 63.95%, and the third line gives the 99% limits as 14.60% to 70.10%.

Interpolate in this table (up to $n = 49$) by dividing L_1^- and L_2^-, the lower and upper confidence limits at the next lower tabled sample size n^-, by desired sample size n, and multiply them by the next lower tabled sample size n^-. Thus, for example, to obtain the confidence limits of the percentage corresponding to 8 individuals showing the given property in a sample of 22 individuals (which corresponds to 36.36% of the individuals showing the property), compute the lower confidence limit $L_1 = L_1^- n^-/n = (19.10)20/22 = 17.36\%$ and the upper confidence limit $L_2 = L_2^- n^-/n = (63.95)20/22 = 58.14\%$.

The second half of the table is for larger sample sizes ($n = 50$, 100, 200, 500, and 1000). The arguments along the left margin of the table are percentages from 0 to 50% in increments of 1%, rather than counts. The 95% and 99% confidence limits corresponding to a given percentage incidence p and sample size n are the functions given in two lines in the body of the table. For instance, the 99% confidence limits of an observed incidence of 12% in a sample of 500 are found to be 8.56–16.19%, in the second of the two lines. Interpolation in this table between the furnished sample sizes can be achieved by means of the following formula for the lower limit:

$$L_1 = \frac{L_1^- n^- (n^+ - n) + L_1^+ n^+ (n - n^-)}{n(n^+ - n^-)}$$

In the above expression, n is the size of the observed sample, n^- and n^+ the next lower and upper tabled sample sizes, respectively, L_1^- and L_1^+ are corresponding tabled confidence limits for these sample sizes, and L_1 is the lower confidence limit to be found by interpolation. The upper confidence limit, L_2, can be obtained by a corresponding formula by substituting 2 for the subscript 1. By way of an example we shall illustrate setting 95% confidence limits to an observed percentage of 25% in a sample size of 80. The tabled 95% limits for $n = 50$ are 13.84–39.27%. For $n = 100$, the corresponding tabled limits are

16.88–34.66%. When we substitute the values for the lower limits in the above formula we obtain

$$L_1 = \frac{(13.84)(50)(100 - 80) + (16.88)(100)(80 - 50)}{80(100 - 50)} = 16.12\%$$

for the lower confidence limit. Similarly, for the upper confidence limit we compute

$$L_2 = \frac{(39.27)(50)(100 - 80) + (34.66)(100)(80 - 50)}{80(100 - 50)} = 35.81\%$$

The tabled values in parentheses are limits for percentages that could not be obtained in any real sampling problem (for example, 25% in 50 items) but are necessary for purposes of interpolation. For percentages greater than 50% look up the complementary percentage as the argument. The complements of the tabled binomial confidence limits are the desired limits.

These tables have been extracted from more extensive ones in D. Mainland, L. Herrera, and M. I. Sutcliffe, *Tables for Use with Binomial Samples* (Department of Medical Statistics, New York University College of Medicine, 1956) with permission of the publisher. The interpolation formulas cited are also due to these authors. Confidence limits of odd percentages up to 13% for $n = 50$ were computed by interpolation. For $Y = 0$, one-sided $(1 - \alpha)100\%$ confidence limits were computed as $L_2 = 1 - \alpha^{1/n}$ with $L_1 = 0$.

TABLE **IX**

Confidence limits for percentages

Y	1 − α	5	10	15	20	25	30	1 − α	Y
	95	0.00-45.07	0.00-25.89	0.00-18.10	0.00-13.91	0.00-11.29	0.00- 9.50	95	
0		0.00	0.00	0.00	0.00	0.00	0.00		0
	99	0.00-60.19	0.00-36.90	0.00-26.44	0.00-20.57	0.00-16.82	0.00-14.23	99	
	95	0.51-71.60	0.25-44.50	0.17-32.00	0.13-24.85	0.10-20.36	0.08-17.23	95	
1		20.00	10.00	6.67	5.00	4.00	3.33		1
	99	0.10-81.40	0.05-54.4	0.03-40.27	0.02-31.70	0.02-26.24	0.02-22.33	99	
	95	5.28-85.34	2.52-55.60	1.66-40.49	1.24-31.70	0.98-26.05	0.82-22.09	95	
2		40.00	20.00	13.33	10.00	8.00	6.67		2
	99	2.28-91.72	1.08-64.80	0.71-48.71	0.53-38.70	0.42-32.08	0.35-27.35	99	
	95		6.67-65.2	4.33-48.07	3.21-37.93	2.55-31.24	2.11-26.53	95	
3			30.00	20.00	15.00	12.00	10.00		3
	99		3.70-73.50	2.39-56.07	1.77-45.05	1.40-37.48	1.16-32.03	99	
	95		12.20-73.80	7.80-55.14	5.75-43.65	4.55-36.10	3.77-30.74	95	
4			40.00	26.67	20.00	16.00	13.33		4
	99		7.68-80.91	4.88-62.78	3.58-50.65	2.83-42.41	2.34-36.39	99	
	95		18.70-81.30	11.85-61.62	8.68-49.13	6.84-40.72	5.64-34.74	95	
5			50.00	33.33	25.00	20.00	16.67		5
	99		12.80-87.20	8.03-68.89	5.85-56.05	4.60-47.00	3.79-40.44	99	
	95			16.33-67.74	11.90-54.30	9.35-45.14	7.70-38.56	95	
6				40.00	30.00	24.00	20.00		6
	99			11.67-74.40	8.45-60.95	6.62-51.38	5.43-44.26	99	
	95			21.29-73.38	15.38-59.20	12.06-49.38	9.92-42.29	95	
7				46.67	35.00	28.00	23.33		7
	99			15.87-79.54	11.40-65.70	8.90-55.56	7.29-48.01	99	
	95				19.10-63.95	14.96-53.50	12.29-45.89	95	
8					40.00	32.00	26.67		8
	99				14.60-70.10	11.36-59.54	9.30-51.58	99	
	95				23.05-68.48	17.97-57.48	14.73-49.40	95	
9					45.00	36.00	30.00		9
	99				18.08-74.30	14.01-63.36	11.43-55.00	99	
	95				27.20-72.80	21.12-61.32	17.29-52.80	95	
10					50.00	40.00	33.33		10
	99				21.75-78.25	16.80-67.04	13.69-58.35	99	
	95					24.41-65.06	19.93-56.13	95	
11						44.00	36.67		11
	99					19.75-70.55	16.06-61.57	99	
	95					27.81-68.69	22.66-59.39	95	
12						48.00	40.00		12
	99					22.84-73.93	18.50-64.69	99	
	95						25.46-62.56	95	
13							43.33		13
	99						21.07-67.72	99	
	95						28.35-65.66	95	
14							46.67		14
	99						23.73-70.66	99	
	95						31.30-68.70	95	
15							50.00		15
	99						26.47-73.53	99	

TABLE **IX**
continued

%	$1 - \alpha$	50	100	*n* 200	500	1000
0	95	.00- 7.11	.00- 3.62	.00- 1.83	.00- 0.74	.00- 0.37
	99	.00-10.05	.00- 5.16	.00- 2.62	.00- 1.05	.00- 0.53
1	95	(.02- 8.88)	.02- 5.45	.12- 3.57	.32- 2.32	.48- 1.83
	99	(.00-12.02)	.00- 7.21	.05- 4.55	.22- 2.80	.37- 2.13
2	95	.05-10.66	.24- 7.04	.55- 5.04	1.06- 3.56	1.29- 3.01
	99	.01-13.98	.10- 8.94	.34- 6.17	.87- 4.12	1.13- 3.36
3	95	(.27-12.19)	.62- 8.53	1.11- 6.42	1.79- 4.81	2.11- 4.19
	99	(.16-15.60)	.34-10.57	.78- 7.65	1.52- 5.44	1.88- 4.59
4	95	.49-13.72	1.10- 9.93	1.74- 7.73	2.53- 6.05	2.92- 5.36
	99	.21-17.21	.68-12.08	1.31- 9.05	2.17- 6.75	2.64- 5.82
5	95	(.88-15.14)	1.64-11.29	2.43- 9.00	3.26- 7.29	3.73- 6.54
	99	(.45-18.76)	1.10-13.53	1.89-10.40	2.83- 8.07	3.39- 7.05
6	95	1.26-16.57	2.24-12.60	3.18-10.21	4.11- 8.43	4.63- 7.64
	99	.69-20.32	1.56-14.93	2.57-11.66	3.63- 9.24	4.25- 8.18
7	95	(1.74-17.91)	2.86-13.90	3.88-11.47	4.96- 9.56	5.52- 8.73
	99	(1.04-21.72)	2.08-16.28	3.17-12.99	4.43-10.42	5.12- 9.31
8	95	2.23-19.25	3.51-15.16	4.70-12.61	5.81-10.70	6.42- 9.83
	99	1.38-23.13	2.63-17.61	3.93-14.18	5.23-11.60	5.98-10.43
9	95	(2.78-20.54)	4.20-16.40	5.46-13.82	6.66-11.83	7.32-10.93
	99	(1.80-24.46)	3.21-18.92	4.61-15.44	6.04-12.77	6.84-11.56
10	95	3.32-21.82	4.90-17.62	6.22-15.02	7.51-12.97	8.21-12.03
	99	2.22-25.80	3.82-20.20	5.29-16.70	6.84-13.95	7.70-12.69
11	95	(3.93-23.06)	5.65-18.80	7.05-16.16	8.41-14.06	9.14-13.10
	99	(2.70-27.11)	4.48-21.42	6.06-17.87	7.70-15.07	8.60-13.78
12	95	4.54-24.31	6.40-19.98	7.87-17.30	9.30-15.16	10.06-14.16
	99	3.18-28.42	5.15-22.65	6.83-19.05	8.56-16.19	9.51-14.86
13	95	(5.18-27.03)	7.11-21.20	8.70-18.44	10.20-16.25	10.99-15.23
	99	(3.72-29.67)	5.77-23.92	7.60-20.23	9.42-17.31	10.41-15.95
14	95	5.82-26.75	7.87-22.37	9.53-19.58	11.09-17.34	11.92-16.30
	99	4.25-30.92	6.46-25.13	8.38-21.40	10.28-18.43	11.31-17.04
15	95	(6.50-27.94)	8.64-23.53	10.36-20.72	11.98-18.44	12.84-17.37
	99	(4.82-32.14)	7.15-26.33	9.15-22.58	11.14-19.55	12.21-18.13
16	95	7.17-29.12	9.45-24.66	11.22-21.82	12.90-19.50	13.79-18.42
	99	5.40-33.36	7.89-27.49	9.97-23.71	12.03-20.63	13.14-19.19
17	95	(7.88-30.28)	10.25-25.79	12.09-22.92	13.82-20.57	14.73-19.47
	99	(6.00-34.54)	8.63-28.65	10.79-24.84	12.92-21.72	14.07-20.25
18	95	8.58-31.44	11.06-26.92	12.96-24.02	14.74-21.64	15.67-20.52
	99	6.60-35.73	9.37-29.80	11.61-25.96	13.81-22.81	14.99-21.32
19	95	(9.31-32.58)	11.86-28.06	13.82-25.12	15.66-22.71	16.62-21.57
	99	(7.23-36.88)	10.10-30.96	12.43-27.09	14.71-23.90	15.92-22.38
20	95	10.04-33.72	12.66-29.19	14.69-26.22	16.58-23.78	17.56-22.62
	99	7.86-38.04	10.84-32.12	13.26-28.22	15.60-24.99	16.84-23.45
21	95	(10.79-34.84)	13.51-30.28	15.58-27.30	17.52-24.83	18.52-23.65
	99	(8.53-39.18)	11.63-33.24	14.11-29.31	16.51-26.05	17.78-24.50
22	95	11.54-35.95	14.35-31.37	16.48-28.37	18.45-25.88	19.47-24.69
	99	9.20-40.32	12.41-34.35	14.97-30.40	17.43-27.12	18.72-25.55
23	95	(12.30-37.06)	15.19-32.47	17.37-29.45	19.39-26.93	20.43-25.73
	99	(9.88-41.44)	13.60-34.82	15.83-31.50	18.34-28.18	19.67-26.59
24	95	13.07-38.17	16.03-33.56	18.27-30.52	20.33-27.99	21.39-26.77
	99	10.56-42.56	13.98-36.57	16.68-32.59	19.26-29.25	20.61-27.64
25	95	(13.84-39.27)	16.88-34.66	19.16-31.60	21.26-29.04	22.34-27.81
	99	(11.25-43.65)	14.77-37.69	17.54-33.68	20.17-30.31	21.55-28.69

TABLE **IX**
continued

%	$1 - \alpha$	50	100	n 200	500	1000
26	95	14.63-40.34	17.75-35.72	20.08-32.65	22.21-30.08	23.31-28.83
	99	11.98-44.73	15.59-38.76	18.43-34.75	21.10-31.36	22.50-29.73
27	95	(15.45-41.40)	18.62-36.79	20.99-33.70	23.16-31.11	24.27-29.86
	99	(12.71-45.79)	16.42-39.84	19.31-35.81	22.04-32.41	23.46-30.76
28	95	16.23-42.48	19.50-37.85	21.91-34.76	24.11-32.15	25.24-30.89
	99	13.42-46.88	17.25-40.91	20.20-36.88	22.97-33.46	24.41-31.80
29	95	(17.06-43.54)	20.37-38.92	22.82-35.81	25.06-33.19	26.21-31.92
	99	(14.18-47.92)	18.07-41.99	21.08-37.94	23.90-34.51	25.37-32.84
30	95	17.87-44.61	21.24-39.98	23.74-36.87	26.01-34.23	27.17-32.95
	99	14.91-48.99	18.90-43.06	21.97-39.01	24.83-35.55	26.32-33.87
31	95	(18.71-45.65)	22.14-41.02	24.67-37.90	26.97-35.25	28.15-33.97
	99	(15.68-50.02)	19.76-44.11	22.88-40.05	25.78-36.59	27.29-34.90
32	95	19.55-46.68	23.04-42.06	25.61-38.94	27.93-36.28	29.12-34.99
	99	16.46-51.05	20.61-45.15	23.79-41.09	26.73-37.62	28.25-35.92
33	95	(20.38-47.72)	23.93-43.10	26.54-39.97	28.90-37.31	30.09-36.01
	99	(17.23-52.08)	21.47-46.19	24.69-42.13	27.68-38.65	29.22-36.95
34	95	21.22-48.76	24.83-44.15	27.47-41.01	29.86-38.33	31.07-37.03
	99	18.01-53.11	22.33-47.24	25.60-43.18	28.62-39.69	30.18-37.97
35	95	(22.06-49.80)	25.73-45.19	28.41-42.04	30.82-39.36	32.04-38.05
	99	(18.78-54.14)	23.19-48.28	26.51-44.22	29.57-40.72	31.14-39.00
36	95	22.93-50.80	26.65-46.20	29.36-43.06	31.79-40.38	33.02-39.06
	99	19.60-55.13	24.08-49.30	27.44-45.24	30.53-41.74	32.12-40.02
37	95	(23.80-51.81)	27.57-47.22	30.31-44.08	32.76-41.39	34.00-40.07
	99	(20.42-56.12)	24.96-50.31	28.37-46.26	31.49-42.76	33.09-41.03
38	95	24.67-52.81	28.49-48.24	31.25-45.10	33.73-42.41	34.98-41.09
	99	21.23-57.10	25.85-51.32	29.30-47.29	32.45-43.78	34.07-42.05
39	95	(25.54-53.82)	29.41-49.26	32.20-46.12	34.70-43.43	35.97-42.10
	99	(22.05-58.09)	26.74-52.34	30.23-48.31	33.42-44.80	35.04-43.06
40	95	26.41-54.82	30.33-50.28	33.15-47.14	35.68-44.44	36.95-43.11
	99	22.87-59.08	27.63-53.35	31.16-49.33	34.38-45.82	36.02-44.08
41	95	(27.31-55.80)	31.27-51.28	34.12-48.15	36.66-45.45	37.93-44.12
	99	(23.72-60.04)	28.54-54.34	32.11-50.33	35.35-46.83	37.00-45.09
42	95	28.21-56.78	32.21-52.28	35.08-49.16	37.64-46.46	38.92-45.12
	99	24.57-60.99	29.45-55.33	33.06-51.33	36.32-47.83	37.98-46.10
43	95	(29.10-57.76)	33.15-53.27	36.05-50.16	38.62-47.46	39.91-46.13
	99	(25.42-61.95)	30.37-56.32	34.01-52.34	37.29-48.84	38.96-47.10
44	95	30.00-58.74	34.09-54.27	37.01-51.17	39.60-48.47	40.90-47.14
	99	26.27-62.90	31.28-57.31	34.95-53.34	38.27-49.85	39.95-48.11
45	95	(30.90-59.71)	35.03-55.27	37.97-52.17	40.58-49.48	41.89-48.14
	99	(27.12-63.86)	32.19-58.30	35.90-54.34	39.24-50.86	40.93-49.12
46	95	31.83-60.67	35.99-56.25	38.95-53.17	41.57-50.48	42.88-49.14
	99	28.00-64.78	33.13-59.26	36.87-55.33	40.22-51.85	41.92-50.12
47	95	(32.75-61.62)	36.95-57.23	39.93-54.16	42.56-51.48	43.87-50.14
	99	(28.89-65.69)	34.07-60.22	37.84-56.31	41.21-52.85	42.91-51.12
48	95	33.68-62.57	37.91-58.21	40.91-55.15	43.55-52.47	44.87-51.14
	99	29.78-66.61	35.01-61.19	38.80-57.30	42.19-53.85	43.90-52.12
49	95	(34.61-63.52)	38.87-59.19	41.89-56.14	44.54-53.47	45.86-52.14
	99	(30.67-67.53)	35.95-62.15	39.77-58.28	43.18-54.84	44.89-53.12
50	95	35.53-64.47	39.83-60.17	42.86-57.14	45.53-54.47	46.85-53.15
	99	31.55-68.45	36.89-63.11	40.74-59.26	44.16-55.84	45.89-54.11

338

TABLE X
The z transformation of correlation coefficient r

r	z	r	z
0.00	0.0000	0.50	0.5493
0.01	0.0100	0.51	0.5627
0.02	0.0200	0.52	0.5763
0.03	0.0300	0.53	0.5901
0.04	0.0400	0.54	0.6042
0.05	0.0500	0.55	0.6184
0.06	0.0601	0.56	0.6328
0.07	0.0701	0.57	0.6475
0.08	0.0802	0.58	0.6625
0.09	0.0902	0.59	0.6777
0.10	0.1003	0.60	0.6931
0.11	0.1104	0.61	0.7089
0.12	0.1206	0.62	0.7250
0.13	0.1307	0.63	0.7414
0.14	0.1409	0.64	0.7582
0.15	0.1511	0.65	0.7753
0.16	0.1614	0.66	0.7928
0.17	0.1717	0.67	0.8107
0.18	0.1820	0.68	0.8291
0.19	0.1923	0.69	0.8480
0.20	0.2027	0.70	0.8673
0.21	0.2132	0.71	0.8872
0.22	0.2237	0.72	0.9076
0.23	0.2342	0.73	0.9287
0.24	0.2448	0.74	0.9505
0.25	0.2554	0.75	0.9730
0.26	0.2661	0.76	0.9962
0.27	0.2769	0.77	1.0203
0.28	0.2877	0.78	1.0454
0.29	0.2986	0.79	1.0714
0.30	0.3095	0.80	1.0986
0.31	0.3205	0.81	1.1270
0.32	0.3316	0.82	1.1568
0.33	0.3428	0.83	1.1881
0.34	0.3541	0.84	1.2212
0.35	0.3654	0.85	1.2562
0.36	0.3769	0.86	1.2933
0.37	0.3884	0.87	1.3331
0.38	0.4001	0.88	1.3758
0.39	0.4118	0.89	1.4219
0.40	0.4236	0.90	1.4722
0.41	0.4356	0.91	1.5275
0.42	0.4477	0.92	1.5890
0.43	0.4599	0.93	1.6584
0.44	0.4722	0.94	1.7380
0.45	0.4847	0.95	1.8318
0.46	0.4973	0.96	1.9459
0.47	0.5101	0.97	2.0923
0.48	0.5230	0.98	2.2976
0.49	0.5361	0.99	2.6467

TABLE **XI**
Critical values of U, the Mann-Whitney statistic

α

n_1	n_2	0.10	0.05	0.025	0.01	0.005	0.001
3	2	6					
	3	8	9				
4	2	8					
	3	11	12				
	4	13	15	16			
5	2	9	10				
	3	13	14	15			
	4	16	18	19	20		
	5	20	21	23	24	25	
6	2	11	12				
	3	15	16	17			
	4	19	21	22	23	24	
	5	23	25	27	28	29	
	6	27	29	31	33	34	
7	2	13	14				
	3	17	19	20	21		
	4	22	24	25	27	28	
	5	27	29	30	32	34	
	6	31	34	36	38	39	42
	7	36	38	41	43	45	48
8	2	14	15	16			
	3	19	21	22	24		
	4	25	27	28	30	31	
	5	30	32	34	36	38	40
	6	35	38	40	42	44	47
	7	40	43	46	49	50	54
	8	45	49	51	55	57	60
9	1	9					
	2	16	17	18			
	3	22	23	25	26	27	
	4	27	30	32	33	35	
	5	33	36	38	40	42	44
	6	39	42	44	47	49	52
	7	45	48	51	54	56	60
	8	50	54	57	61	63	67
	9	56	60	64	67	70	74
10	1	10					
	2	17	19	20			
	3	24	26	27	29	30	
	4	30	33	35	37	38	40
	5	37	39	42	44	46	49
	6	43	46	49	52	54	57
	7	49	53	56	59	61	65
	8	56	60	63	67	69	74
	9	62	66	70	74	77	82
	10	68	73	77	81	84	90

Note: Critical values are tabulated for two samples of sizes n_1 and n_2, where $n_1 \geq n_2$, up to $n_1 = n_2 = 20$. The upper bounds of the critical values are furnished so that the sample statistic U, has to be greater than a given critical value to be significant. The probabilities at the heads of the columns are based on a one-tailed test and represent the proportion of the area of the distribution of U in one tail beyond the critical value. For a two-tailed test use the same critical values but double the probability at the heads of the columns. This table was extracted from a more extensive one (table 11.4) in D. B. Owen, *Handbook of Statistical Tables* (Addison-Wesley Publishing Co., Reading, Mass., 1962): Courtesy of U.S. Atomic Energy Commission, with permission of the publishers.

TABLE **XI**
continued

n_1	n_2	0.10	0.05	0.025	0.01	0.005	0.001
11	1	11					
	2	19	21	22			
	3	26	28	30	32	33	
	4	33	36	38	40	42	44
	5	40	43	46	48	50	53
	6	47	50	53	57	59	62
	7	54	58	61	65	67	71
	8	61	65	69	73	75	80
	9	68	72	76	81	83	89
	10	74	79	84	88	92	98
	11	81	87	91	96	100	106
12	1	12					
	2	20	22	23			
	3	28	31	32	34	35	
	4	36	39	41	42	45	48
	5	43	47	49	52	54	58
	6	51	55	58	61	63	68
	7	58	63	66	70	72	77
	8	66	70	74	79	81	87
	9	73	78	82	87	90	96
	10	81	86	91	96	99	106
	11	88	94	99	104	108	115
	12	95	102	107	113	117	124
13	1	13					
	2	22	24	25	26		
	3	30	33	35	37	38	
	4	39	42	44	47	49	51
	5	47	50	53	56	58	62
	6	55	59	62	66	68	73
	7	63	67	71	75	78	83
	8	71	76	80	84	87	93
	9	79	84	89	94	97	103
	10	87	93	97	103	106	113
	11	95	101	106	112	116	123
	12	103	109	115	121	125	133
	13	111	118	124	130	135	143
14	1	14					
	2	24	25	27	28		
	3	32	35	37	40	41	
	4	41	45	47	50	52	55
	5	50	54	57	60	63	67
	6	59	63	67	71	73	78
	7	67	72	76	81	83	89
	8	76	81	86	90	94	100
	9	85	90	95	100	104	111
	10	93	99	104	110	114	121
	11	102	108	114	120	124	132
	12	110	117	123	130	134	143
	13	119	126	132	139	144	153
	14	127	135	141	149	154	164

TABLE **XI**
continued

n_1	n_2	α 0.10	0.05	0.025	0.01	0.005	0.001
15	1	15					
	2	25	27	29	30		
	3	35	38	40	42	43	
	4	44	48	50	53	55	59
	5	53	57	61	64	67	71
	6	63	67	71	75	78	83
	7	72	77	81	86	89	95
	8	81	87	91	96	100	106
	9	90	96	101	107	111	118
	10	99	106	111	117	121	129
	11	108	115	121	128	132	141
	12	117	125	131	138	143	152
	13	127	134	141	148	153	163
	14	136	144	151	159	164	174
	15	145	153	161	169	174	185
16	1	16					
	2	27	29	31	32		
	3	37	40	42	45	46	
	4	47	50	53	57	59	62
	5	57	61	65	68	71	75
	6	67	71	75	80	83	88
	7	76	82	86	91	94	101
	8	86	92	97	102	106	113
	9	96	102	107	113	117	125
	10	106	112	118	124	129	137
	11	115	122	129	135	140	149
	12	125	132	139	146	151	161
	13	134	143	149	157	163	173
	14	144	153	160	168	174	185
	15	154	163	170	179	185	197
	16	163	173	181	190	196	208
17	1	17					
	2	28	31	32	34		
	3	39	42	45	47	49	51
	4	50	53	57	60	62	66
	5	60	65	68	72	75	80
	6	71	76	80	84	87	93
	7	81	86	91	96	100	106
	8	91	97	102	108	112	119
	9	101	108	114	120	124	132
	10	112	119	125	132	136	145
	11	122	130	136	143	148	158
	12	132	140	147	155	160	170
	13	142	151	158	166	172	183
	14	153	161	169	178	184	195
	15	163	172	180	189	195	208
	16	173	183	191	201	207	220
	17	183	193	202	212	219	232

TABLE **XI**
continued

		α					
n_1	n_2	0.10	0.05	0.025	0.01	0.005	0.001
18	1	18					
	2	30	32	34	36		
	3	41	45	47	50	52	54
	4	52	56	60	63	66	69
	5	63	68	72	76	79	84
	6	74	80	84	89	92	98
	7	85	91	96	102	105	112
	8	96	103	108	114	118	126
	9	107	114	120	126	131	139
	10	118	125	132	139	143	153
	11	129	137	143	151	156	166
	12	139	148	155	163	169	179
	13	150	159	167	175	181	192
	14	161	170	178	187	194	206
	15	172	182	190	200	206	219
	16	182	193	202	212	218	232
	17	193	204	213	224	231	245
	18	204	215	225	236	243	258
19	1	18	19				
	2	31	34	36	37	38	
	3	43	47	50	53	54	57
	4	55	59	63	67	69	73
	5	67	72	76	80	83	88
	6	78	84	89	94	97	103
	7	90	96	101	107	111	118
	8	101	108	114	120	124	132
	9	113	120	126	133	138	146
	10	124	132	138	146	151	161
	11	136	144	151	159	164	175
	12	147	156	163	172	177	188
	13	158	167	175	184	190	202
	14	169	179	188	197	203	216
	15	181	191	200	210	216	230
	16	192	203	212	222	230	244
	17	203	214	224	235	242	257
	18	214	226	236	248	255	271
	19	226	238	248	260	268	284
20	1	19	20				
	2	33	36	38	39	40	
	3	45	49	52	55	57	60
	4	58	62	66	70	72	77
	5	70	75	80	84	87	93
	6	82	88	93	98	102	108
	7	94	101	106	112	116	124
	8	106	113	119	126	130	139
	9	118	126	132	140	144	154
	10	130	138	145	153	158	168
	11	142	151	158	167	172	183
	12	154	163	171	180	186	198
	13	166	176	184	193	200	212
	14	178	188	197	207	213	226
	15	190	200	210	220	227	241
	16	201	213	222	233	241	255
	17	213	225	235	247	254	270
	18	225	237	248	260	268	284
	19	237	250	261	273	281	298
	20	249	262	273	286	295	312

TABLE **XII**
Critical values of the Wilcoxon rank sum.

	nominal α							
	0.05		0.025		0.01		0.005	
n	T	α	T	α	T	α	T	α
5	0	.0312						
	1	.0625						
6	2	.0469	0	.0156				
	3	.0781	1	.0312				
7	3	.0391	2	.0234	0	.0078		
	4	.0547	3	.0391	1	.0156		
8	5	.0391	3	.0195	1	.0078	0	.0039
	6	.0547	4	.0273	2	.0117	1	.0078
9	8	.0488	5	.0195	3	.0098	1	.0039
	9	.0645	6	.0273	4	.0137	2	.0059
10	10	.0420	8	.0244	5	.0098	3	.0049
	11	.0527	9	.0322	6	.0137	4	.0068
11	13	.0415	10	.0210	7	.0093	5	.0049
	14	.0508	11	.0269	8	.0122	6	.0068
12	17	.0461	13	.0212	9	.0081	7	.0046
	18	.0549	14	.0261	10	.0105	8	.0061
13	21	.0471	17	.0239	12	.0085	9	.0040
	22	.0549	18	.0287	13	.0107	10	.0052
14	25	.0453	21	.0247	15	.0083	12	.0043
	26	.0520	22	.0290	16	.0101	13	.0054
15	30	.0473	25	.0240	19	.0090	15	.0042
	31	.0535	26	.0277	20	.0108	16	.0051
16	35	.0467	29	.0222	23	.0091	19	.0046
	36	.0523	30	.0253	24	.0107	20	.0055
17	41	.0492	34	.0224	27	.0087	23	.0047
	42	.0544	35	.0253	28	.0101	24	.0055
18	47	.0494	40	.0241	32	.0091	27	.0045
	48	.0542	41	.0269	33	.0104	28	.0052
19	53	.0478	46	.0247	37	.0090	32	.0047
	54	.0521	47	.0273	38	.0102	33	.0054
20	60	.0487	52	.0242	43	.0096	37	.0047
	61	.0527	53	.0266	44	.0107	38	.0053

Note: This table furnishes critical values for the one-tailed test of significance of the rank sum T, obtained in Wilcoxon's matched-pairs signed-ranks test. Since the exact probability level desired cannot be obtained with integral critical values of T, two such values and their attendant probabilities bracketing the desired signficance level are furnished. Thus, to find the significant 1% values for $n = 19$ we note the two critical of T, 37 and 38, in the table. The probabilities corresponding to these two values of T are 0.0090 and 0.0102. Clearly a rank sum of $T_s = 37$ would have a probability of less than 0.01 and would be considered significant by the stated criterion. For two-tailed tests in which the alternative hypothesis is that the pairs could differ in either direction, double the probabilities stated at the head of the table. For sample sizes $n > 59$ compute

$$t_{\alpha[\infty]} = \left[T_s - \frac{n(n+1)}{4} \right] \Big/ \sqrt{\frac{n(n+1)(2n+1)}{24}}$$

for a two-tailed test.

TABLE **XII**
continued

	nominal α							
	0.05		0.025		0.01		0.005	
n	T	α	T	α	T	α	T	α
21	67	.0479	58	.0230	49	.0097	42	.0045
	68	.0516	59	.0251	50	.0108	43	.0051
22	75	.0492	65	.0231	55	.0095	48	.0046
	76	.0527	66	.0250	56	.0104	49	.0052
23	83	.0490	73	.0242	62	.0098	54	.0046
	84	.0523	74	.0261	63	.0107	55	.0051
24	91	.0475	81	.0245	69	.0097	61	.0048
	92	.0505	82	.0263	70	.0106	62	.0053
25	100	.0479	89	.0241	76	.0094	68	.0048
	101	.0507	90	.0258	77	.0101	69	.0053
26	110	.0497	98	.0247	84	.0095	75	.0047
	111	.0524	99	.0263	85	.0102	76	.0051
27	119	.0477	107	.0246	92	.0093	83	.0048
	120	.0502	108	.0260	93	.0100	84	.0052
28	130	.0496	116	.0239	101	.0096	91	.0048
	131	.0521	117	.0252	102	.0102	92	.0051
29	140	.0482	126	.0240	110	.0095	100	.0049
	141	.0504	127	.0253	111	.0101	101	.0053
30	151	.0481	137	.0249	120	.0098	109	.0050
	152	.0502	138	.0261	121	.0104	110	.0053
31	163	.0491	147	.0239	130	.0099	118	.0049
	164	.0512	148	.0251	131	.0105	119	.0052
32	175	.0492	159	.0249	140	.0097	128	.0050
	176	.0512	160	.0260	141	.0103	129	.0053
33	187	.0485	170	.0242	151	.0099	138	.0049
	188	.0503	171	.0253	152	.0104	139	.0052
34	200	.0488	182	.0242	162	.0098	148	.0048
	201	.0506	183	.0252	163	.0103	149	.0051
35	213	.0484	195	.0247	173	.0096	159	.0048
	214	.0501	196	.0257	174	.0100	160	.0051

TABLE **XII**
continued

n	nominal α 0.05		0.025		0.01		0.005	
	T	α	T	α	T	α	T	α
36	227	.0489	208	.0248	185	.0096	171	.0050
	228	.0505	209	.0258	186	.0100	172	.0052
37	241	.0487	221	.0245	198	.0099	182	.0048
	242	.0503	222	.0254	199	.0103	183	.0050
38	256	.0493	235	.0247	211	.0099	194	.0048
	257	.0509	236	.0256	212	.0104	195	.0050
39	271	.0493	249	.0246	224	.0099	207	.0049
	272	.0507	250	.0254	225	.0103	208	.0051
40	286	.0486	264	.0249	238	.0100	220	.0049
	287	.0500	265	.0257	239	.0104	221	.0051
41	302	.0488	279	.0248	252	.0100	233	.0048
	303	.0501	280	.0256	253	.0103	234	.0050
42	319	.0496	294	.0245	266	.0098	247	.0049
	320	.0509	295	.0252	267	.0102	248	.0051
43	336	.0498	310	.0245	281	.0098	261	.0048
	337	.0511	311	.0252	282	.0102	262	.0050
44	353	.0495	327	.0250	296	.0097	276	.0049
	354	.0507	328	.0257	297	.0101	277	.0051
45	371	.0498	343	.0244	312	.0098	291	.0049
	372	.0510	344	.0251	313	.0101	292	.0051
46	389	.0497	361	.0249	328	.0098	307	.0050
	390	.0508	362	.0256	329	.0101	308	.0052
47	407	.0490	378	.0245	345	.0099	322	.0048
	408	.0501	379	.0251	346	.0102	323	.0050
48	426	.0490	396	.0244	362	.0099	339	.0050
	427	.0500	397	.0251	363	.0102	340	.0051
49	446	.0495	415	.0247	379	.0098	355	.0049
	447	.0505	416	.0253	380	.0100	356	.0050
50	466	.0495	434	.0247	397	.0098	373	.0050
	467	.0506	435	.0253	398	.0101	374	.0051

TABLE XIII

Critical values of the two-sample Kolmogorov-Smirnov statistic.

$$n_2$$

n_1	α	2	3	4	5	6	7	8	9	10	11	12	13	14	15	16	17	18	19	20	21	22	23	24	25
2	.05	-	-	-	-	-	-	16	18	20	22	24	26	26	28	30	32	34	36	38	38	40	42	44	46
	.025	-	-	-	-	-	-	-	-	-	-	24	26	28	30	32	34	36	38	40	40	42	44	46	48
	.01	-	-	-	-	-	-	-	-	-	-	-	-	-	-	-	-	-	38	40	42	44	46	48	50
3	.05	-	-	-	15	18	21	21	24	27	30	30	33	36	36	39	42	45	45	48	51	51	54	57	60
	.025	-	-	-	-	18	21	24	27	30	30	33	36	39	39	42	45	48	51	51	54	57	60	60	63
	.01	-	-	-	-	-	-	-	27	30	33	36	39	42	42	45	48	51	54	57	57	60	63	66	69
4	.05	-	-	16	20	20	24	28	28	30	33	36	39	42	44	48	48	50	53	60	59	62	64	68	68
	.025	-	-	-	20	24	28	28	32	36	36	40	44	44	45	52	52	54	57	64	68	72	72	76	80
	.01	-	-	-	-	24	28	32	36	36	40	44	48	48	52	56	60	60	64	68	72	72	76	80	84
5	.05	-	15	20	25	24	28	30	35	40	39	43	45	46	55	54	55	60	61	65	69	70	72	76	80
	.025	-	-	20	25	30	30	32	36	40	44	45	47	51	55	59	60	65	66	75	74	78	80	81	90
	.01	-	-	-	25	30	35	35	40	45	45	50	52	56	60	64	68	70	71	80	80	83	87	90	95
6	.05	-	18	20	24	30	30	34	39	40	43	48	52	54	57	60	62	72	70	72	75	78	80	90	88
	.025	-	18	24	30	36	36	42	44	48	54	54	58	63	64	67	72	73	84	76	78	81	86	96	96
	.01	-	-	24	30	36	36	40	45	48	54	60	60	64	69	72	73	84	83	88	90	92	97	102	107
7	.05	-	21	24	28	30	42	40	42	46	48	53	56	63	62	64	68	72	76	79	91	84	89	92	97
	.025	-	21	28	30	35	42	41	45	49	52	56	58	70	68	73	77	80	84	86	98	96	98	102	105
	.01	-	-	28	35	36	42	48	49	53	59	60	65	77	75	77	84	87	91	93	105	103	108	112	115
8	.05	16	21	28	30	34	40	48	46	48	53	60	62	64	67	80	77	80	82	88	89	94	98	104	104
	.025	-	24	28	32	36	41	48	48	54	58	64	65	70	74	80	80	86	90	96	97	102	106	112	112
	.01	-	-	32	35	40	48	56	55	60	64	68	72	76	81	88	88	94	98	104	107	112	115	128	125
9	.05	18	24	28	35	39	42	46	54	53	59	63	65	70	75	78	82	90	89	93	99	101	106	111	114
	.025	-	27	32	36	42	45	48	63	60	63	69	72	76	81	85	90	99	98	100	108	110	115	120	123
	.01	-	27	36	40	45	49	55	63	63	70	75	78	84	90	94	99	108	107	111	117	122	126	132	135
10	.05	20	27	30	40	40	46	48	53	70	60	66	70	74	80	84	89	92	94	110	105	108	114	118	125
	.025	-	30	36	40	44	49	54	60	70	68	72	77	82	90	90	96	100	103	120	116	118	124	128	135
	.01	-	30	36	45	48	53	60	63	80	77	80	84	90	100	100	106	108	113	130	126	130	137	140	150

Note: This table furnishes *upper* critical values of $n_1 n_2 D$, the Kolmogorov-Smirnov test statistic D multiplied by the two sample sizes n_1 and n_2. Sample sizes n_1 are given at the left margin of the table, while sample sizes n_2 are given across its top at the heads of the columns. The three values furnished at the intersection of two samples sizes represent the following three two-tailed probabilities: 0.05, 0.025, and 0.01.

For two samples with $n_1 = 16$ and $n_2 = 10$, the 5% critical value of $n_1 n_2 D$ is 84. Any value of $n_1 n_2 D \geq 84$ will be significant at $P \leq 0.05$.

When a one-sided test is desired, approximate probabilities can be obtained from this table by doubling the nominal α values. However, these are not exact, since the distribution of cumulative frequencies is discrete.

This table was copied from table 55 in E. S. Pearson and H. O. Hartley, *Biometrika Tables for Statisticians,* Vol. II (Cambridge University Press, London 1972) with permission of the publishers.

TABLE XIII
continued

n_2

n_1	α	2	3	4	5	6	7	8	9	10	11	12	13	14	15	16	17	18	19	20	21	22	23	24	25
11	.05	22	30	33	39	43	48	53	59	60	77	72	75	82	84	89	93	97	102	107	112	121	119	124	129
	.025	-	30	36	44	48	52	58	63	68	77	76	84	87	94	96	102	107	111	116	123	132	131	137	140
	.01	-	33	40	45	54	59	64	70	77	88	86	91	96	102	106	110	118	122	127	134	143	142	150	154
12	.05	24	30	36	43	48	53	60	63	66	72	84	81	86	93	96	100	108	108	116	120	124	125	144	138
	.025	24	33	40	45	54	56	64	69	72	76	96	84	94	99	104	108	120	120	124	129	134	137	156	150
	.01	-	36	44	50	60	60	68	75	80	86	96	95	104	108	116	119	126	130	140	141	148	149	168	165
13	.05	26	33	39	45	52	56	62	65	70	75	81	91	89	96	101	105	110	114	120	126	130	135	140	145
	.025	26	36	44	47	54	58	65	72	77	84	84	100	104	104	111	114	120	126	130	137	141	146	151	158
	.01	-	39	48	52	60	65	72	78	84	91	95	117	104	115	121	127	131	138	143	150	156	161	166	172
14	.05	26	36	42	46	54	63	64	70	74	82	86	89	112	98	106	111	116	121	126	140	138	142	146	150
	.025	28	39	44	51	58	70	70	76	82	87	94	100	112	110	116	122	126	133	138	147	148	154	160	166
	.01	-	42	48	56	64	77	76	84	90	96	104	104	126	123	126	134	140	148	152	161	164	170	176	182
15	.05	28	36	44	55	57	62	67	75	80	84	93	96	98	120	114	116	123	127	135	138	144	149	156	160
	.025	30	39	45	55	63	68	74	81	90	94	99	104	110	135	119	129	135	141	150	153	154	163	168	175
	.01	-	42	52	60	69	75	81	90	100	102	108	115	123	135	133	142	147	152	160	168	173	179	186	195
16	.05	30	39	48	54	60	64	80	78	84	89	96	101	106	114	128	124	128	133	140	145	150	157	168	167
	.025	32	42	52	59	64	73	80	85	90	96	104	111	116	119	144	136	140	145	156	157	164	169	184	181
	.01	-	45	56	64	72	77	88	94	100	106	116	121	126	133	160	143	154	160	168	173	180	187	200	199
17	.05	32	42	48	55	62	68	77	82	89	93	100	105	111	116	124	136	133	141	146	151	157	163	168	173
	.025	34	45	52	60	67	77	80	90	96	102	108	114	122	129	136	153	148	151	160	166	170	179	183	190
	.01	-	48	60	68	73	84	88	99	106	110	119	127	134	142	143	170	164	166	175	180	187	196	203	207
18	.05	34	45	50	60	72	72	80	90	92	97	108	110	116	123	128	133	162	142	152	159	164	170	180	180
	.025	36	48	54	65	78	80	86	99	100	107	120	120	126	135	140	148	162	159	166	174	178	184	198	196
	.01	-	51	60	70	84	87	94	108	108	118	126	131	140	147	154	164	180	176	182	189	196	204	216	216
19	.05	36	45	53	61	70	76	82	89	94	102	108	114	121	127	133	141	142	171	160	163	169	177	183	187
	.025	38	51	57	66	76	84	90	98	103	111	120	126	133	141	145	151	159	190	169	180	185	190	199	205
	.01	38	54	64	71	83	91	98	107	113	122	130	138	148	152	160	166	176	190	187	199	204	209	218	224
20	.05	38	48	60	65	72	79	88	93	110	107	116	120	126	135	140	146	152	160	180	173	176	184	192	200
	.025	40	51	64	75	78	86	96	100	120	116	124	130	138	150	156	160	166	169	200	180	192	199	208	215
	.01	40	57	68	80	88	93	104	111	130	127	140	143	152	160	168	175	182	187	220	199	212	219	228	235
21	.05	38	51	59	69	75	91	89	99	105	112	120	126	140	138	145	151	159	163	173	189	183	189	198	202
	.025	40	54	63	74	81	98	97	108	116	123	129	137	147	153	157	166	174	180	180	210	203	206	213	220
	.01	42	57	72	80	90	105	107	117	126	134	141	150	161	168	173	180	189	199	199	231	223	227	237	244
22	.05	40	51	62	70	78	84	94	101	108	121	124	130	138	144	150	157	164	169	176	183	198	194	204	209
	.025	42	57	66	78	86	96	102	110	118	132	134	141	148	154	164	170	178	185	192	203	220	214	222	228
	.01	44	60	72	83	92	103	112	122	130	143	148	156	164	173	180	187	196	204	212	223	242	237	242	250
23	.05	42	54	64	72	80	89	98	106	114	119	125	135	142	149	157	163	170	177	184	189	194	230	205	216
	.025	44	60	69	80	86	98	106	115	124	131	137	146	154	163	169	179	184	190	199	206	214	230	226	237
	.01	46	63	76	87	97	108	115	126	137	142	149	161	170	179	187	196	204	209	219	227	237	253	249	262
24	.05	44	57	68	76	90	92	104	111	118	124	144	140	146	156	168	168	180	183	192	198	204	205	240	225
	.025	46	60	72	81	96	102	112	120	128	137	156	151	160	168	184	183	198	199	208	213	222	226	264	238
	.01	48	66	80	90	102	112	128	132	140	150	168	166	176	186	200	203	216	218	228	237	242	249	288	262
25	.05	46	60	68	80	88	97	104	114	125	129	138	145	150	160	167	173	180	187	200	202	209	216	225	250
	.025	48	63	75	90	96	105	112	123	135	140	150	158	166	175	181	190	196	205	215	220	228	237	238	275
	.01	50	69	84	95	107	115	125	135	150	154	165	172	182	195	199	207	216	224	235	244	250	262	262	300

TABLE XIV
Critical values for Kendall's rank correlation coefficient τ

n	α		
	0.10	0.05	0.01
4	1.000	–	–
5	0.800	1.000	–
6	0.733	0.867	1.000
7	0.619	0.714	0.905
8	0.571	0.643	0.786
9	0.500	0.556	0.722
10	0.467	0.511	0.644
11	0.418	0.491	0.600
12	0.394	0.455	0.576
13	0.359	0.436	0.564
14	0.363	0.407	0.516
15	0.333	0.390	0.505
16	0.317	0.383	0.483
17	0.309	0.368	0.471
18	0.294	0.346	0.451
19	0.287	0.333	0.439
20	0.274	0.326	0.421
21	0.267	0.314	0.410
22	0.264	0.307	0.394
23	0.257	0.296	0.391
24	0.246	0.290	0.377
25	0.240	0.287	0.367
26	0.237	0.280	0.360
27	0.231	0.271	0.356
28	0.228	0.265	0.344
29	0.222	0.261	0.340
30	0.218	0.255	0.333
31	0.213	0.252	0.325
32	0.210	0.246	0.323
33	0.205	0.242	0.314
34	0.201	0.237	0.312
35	0.197	0.234	0.304
36	0.194	0.232	0.302
37	0.192	0.228	0.297
38	0.189	0.223	0.292
39	0.188	0.220	0.287
40	0.185	0.218	0.285

Note: This table furnishes 0.10, 0.05, and 0.01 critical values for Kendall's rank correlation coefficient τ. The probabilities are for a two-tailed test. When a one-tailed test is desired, halve the probabilities at the heads of the columns.

To test the significance of a correlation coefficient, enter the table with the appropriate sample size and find the appropriate critical value. For example, for a sample size of 15, the 5% and 1% critical values of τ are 0.390 and 0.505, respectively. Thus, an observed value of 0.498 would be considered significant at the 5% but not at the 1% level. Negative correlations are considered as positive for purposes of this test. For sample sizes $n > 40$ use the asymptotic approximation given in Box 12.3, step 5.

The values in this table have been derived from those furnished in table XI of J. V. Bradley, *Distribution-Free Statistical Tests* (Prentice-Hall, Englewood Cliffs, N.J., 1968) with permission of the author and publisher.

Bibliography

Allee, W. C., and E. Bowen. 1932. Studies in animal aggregations: Mass protection against colloidal silver among goldfishes. *J. Exp. Zool.*, **61**:185–207.

Allee, W. C., E. S. Bowen, J. C. Welty, and R. Oesting. 1934. The effect of homotypic conditioning of water on the growth of fishes, and chemical studies of the factors involved. *J. Exp. Zool.*, **68**:183–213.

Archibald, E. E. A. 1950. Plant populations. II. The estimation of the number of individuals per unit area of species in heterogeneous plant populations. *Ann. Bot. N.S.*, **14**:7–21.

Banta, A. M. 1939. Studies on the physiology, genetics, and evolution of some Cladocera. Carnegie Institution of Washington, Dept. Genetics, Paper 39. 285 pp.

Blakeslee, A. F. 1921. The globe mutant in the jimson weed (*Datura stramonium*). *Genetics*, **6**:241–264.

Block, B. C. 1966. The relation of temperature to the chirp-rate of male snowy tree crickets, *Oecanthus fultoni* (Orthoptera: Gryllidae). *Ann. Entomol. Soc. Amer.*, **59**: 56–59.

Brower, L. P. 1959. Speciation in butterflies of the *Papilio glaucus* group. I. Morphological relationships and hybridization. *Evolution*, **13**:40–63.

Brown, B. E., and A. W. A. Brown. 1956. The effects of insecticidal poisoning on the level of cytochrome oxidase in the American cockroach. *J. Econ. Entomol.*, **49**:675–679.

Brown, F. M., and W. P. Comstock. 1952. Some biometrics of *Heliconius charitonius* (Linnaeus) (Lepidoptera, Nymphalidae). *Amer. Mus. Novitates*, **1574**, 53 pp.

Burr, E. J. 1960. The distribution of Kendall's score S for a pair of tied rankings. *Biometrika*, **47**:151–171.

Carter, G. R., and C. A. Mitchell. 1958. Methods for adapting the virus of rinderpest to rabbits. *Science*, **128**:252–253.

Cowan, I. M., and P. A. Johnston. 1962. Blood serum protein variations at the species and subspecies level in deer of the genus *Odocoilcus*. *Syst. Zool.*, **11**:131–138.

Davis, E. A., Jr. 1955. Seasonal changes in the energy balance of the English sparrow. *Auk*, **72**:385–411.

French, A. R. 1976. Selection of high temperatures for hibernation by the pocket mouse, *Perignathus longimembris:* Ecological advantages and energetic consequences. *Ecology*, **57**:185–191.

Fröhlich, F. W. 1921. *Grundzüge einer Lehre vom Licht- und Farbensinn. Ein Beitrag zur allgemeinen Physiologie der Sinne.* Fischer, Jena. 86 pp.

Gabriel, K. R. 1964. A procedure for testing the homogeneity of all sets of means in analysis of variance. *Biometrics*, **20**:459–477.

Gartler, S. M., I. L. Firschein, and T. Dobzhansky. 1956. Chromatographic investigation of urinary amino-acids in the great apes. *Am. J. Phys. Anthropol.*, **14**:41–57.

Geissler, A. 1889. Beiträge zur Frage des Geschlechtsverhältnisses der Geborenen. *Z. K. Sächs. Stat. Bur.*, **35**:1–24.

Greenwood, M., and G. U. Yule. 1920. An inquiry into the nature of frequency-distributions of multiple happenings. *J. Roy. Stat. Soc.*, **83**:255–279.

Hunter, P. E. 1959. Selection of *Drosophila melanogaster* for length of larval period. *Z. Vererbungsl.*, **90**:7–28.

Johnson, N. K. 1966. Bill size and the question of competition in allopatric and sympatic populations of dusky and gray flycatchers. *Syst. Zool.*, **15**:70–87.

Kouskolekas, C. A., and G. C. Decker. 1966. The effect of temperature on the rate of development of the potato leafhopper, *Empoasca fabae* (Homoptera: Cicadellidae). *Ann. Entomol. Soc. Amer.*, **59**:292–298.

Lee, J. A. H. 1982. Melanoma and exposure to sunlight. *Epidemiol. Rev.* **4**:110–136.

Leinert, J., I. Simon, and D. Hötze. 1983. Methods and their evaluation to estimate the vitamin B_6 status in human subjects. *Int. J. Vitamin and Nutrition Res.*, **53**:166–178.

Lewontin, R. C., and M. J. D. White. 1960. Interaction between inversion polymorphisms of two chromosome pairs in the grasshopper, *Moraba scurra. Evolution*, **14**:116–129.

Littlejohn, M. J. 1965. Premating isolation in the *Hyla ewingi* complex. *Evolution*, **19**:234–243.

Liu, Y. K., R. E. Kosfeld, and V. Koo. 1983. Marrow neutrophil mass in patients with nonhematological tumor. *J. Lab. and Clinical Med.*, **101**:561–568.

Millis, J., and Y. P. Seng. 1954. The effect of age and parity of the mother on birth weight of the offspring. *Ann. Human Genetics*, **19**:58–73.

Mittler, T. E., and R. H. Dadd. 1966. Food and wing determination in *Myzus persicae* (Homoptera: Aphidae). *Ann. Entomol. Soc. Amer.*, **59**:1162–1166.

Mosimann, J. E. 1968. *Elementary Probability for the Biological Sciences.* Appleton-Century-Crofts, New York. 255 pp.

Nelson, V. E. 1964. The effects of starvation and humidity on water content in *Tribolium confusum* Duval (Coleoptera). Unpublished Ph.D. thesis, University of Colorado. 111 pp.

Newman, K. J., and H. V. Meredith. 1956. Individual growth in skeletal bigonial diameter during the childhood period from 5 to 11 years of age. *Am. J. Anatomy*, **99**:157–187.

Olson, E. C., and R. L. Miller. 1958. *Morphological Integration*. University of Chicago Press, Chicago. 317 pp.

Park, W. H., A. W. Williams, and C. Krumwiede. 1924. *Pathogenic Microörganisms*. Lea & Febiger, Philadelphia and New York. 811 pp.

Pearson, E. S., and H. O. Hartley. 1958. *Biometrika Tables for Statisticians*. Vol. I. 2d ed. Cambridge University Press, London. 240 pp.

Pfändner, K. 1984. Nimesulide and antibiotics in the treatment of acute urinary tract infections. *Drug Res.*, **34**:77–79.

Phillips, J. R., and L. D. Newsom. 1966. Diapause in *Heliothis zea* and *Heliothis virescens* (Lepidoptera: Noctuidae). *Ann. Entomol. Soc. Amer.*, **59**:154–159.

Ruffie, J., and N. Taleb. 1965. *Étude hémotypologique des ethnies libanaises*. Hermann, Paris. 104 pp. (Monographies du Centre d'Hémotypologie du C.H.U. de Toulouse.)

Sinnott, E. W., and D. Hammond. 1935. Factorial balance in the determination of fruit shape in *Cucurbita*. *Amer. Nat.*, **64**:509–524.

Sokal, R. R. 1952. Variation in a local population of *Pemphigus*. *Evolution*, **6**:296–315.

Sokal, R. R. 1967. A comparison of fitness characters and their responses to density in stock and selected cultures of wild type and black *Tribolium castaneum*. *Tribolium Information Bull.*, **10**:142–147.

Sokal, R. R., and P. E. Hunter. 1955. A morphometric analysis of DDT-resistant and non-resistant housefly strains. *Ann. Entomol. Soc. Amer.* **48**:499–507.

Sokal, R. R., and I. Karten. 1964. Competition among genotypes in *Tribolium castaneum* at varying densities and gene frequencies (the black locus). *Genetics*, **49**: 195–211.

Sokal, R. R., and F. J. Rohlf. 1981. *Biometry*. 2d ed. W. H. Freeman and Company, New York. 859 pp.

Sokal, R. R., and P. A. Thomas. 1965. Geographic variation of *Pemphigus populitransversus* in Eastern North America: Stem mothers and new data on alates. *Univ. Kansas Sci. Bull.*, **46**:201–252.

Sokoloff, A. 1955. Competition between sibling species of the *Pseudoobscura* subgroup of *Drosophila*. *Ecol. Monogr.*, **25**:387–409.

Sokoloff, A. 1966. Morphological variation in natural and experimental populations of *Drosophila pseudoobscura* and *Drosophila persimilis*. *Evolution*. **20**:49–71.

Student (W. S. Gossett). 1907. On the error of counting with a haemacytometer. *Biometrika*, **5**:351–360.

Sullivan, R. L., and R. R. Sokal. 1963. The effects of larval density on several strains of the housefly. *Ecology*, **44**:120–130.

Swanson, C. O., W. L. Latshaw, and E. L. Tague. 1921. Relation of the calcium content of some Kansas soils to soil reaction by the electrometric titration. *J. Agr. Res.*, **20**:855–868.

Tate, R. F., and G. W. Klett. 1959. Optimal confidence intervals for the variance of a normal distribution. *J. Am. Stat. Assoc.*, **54**:674–682.

Utida, S. 1943. Studies on experimental population of the Azuki bean weevil, *Callosobruchus chinensis* (L.). VIII. Statistical analysis of the frequency distribution of the emerging weevils on beans. *Mem. Coll. Agr. Kyoto Imp. Univ.*, **54**:1–22.

Vollenweider, R. A., and M. Frei. 1953. Vertikale und zeitliche Verteilung der Leitfähigkeit in einem eutrophen Gewässer während der Sommerstagnation. *Schweiz. Z. Hydrol.*, **15**:158–167.

Wilkinson, L., and G. E. Dallal. 1977. Accuracy of sample moments calculations among widely used statistical programs. *Amer. Stat.*, **31**:128–131.

Williams, D. A. 1976. Improved likelihood ratio tests for complete contingency tables. *Biometrika*, **63**:33–37.

Willis, E. R., and N. Lewis. 1957. The longevity of starved cockroaches. *J. Econ. Entomol.*, **50**:438–440.

Willison, J. T., and L. M. Bufa. 1983. Myocardial infarction—1983. *Clinical Res.*, **31**:364–375.

Woodson, R. E., Jr. 1964. The geography of flower color in butterflyweed. *Evolution*, **18**:143–163.

Young, B. H. 1981. A study of striped bass in the marine district of New York State. Anadromous Fish Act, P.L. 89-304. Annual report, New York State Dept. of Environmental Conservation. 21 pp.

Index